CIVILIZATION'S CRISIS
A Set of Linked Challenges

CIVILIZATION'S CRISIS
A Set of Linked Challenges

John Scales Avery
University of Copenhagen, Denmark

NEW JERSEY · LONDON · SINGAPORE · BEIJING · SHANGHAI · HONG KONG · TAIPEI · CHENNAI · TOKYO

Published by

World Scientific Publishing Co. Pte. Ltd.
5 Toh Tuck Link, Singapore 596224
USA office: 27 Warren Street, Suite 401-402, Hackensack, NJ 07601
UK office: 57 Shelton Street, Covent Garden, London WC2H 9HE

Library of Congress Cataloging-in-Publication Data
Names: Avery, John, 1933–
Title: Civilization's crisis : a set of linked challenges / John Scales Avery,
   University of Copenhagen, Denmark.
Description: New Jersey : World Scientific, 2017. | Includes bibliographical
   references and index.
Identifiers: LCCN 2017003573| ISBN 9789813222441 (hardcover : alk. paper) |
   ISBN 9789813222458 (pbk. : alk. paper)
Subjects: LCSH: Climatic changes--Social aspects. | Science--Social aspects.
  | Science and civilization. | Environmental degradation. | Nuclear warfare--Social aspects.
Classification: LCC QC903 .A886 2017 | DDC 361--dc23
LC record available at https://lccn.loc.gov/2017003573

British Library Cataloguing-in-Publication Data
A catalogue record for this book is available from the British Library.

Copyright © 2017 by World Scientific Publishing Co. Pte. Ltd.

All rights reserved. This book, or parts thereof, may not be reproduced in any form or by any means, electronic or mechanical, including photocopying, recording or any information storage and retrieval system now known or to be invented, without written permission from the publisher.

For photocopying of material in this volume, please pay a copying fee through the Copyright Clearance Center, Inc., 222 Rosewood Drive, Danvers, MA 01923, USA. In this case permission to photocopy is not required from the publisher.

Printed in Singapore

Contents

Preface		vii
1.	Economics, Ethics and Ecology	1
2.	Threats to the Environment and Climate Change	41
3.	Growing Population, Vanishing Resources	87
4.	The Global Food and Refugee Crisis	117
5.	Intolerable Economic Inequality	143
6.	The Threat of Nuclear War	163
7.	Facing a Set of Linked Problems	207
8.	Outlawing War	227
9.	The Evolution of Cooperation	261
10.	Education for Peace	293
11.	The Future of International Law	331
12.	The Choice is Ours to Make	363
Index		383

Preface

In written Chinese, the word "crisis" is represented by two characters, one meaning "danger", and the other "opportunity". Today, at the start of the 21st century, the explosive growth of scientific and technological knowledge has brought us many benefits, but we face dangers as well as opportunities. The problem of achieving a stable, peaceful and sustainable world remains challenging, although rational solutions are available.

The aim of this book is to explore the links between the challenges facing civilization today. Here are some of the tasks which history has given to our generation:

- We must achieve a steady-state economic system. Limitless growth on a finite planet is a logical absurdity.
- We must restore democracy in countries where it has decayed, and create it in countries where it never existed.
- We must decrease economic inequality.
- We must leave fossil fuels in the ground.
- We must stabilize and ultimately reduce global population to a level that can be supported by sustainable agriculture.
- We must abolish the institution of war before modern weapons destroy us.
- And finally, we must develop a mature ethical system to match our new technology.

The tasks just mentioned are difficult, but all of the problems are soluble. The aim of this book is to show how the challenges facing civilization are interlinked and thus to throw some light on the remedies. The dangers are balanced by opportunities.

Present economic problems should be seen as an advance warning of the limits to growth that will be reached by the middle of the 21st century. This offers an opportunity to take steps towards ecological sustainability and towards an economic system that does not depend on growth for its health - a steady-state economic system.

As the world economy reaches the limits of growth for resource-using activities, unemployment will become a threat; but public health work, reforestation, soil conservation, windmill construction, hydrogen technology research, and construction of energy-conserving buildings are all labor-intensive activities that will help to prevent unemployment while at the same time aiding the transition to sustainability.

Poverty causes overpopulation, and overpopulation causes poverty; war causes poverty, and poverty causes war; poverty causes disease, and disease causes poverty.

There is a reciprocal relationship between intolerable economic inequality and war. Military might is used by powerful industrialized nations to maintain economic hegemony over less developed countries. This is true today, even though the colonial era is supposed to be over (as has been amply documented by Professor Michael Klare in his books on "Resource Wars"). But, conversely, intolerable economic inequality is also a cause of war: Abolition of the institution of war will require the replacement of "might makes right" by the rule of international law. It will require development of effective global governance. But reform and strengthening of the United Nations is blocked by wealthy countries because they are afraid of loosing their privileged positions. If global economic inequality were less enormous, the problem of unifying the world would be simplified.

Today's military spending of one and a half trillion US dollars per year would be more than enough to finance safe drinking water for the entire world, and to bring primary health care and family planning advice to all. If used constructively, the money now wasted (or worse than wasted) on the institution of war could also help the world to make the transition from fossil fuel use to renewable energy systems.

The dangers of nuclear weapons are linked to the problem of climate change because of the widespread (but false) belief that civilian nuclear power generation is carbon neutral. On the basis of this false premise, it is argued that nuclear power is an answer to the threat of global warming. But because it is almost impossible to distinguish between civilian and military nuclear programs, the widespread use of nuclear power throughout the world would carry with it serious dangers of nuclear proliferation.

We will attempt to explore these and other links between the problems facing civilization at the start of the 21st century. Many of these problems are due to the astonishingly rapid growth of modern science and technology. 40,000 years ago, our ancestors were hunter-gatherers, living in small, mutually-hostile tribes. A few tens of thousands of years later, only an instant on the scale of evolutionary time, the world of our ancestors has been replaced by a world of supercomputers and quantum theory, which is unfortunately also a world of thermonuclear weapons and nerve gas.

We face the serious problems of the 21st century with an inherited emotional nature that has not changed much since our neolithic ancestors flaked flint into speartips. Our emotions still include a tendency towards tribalism, which once contributed importantly to survival, but which today has become a dangerous anachronism.

Although our social and political institutions change much more rapidly than our genomes, the modification of these institutions is still extremely slow compared with the lightning-like speed of scientific and technological progress. This progress has brought us immense benefits, but its very speed threatens to shake human civilization to pieces. We are thus faced with the urgent task of achieving a global ethic which will have both the social and environmental elements needed to save the planet. We must also build social and political institutions which will be in harmony with the enormous powers over nature which science and technology have given us, powers that must be used with wisdom and restraint if human society and the biosphere are to survive.

The first chapter of this book is devoted to a history of economic ideas during the early phases of the Industrial Revolution, a development that was centered in England. It will be seen that the debate that took place at that time has great relevance to the 21st century. Adam Smith's ideas are dominant today, but deficiencies in the concept of a totally free market are becoming more and more apparent. The warning voice of Thomas Robert Malthus is given much space in the first chapter because of the light he throws on our present situation.

I must apologize for trying to treat such a broad panorama of problems. In doing so, one runs the risk of being superficial. However, I believe that insights into links between problems make this attempt worthwhile.

Finally, I would like to express my gratitude to Professor J.B. Opschoor, Ph.D. Stud. Christina Berg Johansen, Canadian Poet and Artist Heather Spears, and Lektor Birgit Schmidt for their careful reading of the book and for many extremely valuable suggestions.

Fig. 0.1 Small and fragile, drifting through the dark immensity of space, our only home. (Image from Pics About Space, Public domain.)

Chapter 1

Economics, Ethics and Ecology

"Like him who perverts the revenues of some pious foundation to profane purposes, he pays the wages of idleness with those funds which the frugality of his forefathers had, as it were, consecrated to the maintenance of industry."

Adam Smith (describing an industrialist who fails to reinvest his profits)

"That population cannot increase without the means of subsistence is a proposition so evident that it needs no illustration. That population does invariably increase, where there are means of subsistence, the history of every people who have ever existed will abundantly prove. And that the superior power cannot be checked without producing misery and vice, the ample portion of these two bitter ingredients in the cup of human life, and the continuation of the physical causes that seem to have produced them, bear all too convincing a testimony"

Thomas Robert Malthus

Introduction

The history of the epoch that immediately preceded the modern era can cast much light on the challenges facing us today, so we will begin by reviewing it. Until the start of the Industrial Revolution in the 18th and 19th centuries, human society maintained a more or less sustainable relationship with nature. However, with the beginning of the industrial era, traditional ways of life, containing both ethical and environmental elements, were replaced by the money-centered, growth-oriented life of today, from which these vital elements are missing.

Economics without ethics

According to the great classical economist Adam Smith (1723-1790), self-interest (even greed) is a sufficient guide to human economic actions. The passage of time has shown that Smith was right in many respects. The free market, which he advocated, has turned out to be the optimum prescription for economic growth. However, history has also shown that there is something horribly wrong or incomplete about the idea that individual self-interest alone, uninfluenced by ethical and ecological considerations, and totally free from governmental intervention, can be the main motivating force of a happy and just society. There has also proved to be something terribly wrong with the concept of unlimited economic growth. Here is what actually happened:

Industrialism in 18th and 19th centuries

Highland Clearances and Enclosure Acts

In pre-industrial Europe, peasant farmers held a low but nevertheless secure position, protected by a web of traditional rights and duties. Their low dirt-floored and thatched cottages were humble but safe refuges. If a peasant owned a cow, it could be pastured on common land.

With the invention of the steam engine and the introduction of spinning and weaving machines towards the end of the 18th century, the pattern changed, at first in England, and afterwards in other European countries. Land-owners in Scotland and Northern England realized that sheep were more profitable to have on the land than "crofters" (i.e., small tenant farmers), and families that had farmed land for generations were violently driven from their homes with almost no warning. The cottages were afterwards burned to prevent the return of their owners.

The following account of the Highland Clearances has been left by Donald McLeod, a crofter in the district of Sutherland: "The consternation and confusion were extreme. Little or no time was given for the removal of persons or property; the people striving to remove the sick or helpless before the fire should reach them; next struggling to save the most valuable of their effects. The cries of the women and children; the roaring of the affrighted cattle, hunted at the same time by the yelling dogs of the shepherds amid the smoke and fire, altogether presented a scene that completely baffles description - it required to be seen to be believed... The conflagration lasted for six days, until the whole of the dwellings were reduced to ashes and smoking ruins."

Between 1750 and 1860, the English Parliament passed a large number of "Enclosure Acts", abolishing the rights of small farmers to pasture their animals on common land that was not under cultivation. The fabric of traditional rights and duties that once had protected the lives of small tenant farmers was torn to pieces. Driven from the land, poor families flocked to the towns and cities, hoping for employment in the textile mills that seemed to be springing up everywhere.

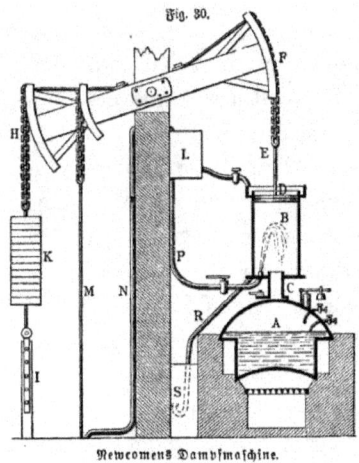

Fig. 1.1 Newcomen's steam engine. (From Hmlopedia.)

Working conditions in 19th century England

According to the new rules by which industrial society began to be governed, traditions were forgotten and replaced by purely economic laws.

Fig. 1.2 A watercolor painting by Vincent van Gogh showing wives of Belgian miners carrying bags of coal. (Public domain.)

Labor was viewed as a commodity, like coal or grain, and wages were paid according to the laws of supply and demand, without regard for the needs of the workers. Wages fell to starvation levels, hours of work increased, and working conditions deteriorated.

John Fielden's book, "The Curse of the Factory System" was written in 1836, and it describes the condition of young children working in the cotton mills. "The small nimble fingers of children being by far the most in request, the custom instantly sprang up of procuring 'apprentices' from the different parish workhouses of London, Birmingham and elsewhere... Overseers were appointed to see to the works, whose interest it was to work the children to the utmost, because their pay was in proportion to the quantity of pay that they could exact."

"Cruelty was, of course, the consequence; and there is abundant evidence on record to show that in many of the manufacturing districts, the most heart-rending cruelties were practiced on the unoffending and friendless creatures... that they were flogged, fettered and tortured in the most exquisite refinements of cruelty, that they were in many cases starved to the bone while flogged to their work, and that they were even in some instances driven to commit suicide... The profits of manufacture were enormous, but this only whetted the appetite that it should have satisfied."

Fig. 1.3 London during the industrial revolution. (Public domain.)

Dr. Peter Gaskell, writing in 1833, described the condition of the English mill workers as follows:

"The vast deterioration in personal form which has been brought about in the manufacturing population during the last thirty years... is singularly impressive, and fills the mind with contemplations of a very painful character... Their complexion is sallow and pallid, with a peculiar flatness of feature caused by the want of a proper quantity of adipose substance to cushion out the cheeks. Their stature is low - the average height of men being five feet, six inches... Great numbers of the girls and women walk lamely or awkwardly... Many of the men have but little beard, and that in patches of a few hairs... (They have) a spiritless and dejected air, a sprawling and wide action of the legs..."

"Rising at or before daybreak, between four and five o'clock the year round, they swallow a hasty meal or hurry to the mill without taking any food whatever... At twelve o'clock the engine stops, and an hour is given for dinner... Again they are closely immured from one o'clock till eight or nine, with the exception of twenty minutes, this being allowed for tea. During the whole of this long period, they are actively and unremittingly engaged in a crowded room at an elevated temperature."

Fig. 1.4 A girl pulling a coaltub through the narrow space left by removal of coal from a seam. (Public domain.)

Dr. Gaskell described the housing of the workers as follows:

"One of the circumstances in which they are especially defective is that of drainage and water-closets. Whole ranges of these houses are either totally undrained, or very partially... The whole of the washings and filth from these consequently are thrown into the front or back street, which, often being unpaved and cut into deep ruts, allows them to collect into stinking and stagnant pools; while fifty, or even more than that number, having only a single convenience common to them all, it is in a very short time choked with excrementous matter. No alternative is left to the inhabitants but adding this to the already defiled street."

"It frequently happens that one tenement is held by several families... The demoralizing effects of this utter absence of domestic privacy must be seen before they can be thoroughly appreciated. By laying bare all the wants and actions of the sexes, it strips them of outward regard for decency - modesty is annihilated - the father and the mother, the brother and the sister, the male and female lodger, do not scruple to commit acts in front of each other which even the savage keeps hid from his fellows."

Adam Smith

The invisible hand

As everyone knows, Adam Smith invented the theory that individual self-interest is, and ought to be, the main motivating force of human economic activity, and that this, in effect, serves the wider social interest. He put forward a detailed description of this concept in an immense book, "The Wealth of Nations" (1776).

Adam Smith (1723-1790) had been Professor of Logic at the University of Glasgow, but in 1764 he withdrew from his position at the university to become the tutor of the young Duke of Buccleuch. In those days a Grand Tour of Europe was considered to be an important part of the education of a young nobleman, and Smith accompanied Buccleuch to the Continent. To while away the occasional dull intervals of the tour, Adam Smith began to write an enormous book on economics which he finally completed twelve years later. He began his "Inquiry into the Nature and Causes of the Wealth of Nations" by praising division of labor. As an example of its benefits, he cited a pin factory, where ten men, each a specialist in his own set of operations, could produce 48,000 pins in a day. In the most complex civilizations, Smith stated, division of labor has the greatest utility.

The second factor in prosperity, Adam Smith maintained, is a competitive market, free from monopolies and entirely free from governmental interference. In such a system, he tells us, the natural forces of competition are able to organize even the most complex economic operations, and are able also to maximize productivity. He expressed this idea in the following words:

"As every individual, therefore, endeavors as much as he can, both to employ his capital in support of domestic industry, and so to direct that industry that its produce may be of greatest value, each individual necessarily labours to render the annual revenue of the Society as great as he can."

"He generally, indeed, neither intends to promote the public interest, nor knows how much he is promoting it. By preferring the support of domestic to that of foreign industry, he intends only his own security; and by directing that industry in such a manner as its produce may be of greatest value, he intends only his own gain; and he is in this, as in many other cases, led by an invisible hand to promote an end that was no part of his intention. Nor is it always the worse for Society that it was no part of it. By pursuing his own interest, he frequently promotes that of Society more effectively than when he really intends to promote it."

For example, a baker does not bake bread out of an unselfish desire to help his fellow humans; he does so in order to earn money; but if he were not performing a useful service, he would not be paid. Thus the "invisible hand" guides him to do something useful. Free competition also regulates prices: If the baker charges too much, he will be undersold. Finally, if there are too many bakers, the trade will become so unprofitable that some bakers will be forced into other trades. Thus highly complex operations are automatically regulated by the mechanisms of the free market. "Observe the accommodation of the most common artificer or day labourer in a civilized and thriving country", Smith continues, "and you will perceive that the number of people of whose industry a part, though but a small part, has been employed in securing him this accommodation, exceeds all computation. The woolen coat, which covers the day-labourer, as coarse and rough as it may seem, is the joint labour of a great multitude of workmen. The shepherd, the sorter of wool, the wool-comber, the carder, the dyer, the scribbler, the spinner, the weaver, the fuller, the dresser, with many others, must all join their different arts to complete even the most homely production. How many merchants and carriers, besides, must have been employed... how much commerce and navigation... how many ship-builders, sailors, sail-makers, rope-makers..."

Reinvestment and growth

An important feature of Adam Smith's economic model is that it is by no means static. The virtuous manufacturer does not purchase pearl necklaces for his wife; he reinvests his profits, buying more machinery or building new factories. An industrialist who ignores the commandment to reinvest is "...like him who perverts the revenues of some pious foundation to profane purposes; he pays the wages of idleness with those funds which the fragility of his forefathers had, as it were, consecrated to the maintenance of industry."

The expansion of the system will not be slowed, Smith maintained, by shortages of labor, because "...the demand for men, like that for any other commodity, necessarily regulates the production of men." Smith did not mean that more births would occur if the demand for workers became greater. He meant that if wages began to rise above the lowest level needed to maintain life, more children of the workers would survive. In those days, the rates of infant and child mortality were horrendous, particularly among the half-starved poor. "It is not uncommon", Smith wrote, "in the Highlands of Scotland, for a mother who has borne twenty children not to have two alive."

Fig. 1.5 Adam Smith (1723-1790). (Public domain.)

Adam Smith's ideas were enthusiastically adopted by the rising class of manufacturers and by their representatives in government. The reverence shown to him can be illustrated by an event that occurred when he visited England's Prime Minister, William Pitt, and his Cabinet. The whole gathering stood up when Smith entered. "Pray be seated, gentlemen", Smith said. "Not until you first are seated Sir", Pitt replied, "for we are all your scholars."

History has shown that Adam Smith was right in many respects. The free market is indeed a dynamo that produces economic growth, and it is capable of organizing even the most complex economic endeavors. Through

Adam Smith's "invisible hand", self interest is capable of guiding the economy so that it will maximize the production of wealth. However, history has also shown the shortcomings of a market that is totally free of governmental regulation.

The landowners of Scotland were unquestionably following self-interest as they burned the cottages of their crofters; and self-interest motivated overseers as they whipped half-starved child workers in England's mills. Adam Smith's "invisible hand" no doubt guided their actions in such a way as to maximize production. But whether a happy and just society was created in this way is questionable. Certainly it was a society with large areas of unhappiness and injustice. Self-interest alone was not enough. A society following purely economic laws - a society where selfishness is exalted as the mainspring for action - lacks both the ethical and ecological dimensions needed for social justice, widespread happiness, and sustainability.[a]

Malthus

A debate between father and son

T.R. Malthus' *Essay on The Principle of Population*, the first edition of which was published in 1798, was one of the the first systematic studies of the problem of population in relation to resources. Earlier discussions of the problem had been published by Boterro in Italy, Robert Wallace in England, and Benjamin Franklin in America. However Malthus' *Essay* was the first to stress the fact that, in general, powerful checks operate continuously to keep human populations from increasing beyond their available food supply. In a later edition, published in 1803, he buttressed this assertion with carefully collected demographic and sociological data from many societies at various periods of their histories.

The publication of Malthus' *Essay* coincided with a wave of disillusionment which followed the optimism of the Enlightenment. The utopian societies predicted by the philosophers of the Enlightenment were compared with reign of terror in Robespierre's France and with the miseries of industrial workers in England; and the discrepancy required an explanation. The optimism which preceded the French Revolution, and the disappointment which followed a few years later, closely paralleled the optimistic expecta-

---

[a] In fact, Adam Smith himself would have accepted this criticism of his enthronement of self-interest as the central principle of society. He believed that his "invisible hand" would not work for the betterment of society except within the context of a certain amount of governmental regulation. His modern Neoliberal admirers, however, forget this aspect of Smith's philosophy, and maintain that market forces alone can achieve a desirable result.

tions of our own century, in the period after the Second World War, when it was thought that the transfer of technology to the less developed parts of the world would eliminate poverty, and the subsequent disappointment when poverty persisted. Science and technology developed rapidly in the second half of the twentieth century, but the benefits which they conferred were just as rapidly consumed by a global population which today is increasing at the rate of one billion people every fourteen years. Because of the close parallel between the optimism and disappointments of Malthus' time and those of our own, much light can be thrown on our present situation by rereading the debate between Malthus and his contemporaries.

Thomas Robert Malthus (1766-1834) came from an intellectual family: His father, Daniel Malthus, was a moderately well-to-do English country gentleman, an enthusiastic believer in the optimistic ideas of the Enlightenment, and a friend of the philosophers Henry Rousseau, David Hume and William Godwin. The famous book on population by the younger Malthus grew out of conversations with his father.

Daniel Malthus attended Oxford, but left without obtaining a degree. He later built a country home near Dorking, which he called "The Rookery". The house had Gothic battlements, and the land belonging to it contained a beech forest, an ice house, a corn mill, a large lake, and serpentine walks leading to "several romantic buildings with appropriate dedications".

Fig. 1.6 The Rookery near Dorking in Surrey. (Public domain.)

Daniel Malthus was an ardent admirer of Rousseau; and when the French philosopher visited England with his mistress, Thérèse le Vasseur, Daniel Malthus entertained him at the Rookery. Rousseau and Thérèse undoubtedly saw Daniel's baby son (who was always called Robert or Bob) and they must have noticed with pity that he had been born with a hare lip. This was later sutured, and apart from a slight scar which marked the operation, he became very handsome.

Robert Malthus was at first tutored at home; but in 1782, when he was 16 years old, he was sent to study at the famous Dissenting Academy at Warrington in Lancashire. Joseph Priestly had taught at Warrington, and he had completed his famous *History of Electricity* there, as well as his *Essay on Government*, which contains the phrase "the greatest good for the greatest number".

Robert's tutor at Warrington Academy was Gilbert Wakefield (who was later imprisoned for his radical ideas). When Robert was 18, Wakefield arranged for him to be admitted to Jesus College, Cambridge University, as a student of mathematics. Robert Malthus graduated from Cambridge in 1788 with a first-class degree in mathematics. He was Ninth Wrangler, which meant that he was the ninth-best mathematician in his graduating class. He also won prizes in declamation, both in English and in Latin, which is surprising in view of the speech defect from which he suffered all his life.

In 1793, Robert Malthus was elected a fellow of Jesus College, and he also took orders in the Anglican Church. He was assigned as Curate to Okewood Chapel in Surrey. This small chapel stood in a woodland region, and Malthus' illiterate parishioners were so poor that the women and children went without shoes. They lived in low thatched huts made of woven branches plastered with mud. The floors of these huts were of dirt, and the only light came from tiny window openings. Malthus' parishioners diet consisted almost entirely of bread. The children of these cottagers developed late, and were stunted in growth. Nevertheless, in spite of the harsh conditions of his parishioners' lives, Malthus noticed that the number of births which he recorded in the parish register greatly exceeded the number of deaths. It was probably this fact which first turned his attention to the problem of population.

By this time, Daniel Malthus had sold the Rookery; and after a period of travel, he had settled with his family at Albury, about nine miles from Okewood Chapel. Robert Malthus lived with his parents at Albury, and it was here that the famous debates between father and son took place.

1793, the year when Robert Malthus took up his position at Okewood, was also the year in which Daniel Malthus friend, William Godwin, published his enormously optimistic book, *Political Justice* [6,14,21]. In this book, Godwin predicted a future society where scientific progress would liberate humans from material want. Godwin predicted that in the future, with the institution of war abolished, with a more equal distribution of property, and with the help of scientific improvements in agriculture and industry, much less labour would be needed to support life. Luxuries are at present used to maintain artificial distinctions between the classes of society, Godwin wrote, but in the future values will change; humans will live more simply, and their efforts will be devoted to self-fulfillment and to intellectual and moral improvement, rather than to material possessions. With the help of automated agriculture, the citizens of a future society will need only a few hours a day to earn their bread.

Godwin went on to say, "The spirit of oppression, the spirit of servility and the spirit of fraud - these are the immediate growth of the established administration of property. They are alike hostile to intellectual improvement. The other vices of envy , malice, and revenge are their inseparable companions. In a state of society where men lived in the midst of plenty, and where all shared alike the bounties of nature, these sentiments would inevitably expire. The narrow principle of selfishness would vanish. No man being obliged to guard his little store, or provide with anxiety and pain for his restless wants, each would lose his own individual existence in the thought of the general good. No man would be the enemy of his neighbor, for they would have nothing to contend; and of consequence philanthropy would resume the empire which reason assigns her. Mind would be delivered from her perpetual anxiety about corporal support, and free to expatiate in the field of thought which is congenial to her. Each man would assist the inquiries of all."

Godwin insisted that there is an indissoluble link between politics, ethics and knowledge. *Political Justice* is an enthusiastic vision of what humans could be like at some future period when the trend towards moral and intellectual improvement has lifted men and women above their their present state of ignorance and vice. Much of the savage structure of the penal system would then be unnecessary, Godwin believed. (At the time when he was writing, there were more than a hundred capital offenses in England, and this number had soon increased to almost two hundred. The theft of any object of greater value than ten shillings was punishable by hanging.)

In its present state, Godwin wrote, society decrees that the majority of its citizens "should be kept in abject penury, rendered stupid with ignorance and disgustful with vice, perpetuated in nakedness and hunger, goaded to the commission of crimes, and made victims to the merciless laws which the rich have instituted to oppress them". But human behavior is produced by environment and education, Godwin pointed out. If the conditions of upbringing were improved, behavior would also improve. In fact, Godwin believed that men and women are subject to natural laws no less than the planets of Newton's solar system. "In the life of every human", Godwin wrote, "there is a chain of causes, generated in that eternity which preceded his birth, and going on in regular procession through the whole period of his existence, in consequence of which it was impossible for him to act in any instance otherwise than he has acted."

Fig. 1.7 Thomas Robert Malthus (1766-1834). (Public domain.)

The chain of causality in human affairs implies that vice and crime should be regarded with the same attitude with which we regard disease. The causes of poverty, ignorance, vice and crime should be removed. Human failings should be cured rather than punished. With this in mind, Godwin wrote, "our disapprobation of vice will be of the same nature as our disapprobation of an infectious distemper."

In France the Marquis de Condorcet had written an equally optimistic book, *Esquisse d'un Tableau Historique des Progrès de l'Esprit Humain*. Condorcet's optimism was unaffected even by the fact that at the time when he was writing he was in hiding, under sentence of death by Robespierre's government. Besides enthusiastically extolling Godwin's ideas to his son, Daniel Malthus also told him of the views of Condorcet.

Condorcet's *Esquisse*, is an enthusiastic endorsement of the idea of infinite human perfectibility which was current among the philosophers of the 18th century, and in this book, Condorcet anticipated many of the evolutionary ideas of Charles Darwin. He compared humans with animals, and found many common traits. Condorcet believed that animals are able to think, and even to think rationally, although their thoughts are extremely simple compared with those of humans. He also asserted that humans historically began their existence on the same level as animals and gradually developed to their present state. Since this evolution took place historically, he reasoned, it is probable, or even inevitable, that a similar evolution in the future will bring mankind to a level of physical, mental and moral development which will be as superior to our own present state as we are now superior to animals.

In his *Esquisse*, Condorcet called attention to the unusually long period of dependency which characterizes the growth and education of human offspring. This prolonged childhood is unique among living beings. It is needed for the high level of mental development of the human species; but it requires a stable family structure to protect the young during their long upbringing. Thus, according to Condorcet, biological evolution brought into existence a moral precept, the sanctity of the family.

Similarly, Condorcet maintained, larger associations of humans would have been impossible without some degree of altruism and sensitivity to the suffering of others incorporated into human behavior, either as instincts or as moral precepts or both; and thus the evolution of organized society entailed the development of sensibility and morality.

Condorcet believed that ignorance and error are responsible for vice; and he listed what he regarded as the main mistakes of civilization: hereditary transmission of power, inequality between men and women, religious bigotry, disease, war, slavery, economic inequality, and the division of humanity into mutually exclusive linguistic groups.

Condorcet believed the hereditary transmission of power to be the source of much of the tyranny under which humans suffer; and he looked forward to an era when republican governments would be established throughout the world. Turning to the inequality between men and women, Condorcet wrote that he could see no moral, physical or intellectual basis for it. He called for complete social, legal, and educational equality between the sexes.

Condorcet predicted that the progress of medical science would free humans from the worst ravages of disease. Furthermore, he maintained that since perfectibility (i.e. evolution) operates throughout the biological world, there is no reason why mankind's physical structure might not gradually improve, with the result that human life in the remote future could be greatly prolonged. Condorcet believed that the intellectual and moral facilities of man are capable of continuous and steady improvement; and he thought that one of the most important results of this improvement will be the abolition of war.

As Daniel Malthus talked warmly about Godwin, Condorcet, and the idea of human progress, the mind of his son, Robert, turned to the unbalance between births and deaths which he had noticed among his parishioners at Okewood Chapel. He pointed out to his father that no matter what benefits science might be able to confer, they would soon be eaten up by population growth. Regardless of technical progress, the condition of the lowest social class would remain exactly the same: The poor would continue to live, as they always had, on the exact borderline between survival and famine, clinging desperately to the lower edge of existence. For them, change for the worse was impossible since it would loosen their precarious hold on life; their children would die and their numbers would diminish until they balanced the supply of food. But any change for the better was equally impossible, because if more nourishment should become available, more of the children of the poor would survive, and the share of food for each of them would again be reduced to the precise minimum required for life.

Observation of his parishioners at Okewood had convinced Robert Malthus that this sombre picture was a realistic description of the condition of the poor in England at the end of the 18th century. Techniques of agriculture and industry were indeed improving rapidly; but among the very poor, population was increasing equally fast, and the misery of society's lowest class remained unaltered.

Publication of the first essay in 1798

Daniel Malthus was so impressed with his son's arguments that he urged him to develop them into a small book. Robert Malthus' first essay on population, written in response to his father's urging, was only 50,000 words in length. It was published anonymously in 1798, and its full title was *An Essay on the Principle of Population, as it affects the future improvement of society, with remarks on the speculations of Mr. Godwin, M. Condorcet, and other writers*. Robert Malthus' *Essay* explored the consequences of his basic thesis: that "the power of population is indefinitely greater than the power in the earth to produce subsistence for man".

"That population cannot increase without the means of subsistence", Robert Malthus wrote, "is a proposition so evident that it needs no illustration. That population does invariably increase, where there are means of subsistence, the history of every people who have ever existed will abundantly prove. And that the superior power cannot be checked without producing misery and vice, the ample portion of these two bitter ingredients in the cup of human life, and the continuance of the physical causes that seem to have produced them, bear too convincing a testimony."

In order to illustrate the power of human populations to grow quickly to enormous numbers if left completely unchecked, Malthus turned to statistics from the United States, where the population had doubled every 25 years for a century and a half. Malthus called this type of growth "geometrical" (today we would call it "exponential"); and, drawing on his mathematical education, he illustrated it by the progression 1,2,4,8,16,32,64,128,256,..etc. In order to show that, in the long run, no improvement in agriculture could possibly keep pace with unchecked population growth, Malthus allowed that, in England, agricultural output might with great effort be doubled during the next quarter century; but during a subsequent 25-year period it could not again be doubled. The growth of agricultural output could at the very most follow an arithmetic (linear) progression, 1,2,3,4,5,6,...etc.

Because of the overpoweringly greater numbers which can potentially be generated by exponential population growth, as contrasted to the slow linear progression of sustenance, Malthus was convinced that at almost all stages of human history, population has not expanded freely, but has instead pressed painfully against the limits of its food supply. He maintained that human numbers are normally held in check either by "vice or misery". (Malthus classified both war and birth control as forms of vice.) Occasionally the food supply increases through some improvement in agriculture, or through the opening of new lands; but population then grows very rapidly, and soon a new equilibrium is established, with misery and vice once more holding the population in check.

Like Godwin's *Political Justice*, Malthus' *Essay on the Principle of Population* was published at exactly the right moment to capture the prevailing mood of England. In 1793, the mood had been optimistic; but by 1798, hopes for reform had been replaced by reaction and pessimism. Public opinion had been changed by Robespierre's Reign of Terror and by the threat of a French invasion. Malthus' clear and powerfully written essay caught the attention of readers not only because it appeared at the right moment, but also because his two contrasting mathematical laws of growth were so striking.

One of Malthus' readers was William Godwin, who recognized the essay as the strongest challenge to his utopian ideas that had yet been published.

Godwin several times invited Malthus to breakfast at his home to discuss social and economic problems. (After some years, however, the friendship between Godwin and Malthus cooled, the debate between them having become more acrimonious.)

In 1801, Godwin published a reply to his critics, among them his former friends James Mackintosh and Samuel Parr, by whom he recently had been attacked. His *Reply to Parr* also contained a reply to Malthus: Godwin granted that the problem of overpopulation raised by Malthus was an extremely serious one. However, Godwin wrote, all that is needed to solve the problem is a change of the attitudes of society. For example we need to abandon the belief "that it is the first duty of princes to watch for (i.e. encourage) the multiplication of their subjects, and that a man or woman who passes the term of life in a condition of celibacy is to be considered as having failed to discharge the principal obligations owed to the community".

"On the contrary", Godwin continued, "it now appears to be rather the man who rears a numerous family that has to some degree transgressed the consideration he owes to the public welfare". Godwin suggested that each marriage should be allowed only two or three children or whatever number might be needed to balance the current rates of mortality and celibacy. This duty to society, Godwin wrote, would surely not be too great a hardship to be endured, once the reasons for it were thoroughly understood.

The second essay, published in 1803

Malthus' small essay had captured public attention in England, and he was anxious to expand it with empirical data which would show his principle of population to be valid not only in England in his own day, but in all societies and all periods. He therefore traveled widely, collecting data. He also made use of the books of explorers, such as Cook and Vancouver.

Malthus second edition - more than three times the length of his original essay on population - was ready in 1803. Book I and Book II of the 1803 edition of Malthus' *Essay* are devoted to a study of the checks to population growth which have operated throughout history in all the countries of the world for which he possessed facts.

In his first chapter, Malthus stressed the potentially enormous power of population growth contrasted the slow growth of the food supply. He concluded that strong checks to the increase of population must almost always be operating to keep human numbers within the bounds of sustenance. He classified the checks as either preventive or positive, the preventive checks being those which reduce fertility, while the positive checks are those which increase mortality. Among the positive checks, Malthus listed "unwholesome occupations, severe labour and exposure to the seasons, extreme poverty, bad nursing of children, great towns, excesses of all kinds, the whole train of common diseases and epidemics, wars, plague, and famine".

In the following chapters of Books I, Malthus showed in detail the mechanisms by which population is held at the level of sustenance in various cultures. He first discussed primitive hunter-gatherer societies, such as the inhabitants of Tierra del Fuego, Van Diemens Land and New Holland, and those tribes of North American Indians living predominantly by hunting. In hunting societies, he pointed out, the population is inevitably very sparse: "The great extent of territory required for the support of the hunter has been repeatedly stated and acknowledged", Malthus wrote, "...The tribes of hunters, like beasts of prey, whom they resemble in their mode of subsistence, will consequently be thinly scattered over the surface of the earth. Like beasts of prey, they must either drive away or fly from every rival, and be engaged in perpetual contests with each other...The neighboring nations live in a perpetual state of hostility with each other. The very act of increasing in one tribe must be an act of aggression against its neighbors, as a larger range of territory will be necessary to support its increased numbers. The contest will in this case continue, either till the equilibrium is restored by mutual losses, or till the weaker party is exterminated or driven from its country... Their object in battle is not conquest but destruction. The life of the victor depends on the death of the enemy". Malthus concluded that among the American Indians of his time, war was the predominant check to population growth, although famine, disease and infanticide each played a part.

In the next chapter, Malthus quoted Captain Cook's description of the natives of the region near Queen Charlotte's Sound in New Zealand, whose way of life involved perpetual war. "If I had followed the advice of all our pretended friends", Cook wrote, "I might have extirpated the whole race; for the people of each hamlet or village, by turns, applied to me to destroy the other". According to Cook, the New Zealanders practiced both ceaseless war and cannibalism; and population pressure provided a motive for both practices.

In later chapters on nomadic societies of the Near East and Asia, war again appears, not only as a consequence of the growth of human numbers, but also as one of the major mechanisms by which these numbers are reduced to the level of their food supply. The studies quoted by Malthus make it seem likely that the nomadic Tartar tribes of central Asia made no use of the preventive checks to population growth. In fact the Tartar tribes may have regarded growth of their own populations as useful in their wars with neighboring tribes.

Malthus also described the Germanic tribes of Northern Europe, whose population growth led them to the attacks which destroyed the Roman Empire. He quoted the following passage from Machiavelli's *History of Florence*: "The people who inhabit the northern parts that lie between the Rhine and the Danube, living in a healthful and prolific climate, often increase to such a degree that vast numbers of them are forced to leave their native country and go in search of new habitations. When any of those provinces begins to grow too populous and wants to disburden itself, the following method is observed. In the first place, it is divided into three parts, in each of which there is an equal portion of the nobility and commonality, the rich and the poor. After this they cast lots; and that division on which the lot falls quits the country and goes to seek its fortune, leaving the other two more room and liberty to enjoy their possessions at home. These emigrations proved the destruction of the Roman Empire". Regarding the Scandinavians in the early middle ages, Malthus wrote: "Mallet relates, what is probably true, that it was their common custom to hold an assembly every spring for the purpose of considering in what quarter they should make war".

In Book II, Malthus turned to the nations of Europe, as they appeared at the end of the 18th century, and here he presents us with a different picture. Although in these societies poverty, unsanitary housing, child labor, malnutrition and disease all took a heavy toll, war produced far less mortality than in hunting and pastoral societies, and the preventive checks, which lower fertility, played a much larger roll.

Malthus had visited Scandinavia during the summer of 1799, and he had made particularly detailed notes on Norway. He was thus able to present a description of Norwegian economics and demography based on his own studies. Norway was remarkable for having the lowest reliably-recorded death rate of any nation at that time: Only 1 person in 48 died each year in Norway. (By comparison, 1 person in 20 died each year in London.) The rate of marriage was also remarkably low, with only 1 marriage each year for every 130 inhabitants; and thus in spite of the low death rate, Norway's population had increased only slightly from the 723,141 inhabitants recorded in 1769.

There were two reasons for late marriage in Norway: Firstly, every man born of a farmer or a labourer was compelled by law to be a soldier in the reserve army for a period of ten years; and during his military service, he could not marry without the permission of both his commanding officer and the parish priest. These permissions were granted only to those who were clearly in an economic position to support a family. Men could be inducted into the army at any age between 20 and 30, and since commanding officers preferred older recruits, Norwegian men were often in their 40's before they were free to marry. At the time when Malthus was writing, these rules had just been made less restrictive; but priests still refused to unite couples whose economic foundations they judged to be insufficient.

The second reason for late marriages was the structure of the farming community. In general, Norwegian farms were large; and the owner's household employed many young unmarried men and women as servants. These young people had no chance to marry unless a smaller house on the property became vacant, with its attached small parcel of land for the use of the "houseman"; but because of the low death rate, such vacancies were infrequent. Thus Norway's remarkably low death rate was balanced by a low birth rate. Other chapters in Book II are devoted to the checks to population growth in Sweden, Russia, Central Europe, Switzerland, France, England, Scotland and Ireland.

Malthus painted a very dark panorama of population pressure and its consequences in human societies throughout the world and throughout history: At the lowest stage of cultural development are the hunter-gatherer societies, where the density of population is extremely low. Nevertheless, the area required to support the hunters is so enormous that even their sparse and thinly scattered numbers press hard against the limits of sustenance. The resulting competition for territory produces merciless intertribal wars.

The domestication of animals makes higher population densities possible; and wherever this new mode of food production is adopted, human numbers rapidly increase; but very soon a new equilibrium is established, with the population of pastoral societies once more pressing painfully against the limits of the food supply, growing a little in good years, and being cut back in bad years by famine, disease and war.

Finally, agricultural societies can maintain extremely high densities of population; but the time required to achieve a new equilibrium is very short. After a brief period of unrestricted growth, human numbers are once more crushed against the barrier of limited resources; and if excess lives are produced by overbreeding, they are soon extinguished by deaths among the children of the poor.

Malthus was conscious that he had drawn an extremely dark picture of the human condition. He excused himself by saying that he has not done it gratuitously, but because he was convinced that the dark shades really are there, and that they form an important part of the picture. He did allow one ray of light, however: By 1803, his own studies of Norway, together with personal conversations with Godwin and the arguments in Godwin's *Reply to Parr*, had convinced Malthus that "moral restraint" should be included among the possible checks to population growth. Thus he concluded Book II of his 1803 edition by saying that the checks which keep population down to the level of the means of subsistence can all be classified under the headings of "moral restraint, vice and misery". (In his first edition he had maintained that vice and misery are the only possibilities).

Systems of Equality

In the 1803 edition of Malthus' *Essay*, Books III and IV form a second volume. The ideas which he put forward in this second volume are much more open to dispute than are the solidly empirical demographic studies of Books I and II. Malthus excused himself at the beginning of the second volume, saying that he realized that the ideas which he was about to put forward were less solidly based than those in his first volume. However, he said that he wished to explore all the consequences of his principle of population: "..Even the errors into which I may have fallen", he wrote, "by affording a handle to argument, and an additional excitement to examination, may be subservient to the important end of bringing a subject so nearly connected with the happiness of society into more general notice".

Malthus began Book III by discussing the systems of equality proposed by Condorcet and Godwin; and he tried to show that such utopian societies would prove impossible in practice, because they would rapidly drown in a flood of excess population. Condorcet himself had recognized this difficulty. He realized that improved living conditions for the poor would lead to a rapid growth of population. "Must not a period then arrive", Condorcet had written, "... when the increase of the number of men surpassing their means of subsistence, the necessary result must be either a continual diminution of happiness and population... or at least a kind of oscillation between good and evil?"

Condorcet believed the serious consequences of population pressure to be far in the future, but Malthus disagreed with him on exactly that point: "M. Condorcet's picture of what may be expected to happen when the number of men shall surpass subsistence is justly drawn... The only point in which I differ from M. Condorcet in this description is with regard to the period when it may be applied to the human race... This constantly subsisting cause of periodical misery has existed in most countries ever since we have had any histories of mankind, and continues to exist at the present moment."

"M. Condorcet, however, goes on to say", Malthus continued, "that should the period, which he conceives to be so distant, ever arrive, the human race, and the advocates of the perfectibility of man, need not be alarmed at it. He then proceeds to remove the difficulty in a manner which I profess not to understand. Having observed that the ridiculous prejudices of superstition would by that time have ceased to throw over morals a corrupt and degrading austerity, he alludes either to a promiscuous concubinage, which would prevent breeding, or to something else as unnatural. To remove the difficulty in this way will surely, in the opinion of most men, be to destroy that virtue and purity of manners which the advocates of equality and of the perfectibility of man profess to be the end and object of their views." When Malthus referred to "something else as unnatural", he of course meant birth control, some forms of which existed at the time when he was writing; and in this passage we see that he was opposed to the practice. He preferred late marriage or "moral restraint" as a means of limiting excessive population growth.

After his arguments against Condorcet, Malthus discussed William Godwin's egalitarian utopia, which, he said, would be extremely attractive if only it could be achieved: "The system of equality which Mr. Godwin proposes", Malthus wrote, "is, on the first view of it, the most beautiful and engaging which has yet appeared. A melioration of society to be produced merely by reason and conviction gives more promise of permanence than than any change effected and maintained by force. The unlimited exercise of private judgement is a doctrine grand and captivating, and has a vast superiority over those systems where every individual is in a manner the slave of the public. The substitution of benevolence, as a master-spring and moving principle of society, instead of self-love, appears at first sight to be a consummation devoutly to be wished. In short, it is impossible to contemplate the whole of this fair picture without emotions of delight and admiration, accompanied with an ardent longing for the period of its accomplishment."

"But alas!" Malthus continued, "That moment can never arrive.... The great error under which Mr. Godwin labours throughout his whole work is the attributing of almost all the vices and misery that prevail in civil society to human institutions. Political regulations and the established administration of property are, with him, the fruitful sources of all evil, the hotbeds of all the crimes that degrade mankind. Were this really a true state of the case, it would not seem a completely hopeless task to remove evil completely from the world; and reason seems to be the proper and adequate instrument for effecting so great a purpose. But the truth is, that though human institutions appear to be, and indeed often are, the obvious and obtrusive causes of much misery in society, they are, in reality, light and superficial in comparison with those deeper-seated causes of evil which result from the laws of nature and the passions of mankind."

The passions of mankind drive humans to reproduce, while the laws of nature set limits to the carrying capacity of the environment. Godwin's utopia, if established, would be very favorable to the growth of population; and very soon the shortage of food would lead to its downfall: Because of the overpowering force of population growth, "Man cannot live in the midst of plenty. All cannot share alike the bounties of nature. Were there no established administration of property, every man would be obliged to guard with his force his little store. Selfishness would be triumphant. The subjects of contention would be perpetual. Every individual would be under constant anxiety about corporal support, and not a single intellect would be left free to expatiate in the field of thought."

Malthus believed that all systems of equality are doomed to failure, not only because of the powerful pressure of population growth, but also because differences between the upper, middle, and lower classes serve the useful purpose of providing humans with an incentive for hard work. He thought that fear of falling to a lower social status, and hope of rising to a higher one, provide a strong incentive for constructive activity. However, he believed that happiness is most often found in the middle ranks of society, and that therefore the highest and lowest classes ought not to be large. Malthus advocated universal education and security of property as means by which the lowest classes of society could be induced to adopt more virtuous and prudent patterns of behavior.

The Poor Laws

Among the most controversial chapters of Malthus' second volume are those dealing with the Poor Laws. During the reign of Queen Elisabeth I, a law had been enacted according to which justices were authorized to collect taxes in order to set to work "...the children of all such, whose parents shall not by the said persons be thought able to keep and maintain their children; and also such persons, married or unmarried, as, having no means to maintain them, use no ordinary or daily trade to get their living by..". Malthus commented: "What is this but saying that the funds for the maintenance of labour in this country may be increased without limit by a *fiat* of government...? Strictly speaking, this clause is as arrogant and absurd as if it had enacted that two ears of wheat should in the future grow where one had grown before. Canute, when he commanded the waves not to wet his princely foot, did not assume a greater power over the laws of nature." Malthus pointed out that if we believe that every person has a right to have as many children as he or she wishes, and if we enact a law, according to which every person born has a right to sustenance, then we implicitly assume that the supply of food can be increased without limit, which of course is impossible.

During the first few years of the nineteenth century there was a severe shortage of food in England, partly because of war with France, and partly because of harvest failures. As a result, the price of wheat tripled, causing great distress among the poor. By 1803, 3,000,000 pounds sterling were being distributed to make up the difference between the wages of poor workers and the amount which they needed to pay for food. Malthus regarded the supply of grain as constant, i.e. independent of the price; and he therefore believed that distribution of money under the Poor Laws merely raised the price of grain still further in relation to wages, forcing a larger number of independent workers to seek help. He thought that the distributed money helped to relieve suffering in some cases, but that it spread the suffering over a wider area.

In some parishes, the amount of money distributed under the Poor Laws was proportional to the number of children in a family, and Malthus believed that this encouraged the growth of population, further aggravating the shortage of food. "A poor man may marry with little or no prospect of being able to support a family in independence", he wrote, "...and the Poor Laws may be said therefore in some measure to create the poor which they maintain; and as the provisions of the country must, in consequence of the increased population, be distributed to every man in smaller proportions, it is evident that the labour of those who are not supported by parish assistance, will purchase a smaller quantity of provisions than before, and consequently more of them must be driven to ask for support." Malthus advocated a very gradual abolition of the Poor Laws, and he believed that while this change was being brought about, the laws ought to be administered in such a way that the position of least well-off independent workers should not be worse than the position of those supported by parish assistance.

Replies to Malthus

The second edition of Malthus' *Essay* was published in 1803. It provoked a storm of controversy, and a flood of rebuttals. In 1803 England's political situation was sensitive. Revolutions had recently occurred both in America and in France; and in England there was much agitation for radical change, against which Malthus provided counter-arguments. Pitt and his government had taken Malthus' first edition seriously, and had abandoned their plans for extending the Poor Laws. Also, as a consequence of Malthus' ideas, England's first census was taken in 1801. This census, and subsequent ones, taken in 1811, 1821 and 1831, showed that England's population was indeed increasing rapidly, just as Malthus had feared. (The population of England and Wales more than doubled in 80 years, from an estimated 6.6 million in 1750 to almost 14 million in 1831.) In 1803, the issues of poverty and population were at the center of the political arena, and articles refuting Malthus began to stream from the pens of England's authors.

William Coleridge planned to write an article against Malthus, and he made extensive notes in the margins of his copy of the *Essay*. In one place he wrote: "Are Lust and Hunger both alike Passions of physical Necessity, and the one equally with the other independent of the Reason and the Will? Shame upon our race that there lives an individual who dares to ask the Question." In another place Coleridge wrote: "Vice and Virtue subsist in the agreement of the habits of a man with his Reason and Conscience, and these can have but one moral guide, Utility, or the virtue and Happiness of Rational Beings". Although Coleridge never wrote his planned article, his close friend Robert Southey did so, using Coleridge's notes almost verbatim. Some years later Coleridge remarked: "Is it not lamentable - is it not even marvelous - that the monstrous practical sophism of Malthus should now have gained complete possession of the leading men of the kingdom! Such an essential lie in morals - such a practical lie in fact it is too! I solemnly declare that I do not believe that all the heresies and sects and factions which ignorance and the weakness and wickedness of man have ever given birth to, were altogether so disgraceful to man as a Christian, a philosopher, a statesman or citizen, as this abominable tenet."

In 1812, Percy Bysshe Shelley, who was later to become William Godwin's son-in-law, wrote: "Many well-meaning persons... would tell me not to make people happy for fear of over-stocking the world... War, vice and misery are undoubtedly bad; they embrace all that we can conceive of temporal and eternal evil. Are we to be told that these are remedyless, because the earth would in case of their remedy, be overstocked?" A year later, Shelley called Malthus a "priest, eunuch, and tyrant", and accused him, in a pamphlet, of proposing that ".. after the poor have been stript naked by the tax-gatherer and reduced to bread and tea and fourteen hours of hard labour by their masters.. the last tie by which Nature holds them to benignant earth (whose plenty is garnered up in the strongholds of their tyrants) is to be divided... They are required to abstain from marrying under penalty of starvation... whilst the rich are permitted to add as many mouths to consume the products of the poor as they please"

Godwin himself wrote a long book (which was published in 1820) entitled *Of Population, An Enquiry Concerning the Power and Increase in the Number of Mankind, being an answer to Mr. Malthus*. One can also view many of the books of Charles Dickens as protests against Malthus' point of view. For example, *Oliver Twist* gives us a picture of a workhouse "administered in such a way that the position of least well-off independent workers should not be worse than the position of those supported by parish assistance."

Among the 19th century authors defending Malthus was Harriet Martineau, who wrote: "The desire of his heart and the aim of his work were that domestic virtue and happiness should be placed within the reach of all... He found that a portion of the people were underfed, and that one consequence of this was a fearful mortality among infants; and another consequence the growth of a recklessness among the destitute which caused infanticide, corruption of morals, and at best, marriage between pauper boys and girls; while multitudes of respectable men and women, who paid rates instead of consuming them, were unmarried at forty or never married at all. Prudence as to time of marriage and for making due provision for it was, one would think, a harmless recommendation enough, under the circumstances."

At the end of the 19th century, the founders of neoclassical economic theory looked back on the problems raised by Malthus and concluded that they had been overcome because of improvements in agriculture and transportation, and the opening up of new "unclaimed" lands in other parts of the world. However Alfred Marshall pointed out that in the very long run,the problems that Malthus raised would return. In his "Principles of Economics" (1890), Chapter IV, he wrote: "... it was not Malthus' fault that he could not foresee the great developments of steam transport by land and by sea, which have enabled Englishmen of the present generation to obtain the products of the richest lands of the earth at comparatively small cost... But... [i]t remains true that unless the checks on the growth of population in force at the end of the nineteenth century are on the whole increased... it will be impossible for the habits of comfort prevailing in Western Europe to spread themselves over the whole world and maintain themselves for many hundred years". Then there follows a footnote in which Malthus first extrapolates the then population size to find that before 2120 it will be 6 billion; he then assumes further innovation, but eventually says that: "... the pressure of population on the means of subsistence may be held in check for about two hundred years, but not longer."

The Irish Potato Famine of 1845

Meanwhile, in Ireland, a dramatic series of events had occurred, confirming the ideas of Malthus. Anti-Catholic laws prevented the Irish cottagers from improving their social position; and instead they produced large families, fed almost exclusively on a diet of milk and potatoes. The potato and milk diet allowed a higher density of population to be supported in Ireland than would have been the case if the Irish diet had consisted primarily of wheat. As a result, the population of Ireland grew rapidly: In 1695 it had been approximately one million, but by 1821 it had reached 6,801,827. By 1845, the population of Ireland was more than eight million; and in that year the potato harvest failed because of blight. All who were able to do so fled from the country, many emigrating to the United States; but two million people died of starvation. As the result of this shock, Irish marriage habits changed, and late marriage became the norm, just as Malthus would have wished. After the Potato Famine of 1845, Ireland maintained a stable population of roughly four million.

Malthus continued a life of quiet scholarship, unperturbed by the heated public debate which he had caused. At the age of 38, he married a second cousin. The marriage produced only three children, which at that time was considered to be a very small number. Thus he practiced the pattern of late marriage which he advocated. Although he was appointed rector of a church in Lincolnshire, he never preached there, hiring a curate to do this in his place. Instead of preaching, Malthus accepted an appointment as Professor of History and Political Economy at the East India Company's College at Haileybury. This appointment made him the first professor of economics in England, and probably also the first in the world. Among the important books which he wrote while he held this post was *Principles of Political Economy, Considered with a View to their Practical Application*. Malthus also published numerous revised and expanded editions of his *Essay on the Principle of Population*. The third edition was published in 1806, the fourth in 1807, the fifth in 1817, and the sixth in 1826.

In the societies that Malthus describes, we can see a clear link not only between population pressure and poverty, but also between population pressure and war. Undoubtedly this is why the suffering produced by poverty and war saturates so much of human history. Stabilization of population through birth control offers a key to eliminating this suffering.

We will return to the ideas of Malthus in Chapter 2, since they are extremely relevant to the problems of the 21st century.

Ricardo's theory of rent

Among Malthus' closest friends was the financier David Ricardo (1772-1823). Ricardo had been born into a Jewish family that had moved to London from Portugal. However, at the age of 21 he had broken relations with his family and rejected his orthodox Jewish faith in order to marry a Quaker girl. Ricardo, who had worked with his father on the London Sock Exchange since the age of 14, then proceeded to become a financier in his own right, amassing a fortune worth over a million pounds, in those days an immense sum.

Having read a copy of Adam Smith's *Wealth of Nations*, Ricardo became interested in theoretical economics, and at the age of 37 he began to write about this subject. His articles and books were admired by Malthus, and the two became close friends, although they disagreed on many issues.

Malthus had been brought up as a member of the British landowning class. He valued the beauty of the countryside, and was disturbed by the growth of industrialism. By contrast, Ricardo's sympathies lay with the rising and vigorous class of industrialists. The theory of rent, developed by Ricardo, showed that there is an inevitable conflict between these two classes.

Ricardo's theory of rent dealt with the effect of economic growth on prices, wages and profits. He and Malthus both agreed with Adam Smith's picture of growth: The virtuous industrialist does not spend his profits on luxuries, but instead reinvests them. New factories are built, the demand for workers increases, wages rise, and more workers are "produced" in response to the demand, i.e., more of the worker's children survive, and their numbers grow.

With each turn of the spiral of economic growth, there is an increased demand for food, since the population of workers increases. The most fertile land is already in use, but to meet the larger demand for food, marginal land is tilled, for example land on steep hillside slopes. It costs more to grow grain on marginal land, and therefore grain prices rise. According to Ricardo, the only people who benefit from economic growth are the owners of especially fertile land. The factory owners do not benefit, because they must pay higher wages to meet the increased price of food for their workers, and their profits remain the same. The workers do not benefit, because regardless of the price of grain, each of them is given only enough food to survive. The true beneficiaries of economic growth, according to Ricardo, are the owners of the most fertile land, i.e., the landowning aristocracy.

Ricardo defines "rent" to be the difference, per acre, between the cost of growing grain on good land, and the cost on marginal land. This difference is pocketed by the owners of good land. They do not really deserve it because ownership of fertile land is something that they inherited, rather than something that they produced by their own efforts.

The Corn Laws

At the time when Ricardo was writing, imports of cheap foreign grain were effectively blocked by the Corn Laws, a series of acts of Parliament which were in force between 1815 and 1846. These laws imposed prohibitively high tariffs on the import of foreign grain. Ricardo's theory of rents showed that the Corn Laws benefited the landowning aristocracy at the expense of the industrialists. His sympathies were with the industrialists, because he felt that the Corn Laws were forcing England back into feudalism and economic stagnation. By contrast, Malthus favored the Corn laws because he felt that it was dangerous for England to become dependent on imports of foreign grain. What would the country do in case of war?, Malthus asked. What would England do if it lost its industrial edge and became unable to export its manufactured products? How would the country then support its overgrown population?

In the end, the aristocracy lost its control of Parliament, the Corn Laws were repealed, and the population of England continued to grow. It has grown from 8.3 million in 1801, the year of the first census, to 50.7 million in 2006. Today, England could not possibly support its population on home-grown food. Like the Netherlands and Japan, Britain is dependent on exports of manufactured goods and imports of grain.

The Iron Law of Wages

Ricardo believed that the "natural price" of any commodity is the lowest possible cost of its production, and that in the long run, prices of any commodity would approach this natural value. When he applied this idea to labor, the result was his "Iron Law of Wages". Since the lowest cost of "producing" workers is the cost of keeping them alive at the subsistence level, he reasoned, the natural price of labor is determined by the lowest possible cost of sustenance. If workers are paid less than this, they will die, their numbers will decrease, the demand for workers will increase, and the price of labor will rise. If they are paid more, a greater number of their children will survive, the number of workers will increase above demand, and wages will fall. According to this argument, starvation wages are inevitable.

Ricardo's reasoning assumes industrialists to be completely without social conscience or governmental regulation; it fails to anticipate the development of trade unionism; and it assumes that the working population will multiply without restraint as soon as their wages rise above the starvation level. This was an accurate description of what was happening in England during Ricardo's lifetime, but it obviously does not hold for all times and all places.

The Reform Movement

The slow acceptance of birth control in England

With the gradual acceptance of birth control in England, the growth of trade unions, the passage of laws against child labor and finally minimum wage laws, conditions of workers gradually improved, and the benefits of industrialization began to spread to the whole of society.

One of the arguments which was used to justify the abuse of labor was that the alternative was starvation. The population of Europe had begun to grow rapidly for a variety of reasons: - because of the application of scientific knowledge to the prevention of disease; because the potato had been introduced into the diet of the poor; and because bubonic plague had become less frequent after the black rat had been replaced by the brown rat, accidentally imported from Asia.

It was argued that the excess population could not be supported unless workers were employed in the mills and factories to produce manufactured goods, which could be exchanged for imported food. In order for the manufactured goods to be competitive, the labor which produced them had to be cheap: hence the abuses. (At least, this is what was argued).

Industrialization benefited England, but in a very uneven way, producing great wealth for some parts of society, but also extreme misery in other social classes. For many, technical progress by no means led to an increase of happiness. The persistence of terrible poverty in 19th-century England, and the combined pessimism of Ricardo and Malthus, caused Thomas Carlyle to call economics "the Dismal Science".

Among the changes which were needed to insure that the effects of technical progress became beneficial rather than harmful, the most important were the abolition of child labor, the development of unions, the minimum wage law, and the introduction of birth control.

Francis Place (1771-1854), a close friend of William Godwin and James Mill, was one of the earliest and most courageous pioneers of these needed changes. Place had known extreme poverty as a child, but he had risen to become a successful businessman and a leader of the trade union movement.

Place and Mill were Utilitarians, and like other members of this movement they accepted the demographic studies of Malthus while disagreeing with Malthus' rejection of birth control. They reasoned that since abortion and infanticide were already widely used by the poor to limit the size of their families, it was an indication that reliable and humane methods of birth control would be welcome. If marriage could be freed from the miseries which resulted from excessive numbers of children, the Utilitarians believed, prostitution would become less common, and the health and happiness of women would be improved.

Francis Place and James Mill decided that educational efforts would be needed to make the available methods of birth control more widely known and accepted. In 1818, Mill cautiously wrote "The great problem of a real check to population growth has been miserably evaded by all those who have meddled with the subject... And yet, if the superstitions of the nursery were discarded, and the principle of utility kept steadily in view, a solution might not be very difficult to be found."

Fig. 1.8 The Utilitarian philosopher and economist James Mill (1773-1836) was an early advocate of birth control. (He was the father of John Stuart Mill.) (Public domain.)

A few years later, Mill dared to be slightly more explicit: "The result to be aimed at", he wrote in his *Elements of Political Economy* (1821), "is to secure to the great body of the people all the happiness which is capable of being derived from the matrimonial union, (while) preventing the evils which the too rapid increase of their numbers would entail. The progress of legislation, the improvement of the education of the people, and the decay of superstition will, in time, it may be hoped, accomplish the difficult task of reconciling these important objects."

In 1822, Francis Place took the considerable risk of publishing a four-page pamphlet entitled *To the Married of Both Sexes of the Working People*, which contained the following passages:

"It is a great truth, often told and never denied, that when there are too many working people in any trade or manufacture, they are worse paid than they ought to be paid, and are compelled to work more hours than they ought to work. When the number of working people in any trade or manufacture has for some years been too great, wages are reduced very low, and the working people become little better than slaves."

"When wages have thus been reduced to a very small sum, working people can no longer maintain their children as all good and respectable people wish to maintain their children, but are compelled to neglect them; - to send them to different employments; - to Mills and Manufactories, at a very early age. The miseries of these poor children cannot be described, and need not be described to you, who witness them and deplore them every day of your lives."

"The sickness of yourselves and your children, the privation and pain and premature death of those you love but cannot cherish as you wish, need only be alluded to. You know all these evils too well."

"And what, you will ask, is the remedy? How are we to avoid these miseries? The answer is short and plain: the means are easy. Do as other people do, to avoid having more children than they wish to have, and can easily maintain."

"What is to be done is this. A piece of soft sponge is tied by a bobbin or penny ribbon, and inserted just before the sexual intercourse takes place, and is withdrawn again as soon as it has taken place. Many tie a sponge to each end of the ribbon, and they take care not to use the same sponge again until it has been washed. If the sponge be large enough, that is, as large as a green walnut, or a small apple, it will prevent conception... without diminishing the pleasures of married life..."

"You cannot fail to see that this address is intended solely for your good. It is quite impossible that those who address you can receive any benefit from it, beyond the satisfaction which every benevolent person and true Christian, must feel, at seeing you comfortable, healthy and happy."

The publication of Place's pamphlet in 1822 was a landmark in the battle for the acceptance of birth control in England. Another important step was taken in 1832, when a small book entitled *The Fruits of Philosophy or, the Private Companion of Young Married People* was published by a Boston physician named Dr. Charles Knowlton. The book contained simple contraceptive advice. It reviewed the various methods of birth control available at the time. In order for the sponge method to be reliable, Knowlton's book pointed out, use of a saline douching solution was necessary.

The battle for these social reforms was not easily won. For example, in 1876, "The Fruits of Philosophy" was ruled by an English court to be obscene, and a bookseller was sentenced to two years imprisonment for distributing it. The liberal politician Charles Bradlaugh and his friend, the feminist author Annie Besant then decided to provoke a new trial by selling the book themselves. They wrote polite letters to the Chief Clerk of the Magistrates, the Detective Department, and the City Solicitor announcing the time and the place at which they intended to sell the book, and they asked to be arrested. The result was a famous trial in which the two reformers were acquitted, but the jury again ruled " The Fruits of Philosophy" to be obscene.

As the nineteenth century progressed, birth control gradually came to be accepted in England, and the average number of children per marriage fell from 6.16 in 1860 to 4.13 in 1890. By 1915 this figure had fallen to 2.43. Because of lowered population pressure, combined with the growth of trade unions and better social legislation, the condition of England's industrial workers improved; and under the new conditions, Ricardo's Iron Law of Wages fortunately no longer seemed to hold.

Trade unions and child labor laws

Nor was the battle to establish trade unions easily won. At the start of the 19th century, many countries had laws prohibiting organizing unions, and these invoked penalties up to and including death. In England, the Reform Act of 1832 made unions legal, but nevertheless in 1834, six men from Dorset who had formed the "Friendly Society of Agricultural Workers" were arrested and sentenced to a seven years' transportation to Australia. An obscure law from 1797 was invoked, which prohibited swearing secret oaths. This they had in fact done, but their main crime seems to have been refusing to work for less than 10 shillings a week. Despite bitter opposition, trade unions gradually developed both in England and in other industrial countries.

One of the important influences for reform was the Fabian Society, founded in London in 1884. The group advocated gradual rather than revolutionary reform (and took its name from Quintus Fabius Maximus, the Roman general who defeated Hannibal's Carthaginian army by using harassment and attrition rather than head-on battles). The Fabian Society came to include a number of famous people, including Sydney and Beatrice Webb, George Bernard Shaw, H.G. Wells, Annie Besant, Leonard Woolf, Emaline Pankhurst, Bertrand Russell, John Maynard Keynes, Harold Laski, Ramsay MacDonald, Clement Attlee, Tony Benn and Harold Wilson. Jawaharlal Nehru, India's first Prime Minister, was greatly influenced by Fabian economic ideas.

The group was instrumental in founding the British Labour Party (1900), the London School of Economics and the New Statesman. In 1906, Fabians lobbied for a minimum wage law, and in 1911 they lobbied for the establishment of a National Health Service.

Adam Smith had praised division of labor as one of the main elements in industrial efficiency, but precisely this aspect of industrialism was criticized by Thomas Carlyle (1795-1891), John Ruskin (1819-1900) and William Morris (1834-1896). They considered the numbingly repetitive work of factory laborers to be degrading, and they rightly pointed out that important traditions of design were being lost and replaced by ugly mass produced

Fig. 1.9 Beatrice Webb (1858-1943). Together with her husband Sidney Webb, Graham Wallace and George Bernard Shaw, she founded the London School of Economics using money left to the Fabian Society by Henry Hutchinson. The Fabians also founded the British Labour Party, and they lobbied for a minimum wage law and National Health Service. (Public domain.)

artifacts. The Arts and Crafts movement founded by Ruskin and Morris advocated cooperative workshops, where creative freedom and warm human relationships would make work rewarding and pleasant. In several Scandinavian countries, whose industrialization came later than England's, efforts were made to preserve traditions of design. Hence the present artistic excellence of Scandinavian furniture and household articles.

Through the influence of reformers, the more brutal aspects of Adam Smith's economic model began to be moderated. Society was learning that free market mechanisms alone do not lead to a happy and just society. In addition, ethical and ecological considerations and some degree of governmental regulation are also needed.

The Reform Movement aimed at social goals, but left ecological problems untreated. Thus our economic system still does not reflect the true price to society of environmentally damaging activities. For example, the price of coal does not the reflect the cost of the environmental damage done by burning it. This being so, our growth-worshiping economic system of today thunders ahead towards an environmental mega-catastrophe, as we will see in the next chapter.

Suggestions for further reading

(1) John Fielden, *The Curse of the Factory System*, (1836).
(2) A. Smith, *The Theory of Moral Sentiments...* (1759), ed. D.D. Raphael and A.L. MacPhie, Clarendon, Oxford, (1976).
(3) A. Smith, *An Inquiry into the Nature and Causes of the Wealth of Nations* (1776), Everyman edn., 2 vols., Dent, London, (1910).
(4) Charles Knowlton *The Fruits of Philosophy, or The Private Companion of Young Married People*, (1832).
(5) John A. Hobson, *John Ruskin, Social Reformer*, (1898).
(6) E. Pease, *A History of the Fabian Society*, Dutton, New York, (1916).
(7) G. Claeys, ed., *New View of Society, and other writings by Robert Owen*, Penguin Classics, (1991).
(8) W. Bowden, *Industrial Society in England Towards the End of the Eighteenth Century*, MacMillan, New York, (1925).
(9) G.D. Cole, *A Short History of the British Working Class Movement*, MacMillan, New York, (1927).
(10) P. Deane, *The First Industrial Revolution*, Cambridge University Press, (1969).
(11) Marie Boaz, *Robert Boyle and Seventeenth Century Chemistry*, Cambridge University Press (1958).
(12) J.G. Crowther, *Scientists of the Industrial Revolution*, The Cresset Press, London (1962).
(13) R.E. Schofield, *The Lunar Society of Birmingham*, Oxford University Press (1963).
(14) L.T.C. Rolt, *Isambard Kingdom Brunel*, Arrow Books, London (1961).
(15) J.D. Bernal, *Science in History*, Penguin Books Ltd. (1969).
(16) Bertrand Russell, *The Impact of Science on Society*, Unwin Books, London (1952).
(17) Wilbert E. Moore, *The Impact of Industry*, Prentice Hall (1965).
(18) Charles Morazé, *The Nineteenth Century*, George Allen and Unwin Ltd., London (1976).
(19) Carlo M. Cipolla (editor), *The Fontana Economic History of Europe*, Fontana/Collins, Glasgow (1977).
(20) Martin Gerhard Geisbrecht, *The Evolution of Economic Society*, W.H. Freeman and Co. (1972).
(21) P.N. Stearns, *The Industrial Revolution in World History*, Westvieiw Press, (1998).
(22) E.P. Thompson, *The Making of the English Working Class*, Pennguin Books, London, (1980).
(23) N.J. Smelser, *Social Change and the Industrial Revolution: An Application of Theory to the British Cotton Industry*, University of Chicago Press, (1959).

(24) D.S. Landes, *The Unbound Prometheus: Technical Change and Industrial Development in Western Europe from 1750 to the Present*, 2nd ed., Cambridge University Press, (2003).
(25) S. Pollard, *Peaceful Conquest: The Industrialization of Europe, 1760-1970*, Oxford University Press, (1981).
(26) M. Kranzberg and C.W. Pursell, Jr., eds., *Technology in Western Civilization*, Oxford University Press, (1981).
(27) M.J. Daunton, *Progress and Poverty: An Economic and Social History of Britain, 1700-1850*, Oxford University Press, (1990).
(28) L.R. Berlanstein, *The Industrial Revolution and Work in 19th Century Europe*, Routledge, (1992).
(29) J.D. Bernal, *Science and Industry in the 19th Century*, Indiana University Press, Bloomington, (1970).
(30) P.A. Brown, *The French Revolution in English History*, 2nd edn., Allen and Unwin, London, (1923).
(31) E. Burke, *Reflections on the Revolution in France and on the Proceedings of Certain Societies in London Relative to that Event...*, Dent, London, (1910).
(32) J.B. Bury, *The Idea of Progress*, MacMillan, New York, (1932).
(33) I.R. Christie, *Stress and Stability in Late Eighteenth Century Britain; Reflections on the British Avoidance of Revolution* (Ford Lectures, 1983-4), Clarendon, Oxford, (1984).
(34) H.T. Dickenson, *Liberty and Property, Political Ideology in Eighteenth Century Britain*, Holmes and Meier, New York, (1977).
(35) W. Eltis, *The Classical Theory of Economic Growth*, St. Martin's, New York, (1984).
(36) E. Halévy, *A History of the English People in the Nineteenth Century*, (transl. E.I. Watkin), 2nd edn., Benn, London, (1949).
(37) E. Halévy, *The Growth of Philosophic Radicalism*, (transl. M. Morris), new edn., reprinted with corrections, Faber, London, (1952).
(38) W. Hazlitt, *The Complete Works of William Hazlitt*, ed. P.P. Howe, after the edition of A.R. Walker and A. Glover, 21 vols., J.M. Dent, London, (1932).
(39) W. Hazlitt, *A Reply to the Essay on Population by the Rev. T.R. Malthus...*, Longman, Hurst, Rees and Orme, London, (1807).
(40) R. Heilbroner, *The Worldly Philosophers: The Lives, Times and Ideas of the Great Economic Thinkers*, 5th edn., Simon and Schuster, New York, (1980).
(41) R.K. Kanth, *Political Economy and Laissez-Faire: Economics and Ideology in the Ricardian Era*, Rowman and Littlefield, Totowa N.J., (1986).
(42) J.M. Keynes, *Essays in Biography*, in *The Collected Writings of John Maynard Keynes*, MacMillan, London, (1971-82).

(43) F. Knight, *University Rebel: The Life of William Frend, 1757-1841*, Gollancz, London (1971).
(44) M. Lamb, and C. Lamb, *The Works of Charles and Mary Lamb*, ed. E.V. Lucas, 7 vols., Methuen, London, (1903).
(45) A. Lincoln, *Some Political and Social Ideas of English Dissent, 1763-1800*, Cambridge University Press, (1938).
(46) D. Locke, *A Fantasy of Reason: The Life and Thought of William Godwin*, Routledge, London, (1980).
(47) J. Locke, *Two Treatises on Government. A Critical Edition with an Introduction and Apparatus Criticus*, ed. P. Laslett, Cambridge University Press, (1967).
(48) J. Macintosh, *Vindicae Gallicae. Defense of the French Revolution and its English Admirers against the Accusations of the Right Hon. Edmund Burke...*, Robinson, London, (1791).
(49) J. Macintosh, *A Discourse on the Study of the Law of Nature and of Nations*, Caldell, London, (1799).
(50) T. Paine, *The Rights of Man: being an Answer to Mr. Burke's Attack on The French Revolution*, Jordan, London, part I (1791), part II (1792).
(51) H.G. Wells, *Anticipations of the Reaction of Mechanical and Scientific Progress on Human Life and Thought*, Chapman and Hall, London, (1902).
(52) B. Wiley, *The Eighteenth Century Background: Studies of the Idea of Nature in the Thought of the Period*, Chatto and Windus, London, (1940).
(53) G.R. Morrow, *The Ethical and Economic Theories of Adam Smith: A Study in the Social Philosophy of the 18th Century*, Cornell Studies in Philosophy, **13**, 91-107, (1923).
(54) H.W. Schneider, ed., *Adam Smith's Moral and Political Philosophy*, Harper Torchbook edition, New York, (1948).
(55) F. Rosen, *Classical Utilitarianism from Hume to Mill*, Routledge, (2003).
(56) J.Z. Muller, *The Mind and the Market: Capitalism in Western Thought*, Anchor Books, (2002).
(57) J.Z. Muller, *Adam Smith in His Time and Ours: Designing the Decent Society*, Princeton University Press, (1995).
(58) S. Hollander, *The Economics of Adam Smith*, University of Toronto Press, (19773).
(59) K. Haakonssen, *The Cambridge Companion to Adam Smith*, Cambridge University Press, (2006).
(60) K. Haakonssen, *The Science of a Legeslator: The Natural Jurisprudence of David Hume and Adam Smith*, Cambridge University Press, (1981).

(61) I. Hont and M. Ignatieff, *Wealth and Virtue: The Shaping of Political Economy in the Scottish Enlightenment*, Cambridge University Press, (1983).
(62) I.S. Ross, *The Life of Adam Smith*, Clarendon Press, Oxford, (1976).
(63) D. Winch, *Adam Smith's Politics: An Essay in Historiographic Revision*, Cambridge University Press, (1979).

Chapter 2

Threats to the Environment and Climate Change

"Some of the potential risks could be irreversible and could accelerate the process of global warming. Melting of permafrost in the Arctic could lead to the release of huge quantities of methane. Dieback of the Amazon forest could mean that the region starts to emit rather than to absorb greenhouse gases. These feedbacks could lead to warming that is at least twice as fast as current high-emission projections, leading to temperatures higher than seen in the last 50 million years. There are still uncertainties about how much warming would be needed to trigger these abrupt changes. Nevertheless, the consequences would be catastrophic if they do occur."

Stern Report Discussion Paper, January 31, 2006

Introduction

The worst dangers from a disastrous increase in global temperatures lie in the distant future; but to avoid them, action must be taken immediately. The huge subsidies currently given to fossil fuel companies must be abolished, or, better yet, shifted to the support of renewable energy.

In the long-term future (in several hundred years) climate change threatens to produce ocean level rises which will drown most of the world's coastal cities, and which will wipe out countries such as Bangladesh and Holland. At the same time, increases in temperature will make large parts of the Middle East, India and Africa uninhabitable.

Hope that catastrophic climate change can be avoided comes from the exponentially growing world-wide use of renewable energy and from the fact prominent public figures, such as Pope Francis, Leonardo DiCaprio, Elon Musk, Bill McKibben, Naomi Klein and Al Gore, are making the public increasingly aware of the long-term dangers. Short-term disasters due to climate change may also be sufficiently severe to wake us up.

We must work with dedication to save the future for our grandchildren and their grandchildren, a future, which we share with all other living creatures on earth. We must accept our responsibility for the long-term future of human civilization and the biosphere.

Malthus revisited

Avoiding the grim Malthusian forces

Malthus died in Bath in 1834, but debate on his ideas continued to rage, both in his own century and our own. Each year he is refuted, and each year revived. Despite impressive scientific progress since his time, the frightful Malthusian forces - poverty, famine, disease, and war - cast as dark a shadow in our own times as they did in the nineteenth century. Indeed, the enormous power of modern weapons has greatly intensified the dangers posed by war; and the rapid growth of global population has given new dimensions to the problems of poverty and famine.

Looking at the world today, we can see regions where Malthus seems to be a truer prophet than Condorcet and Godwin. In most developing countries, poverty and disease are still major problems. In other parts of the world, the optimistic prophecies of Condorcet and Godwin have been at least partially fulfilled. In the industrialized nations, Godwin's prophecy of automated agriculture has certainly come true. In the nations of the North, only a small percentage of the population is engaged in agriculture, while most of the citizens are free to pursue other goals than food production.

Scandinavia is an example of an area where poverty and war have both been eliminated locally, and where death from infectious disease is a rarity. These achievements would have been impossible without the low birth rates which also characterize the region. In Scandinavia, and in other similar regions, low birth rates and death rates, a stable population, high educational levels, control of infectious disease, equal status for women, democratic governments, and elimination of poverty and war are linked together in a mutually re-enforcing circle of cause and effect. By contrast, in many large third-world cities, overcrowding, contaminated water, polluted air, dense population without adequate sanitation, low status of women, high birth rates, rapidly increasing population, high unemployment levels, poverty, crime, ethnic conflicts, and resurgence of infectious disease are also linked in a self-perpetuating causal loop - in this case a vicious circle.

Population stabilization and sustainability

Does the contrast between the regions of our contemporary world mean that Malthus has been "proved wrong" in some regions and "proved right" in others? To answer this question, let us re-examine the basic assertion which Malthus puts forward in Books I and II of the 1803 version of his *Essay*. His basic thesis is that the maximum natural fertility of human populations is greatly in excess of replacement fertility. This being so, Malthus points out, human populations would always increase exponentially if they were not prevented from doing so by powerful and obvious checks.

In general, Malthus tells us, populations cannot increase exponentially because the food supply increases slowly, or is constant. Therefore, he concludes, in most societies and almost all periods of history, checks to population growth are operating. These checks may be positive, or they may be preventive, the positive checks being those which raise the death rate, while the preventive checks lower the birth rate. There are, however, Malthus says, exceptional periods of history when the populations of certain societies do actually increase exponentially because of the opening of new lands or because of the introduction of new methods of food production. As an example, he cites the growth of the population of the United States, which doubled every 25 years over a period of 150 years.

We can see, from this review of Malthus' basic thesis, that his demographic model is flexible enough to describe all of the regions of our contemporary world: If Malthus were living today, he would say that in countries with low birth and death rates and stable populations, the checks to population growth are primarily preventive, while in countries with high death rates, the positive checks are important. Finally, Malthus would describe our rapidly-growing global population as the natural result of the introduction of improved methods of food production in the developing countries. We should notice, however, that the flexibility of Malthus' demographic model first appears in the 1803 version of his *Essay*: In the 1798 version, he maintained "..that population does invariably increase, where there are means of subsistence.." and "that the superior power (of population) cannot be checked without producing misery and vice.." This narrower model of population did not agree with Malthus' own observations in Norway in 1799, and therefore in his 1803 *Essay* he allowed more scope for preventive checks, which included late marriage and moral restraint as well as birth control (which he classified under the heading of "vice").

Today we are able to estimate the population of the world at various periods in history, and we can also make estimates of global population in prehistoric times. Looking at the data, we can see that the global population of humans has not followed an exponential curve as a function of time, but has instead followed a hyperbolic trajectory. At the time of Christ, the population of the world is believed to have been approximately 220 million. By 1500, the earth contained 450 million people, and by 1750, the global population exceeded 700 million. As the industrial and scientific revolution has accelerated, global population has responded by increasing at a breakneck speed: In 1930, the population of the world reached two billion; in 1958 three billion; in 1974 four billion; in 1988 five billion, and in 1999, six billion. Today, roughly a billion people are being added to the world's population every fourteen years.

As the physicist Murry Gell-Mann has pointed out, a simple mathematical curve which closely approximates the global population of humans over a period of several thousand years is an hyperbola of the form

$$P = \frac{190,000,000,000}{2025 - t}$$

Here P is the population and t is the year. How are we to explain the fact that the population curve is not an exponential? We can turn to Malthus for an answer: According to his model, population does not increase exponentially, except under special circumstances, when the food supply is so ample that the increase of population is entirely unchecked. Malthus gives us a model of culturally-driven population growth. He tells us that population increase tends to press against the limits of the food supply, and since these limits are culturally determined, population density is also culturally-determined. Hunter-gatherer societies need large tracts of land for their support; and in such societies, the population density is necessarily low. Pastoral methods of food production can support populations of a higher density. Finally, extremely high densities of population can be supported by modern agriculture. Thus, the hyperbolic curve, $P=C/(2025-t)$,

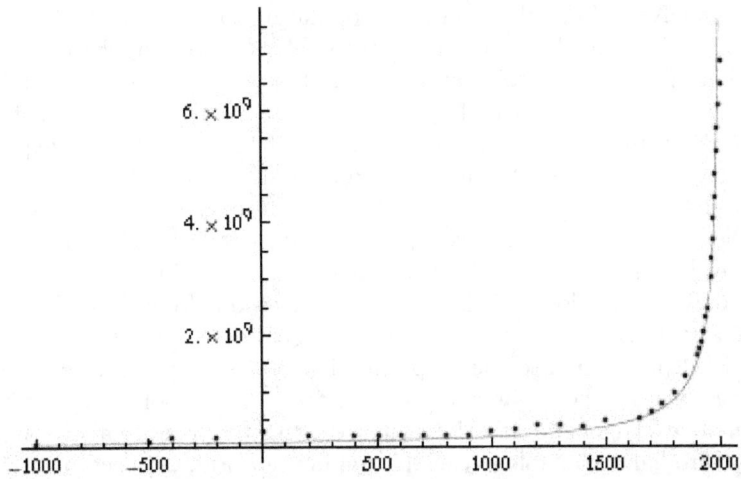

Fig. 2.1 The hyperbola C/(2025-t) compared with global population estimates from the U.S. Census Bureau. Here we choose $C = 190,000,000,000$. (Author's own graph.)

where C is a constant, should be seen as describing the rapidly-accelerating growth of human culture, this being understood to include methods of food production.

If we look at the curve, P=C/(2025-t), it is obvious that human culture has reached a period of crisis. The curve predicts that the world's population will rise to infinity in the year 2025, which of course is impossible. Somehow the actual trajectory of global population as a function of time must deviate from the hyperbolic curve, and in fact, the trajectory has already begun to fall away from the hyperbola. Because of the great amount of human suffering which may be involved, and the potentially catastrophic damage to the earth's environment, the question of how the actual trajectory of human population will come to deviate from the hyperbola is a matter of enormous importance. Will population overshoot the sustainable limit, and crash? Or will it gradually approach a maximum? In the case of the second alternative, will the checks which slow population growth be later marriage and family planning? Or will the grim Malthusian forces - famine, disease and war - act to hold the number of humans within the carrying capacity of their environment?

We can anticipate that as the earth's human population approaches 10 billion, severe famines will occur in many developing countries. The beginnings of this tragedy can already be seen. It is estimated that roughly 40,000 children now die every day from starvation, or from a combination of disease and malnutrition.

An analysis of the global ratio of population to cropland shows that we have probably already exceeded the sustainable limit of population through our dependence on petroleum: Between 1950 and 1982, the use of cheap synthetic fertilizers increased by a factor of 8. Much of our present agricultural output depends on their use, but their production is expensive in terms of energy. Furthermore, petroleum-derived synthetic fibers have reduced the amount of cropland needed for growing natural fibers, and petroleum-driven tractors have replaced draft animals which required cropland for pasturage. Also, petroleum fuels have replaced fuelwood and other fuels derived for biomass. The reverse transition, from fossil fuels back to renewable energy sources, will require a considerable diversion of land from food production to energy production. For example, 1.1 hectares are needed to grow the sugarcane required for each alcohol-driven Brazilian automobile. This figure may be compared with the steadily falling average area of cropland available to each person in the world: .24 hectares in 1950, .16 hectares in 1982.

As population increases, the cropland per person will continue to fall, and we will be forced to make still heavier use of fertilizers to increase output per hectare. Also marginal land will be used in agriculture, with the probable result that much land will be degraded through erosion and salination. Climate change will reduce agricultural output. The Hubbert peaks for oil and natural gas will occur within one or two decades, and the fossil fuel era will be over by the end of 21st century. Thus there is a danger that just as global population reaches the unprecedented level of 10 billion or more, the agricultural base for supporting it may suddenly collapse. Ecological catastrophe, possibly compounded by war and other disorders, could produce famine and death on a scale unprecedented in history - a disaster of unimaginable proportions, involving billions rather than millions of people, as will be discussed in Chapter 4.

The resources of the earth and the techniques of modern science can support a global population of moderate size in comfort and security; but the optimum size is undoubtedly smaller than the world's present population (see Chapter 4). Given a sufficiently small global population, renewable sources of energy can be found to replace disappearing fossil fuels. Technology may also be able to find renewable substitutes for many disappearing mineral resources for a global population of a moderate size. What technology cannot do, however, is to give a global population of 10 billion people the standard of living which the industrialized countries enjoy today.

What would Malthus tell us if he were alive today? Certainly he would say that we have reached a period of human history where it is vital to stabilize the world's population if catastrophic environmental degradation and famine are to be avoided. He would applaud efforts to reduce suffering by eliminating poverty, widespread disease, and war; but he would point out that, since it is necessary to stop the rapid increase of human numbers, it follows that whenever the positive checks to population growth are removed, it is absolutely necessary to replace them by preventive checks. Malthus' point of view became more broad in the successive editions of his *Essay*; and if he were alive today, he would probably agree that family planning is the most humane of the preventive checks.

In Malthus' *Essay on the Principle of Population*, population pressure appears as one of the main causes of war; and Malthus also discusses many societies in which war is one of the the principle means by which population is reduced to the level of the food supply. Thus, his *Essay* contains another important message for our own times: If he were alive today, Malthus would also say that there is a close link between the two most urgent tasks which history has given to the 21st century - stabilization of the global population, and abolition of the institution of war.

In most of the societies which Malthus described, a clear causal link can be seen, not only between population pressure and poverty, but also between population pressure and war. As one reads his *Essay*, it becomes clear why both these terrible sources of human anguish saturate so much of history, and why efforts to eradicate them have so often met with failure: The only possible way to eliminate poverty and war is to reduce the pressure of population by preventive checks, since the increased food supply produced by occasional cultural advances can give only very temporary relief. Today, the links between population pressure, poverty, and war are even more pronounced than they were in the past, because the growth of human population has brought us to the absolute limits imposed by ecological constraints.

Biology and economics

Classical economists like Smith and Ricardo pictured the world as largely empty of human activities. According to the "empty-world" picture of economics, the limiting factors in the production of food and goods are shortages of capital and labor. The land, forests, fossil fuels, minerals, oceans filled with fish, and other natural resources upon which human labor and capital operate, are assumed to be present in such large quantities that they are not limiting factors. In this picture, there is no naturally-determined upper limit to the total size of the human economy. It can continue to grow as long as new capital is accumulated, as long as new labor is provided by population growth, and as long as new technology replaces labor by automation.

Biology, on the other hand, presents us with a very different picture. Biologists remind us that if any species, including our own, makes demands on its environment which exceed the environment's carrying capacity, the result is a catastrophic collapse both of the environment and of the population which it supports. Only demands which are within the carrying capacity are sustainable. For example, there is a limit to regenerative powers of a forest. It is possible to continue to cut trees in excess of this limit, but only at the cost of a loss of forest size, and ultimately the collapse and degradation of the forest. Similarly, cattle populations may for some time exceed the carrying capacity of grasslands, but the ultimate penalty for overgrazing will be degradation or desertification of the land. Thus, in biology, the concept of the carrying capacity of an environment is extremely important; but in economic theory this concept has not yet been given the weight that it deserves.

The terminology of economics can be applied to natural resources: For example, a forest can be thought of as natural capital, and the sustainable yield from the forest as interest. Exceeding the biological carrying capacity then corresponds, in economic terms, to spending one's capital.

If it is to be prevented from producing unacceptable contrasts of affluence and misery within a society, the free market advocated by Adam Smith needs the additional restraints of ethical principles, as well as a certain amount of governmental regulation. Furthermore, in the absence of these restraints, it will destroy the natural environment of our planet.

There is much evidence to indicate that the total size of the human economy is rapidly approaching the absolute limits imposed by the carrying capacity of the global environment. For example, a recent study by Vitousek et. al. showed that 40 percent of the net primary product of land-based photosynthesis is appropriated, directly or indirectly, for human use. (The net primary product of photosynthesis is defined as the total quantity of solar energy converted into chemical energy by plants, minus the energy used by the plants themselves). Thus we are only a single doubling time away from 80 percent appropriation, which would imply a disastrous environmental degradation.

Another indication of our rapid approach to the absolute limits of environmental carrying capacity can be found in the present rate of loss of biodiversity. Biologists estimate that between 10,000 and 50,000 species are being driven into extinction each year as the earth's rainforests are destroyed.

The burning of fossil fuels and the burning of tropical rainforests have released so much carbon dioxide that the atmospheric concentration of this greenhouse gas has increased from a preindustrial value of 260 ppm to its present value: 380 ppm. Most scientists agree that unless steps are taken to halt the burning of rainforests and to reduce the use of fossil fuels, the earth's temperature will steadily rise during the coming centuries. This gradual long-term climate change will threaten future agricultural output by changing patterns of rainfall. Furthermore, the total melting of the Arctic and Antarctic icecaps, combined with the thermal expansion of the oceans, threatens to produce a sea level rise of up to 12 meters. Although these are slow, long-term effects, we owe it to future generations to take steps now to halt global warming.

The switch from fossil fuels to renewable energy sources is vital not only because of the need to reduce global warming, but also because the earth's supply of fossil fuels is limited. A peak in the production and consumption of conventional petroleum is predicted within one or two decades. Such a peak in the use of any non-renewable natural resource is called a "Hubbert peak" after the oil expert Dr. M. King Hubbert. It occurs when reserves of the resource are approximately half exhausted. After that point, the resource does not disappear entirely, but its price increases steadily because supply fails to meet demand, and because of rising extraction costs. It is predicted that the Hubbert peak for both oil and natural gas will also occur within a few decades. The peak for oil may occur within the present decade. Thus, halfway through the 21st Century, oil and natural gas will become very expensive - perhaps so expensive that they will not be burned but will instead be reserved as starting points for chemical synthesis.

The reserves of coal are much larger, and at the present rate of use they would last for slightly more than two centuries. However, it seems likely that as petroleum is exhausted, coal will be converted into liquid fuels, as was done in Germany during World War II, and in South Africa during the oil embargo. Thus, in predicting a date for the end of the fossil fuel era, we ought to lump oil, natural gas and coal together. If we do so, we find the total supply has an energy content of 1260 terawatt-years. (1 terawatt is equal to 1,000,000,000,000 Watts). One finds in this way that if they are used at the present rate of 13 terawatts, fossil fuels will last about 100 years.

Resolute government intervention is needed to promote energy conservation measures and to bring about the switch from fossil fuels to renewable energy sources, such as biomass, photovoltaics, solar thermal power, wind and wave power, and hydropower. Both subsidies for renewable energy technologies, to help them get started, and taxes on fossil fuels will be needed. Changes in tax structure could also encourage smaller families, encourage resource conservation, or diminish pollution. In general, taxation should be used, not merely to raise money, but, more importantly, to guide the evolution of society towards humane and sustainable goals.

Fossil fuel use and climate change

Melting of the polar ice caps

At present the amount of carbon in the atmosphere is increasing by about 6 gigatons per year because of human activities; and projections estimate that the CO_2 concentration will reach about 600 ppm by 2050 (more than double the preindustrial concentration). In addition to CO_2, methane, CH_4, and nitrous oxide, NO_2, are also released into the atmosphere by human activities. Anthropogenic methane comes from the production and transportation of coal, natural gas and oil, decomposition of organic wastes in municipal landfills, cultivation of rice paddies, and the raising of livestock.

The greenhouse gases (which include water vapor, carbon dioxide, methane, ozone, nitrous oxide, sulfur hexafluoride, hydroflurocarbons, perflurocarbons and many other gases) absorb a part of the infrared radiation from the earth's surface, which otherwise would have been sent directly into outer space. Part of this energy is re-radiated into space, but a part is sent downward to the earth, where it is absorbed. The result is that the earth's surface is much warmer than it otherwise would be. The mechanism is much the same as that of a greenhouse, where the glass absorbs and re-radiates infrared radiation. A moderate greenhouse effect on earth is helpful to life, but climatologists believe that anthropogenic CO_2 and CH_4 emissions may produce a dangerous amount of global warming during the next few centuries.

According to the Intergovernmental Panel on Climate Change the percentages of greenhouse gas emissions contributed by various human activities are as follows:

$$\text{Energy use} \begin{cases} \text{Transportation} & 13.5\% \\ \text{Electricity and heat} & 24.6\% \\ \text{Other fuel combustion} & 9.0\% \\ \text{Industry} & 10.4\% \\ \text{Fugitive emissions} & 3.9\% \end{cases}$$

$$\text{Other sources} \begin{cases} \text{Industrial processes} & 3.4\% \\ \text{Land use change} \\ \text{(deforestation)} & 18.2\% \\ \text{Agriculture} & 13.5\% \\ \text{Waste} & 3.6\% \end{cases}$$

In thinking about global warming, it is important to remember that it is a very slow and long-term phenomenon. Stephen H. Schneider and Janica Lane of Stanford University, in an article entitled *An Overview of 'Dangerous' Climate Change* include a figure that emphasizes the long-term nature of global warming. The figure presupposes that CO_2 emissions will peak within 50 years and will thereafter be reduced. According to the figure, it will still take more than a century for the level of CO_2 in the atmosphere to stabilize. The establishment of temperature equilibrium will require several centuries. Sea level rises due to thermal expansion of ocean water will not be complete before the end of the millennium, while sea level rises due to melting of the polar icecaps might not be complete for several millennia!

It is worrying to think that total melting of the Greenland ice cap, which some authors think might begin in earnest during the 22nd century, would result in a sea level rise of up to 7 meters. Of course, society would have some time to adjust to this event. But a glance at maps and elevations makes one realize the extent of such a catastrophe and the importance of preventing it.

The IPCC and Stern reports

Models put forward by the Intergovernmental Panel on Climate Change (IPCC, 2007 Report) suggest that if no steps are taken to reduce carbon emissions, a temperature increase of 1.4-5.6°C will occur by 2100.[a] Global warming may have some desirable effects, such as increased possibilities for agriculture in Canada, Sweden and Siberia. However, most of the expected effects of global warming will be damaging. These unwanted effects include ocean level rises, extreme weather conditions (such as heat waves, hurricanes and tropical cyclones), changes in the patterns of ocean currents, melting of polar ice and glaciers, abnormal spread of diseases, extinctions

---
[a]Relative to 1990 temperatures.

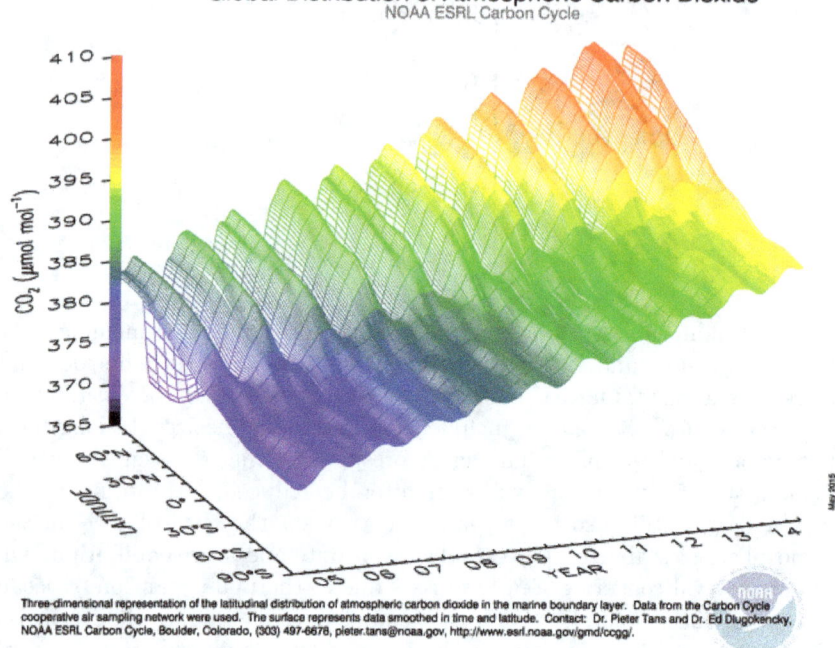

Fig. 2.2 Carbon Dioxide observations from 2005 to 2014 showing the seasonal variations and the difference between northern and southern hemispheres. (Public domain.)

of plant and animal species, together with aridity and crop failures in some areas of the world which are now able to produce and export large quantities of grain.

According to a report presented to the Oxford Institute of Economic Policy by Sir Nicholas Stern on 31 January, 2006, areas likely to lose up to 30% of their rainfall by the 2050's because of climate change include much of the United States, Brazil, the Mediterranean region, Eastern Russia and Belarus, the Middle East, Southern Africa and Southern Australia. Meanwhile rainfall is predicted to increase up to 30% in Central Africa, Pakistan, India, Bangladesh, Siberia, and much of China.

Stern and his team point out that "We can... expect to see changes in the Indian monsoon, which could have a huge impact on the lives of hundreds of millions of people in India, Pakistan and Bangladesh. Most climate models suggest that the monsoon will change, although there is still uncertainty about exactly how. Nevertheless, small changes in the

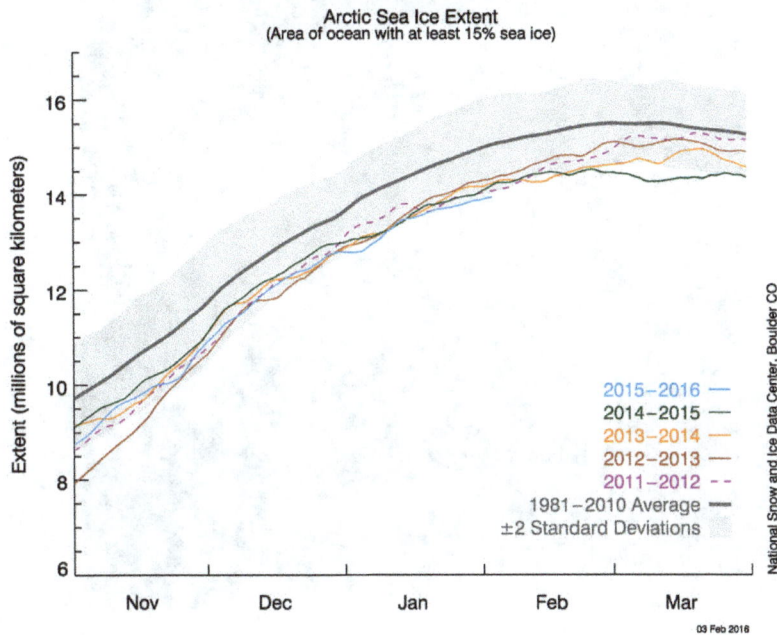

Fig. 2.3 Arctic sea ice extent as of February 3, 2016. January Arctic sea ice extent was the lowest in the satellite record. credit: NSIDC. The rapid nonlinear loss of Arctic sea ice has surprised IPCC scientists. (Public domain.)

monsoon could have a huge impact. Today, a fluctuation of just 10% in either direction from average monsoon rainfall is known to cause either severe flooding or drought. A weak summer monsoon, for example, can lead to poor harvests and food shortages among the rural population - two-thirds of India's almost 1.1 billion people. Heavier-than-usual monsoon downpours can also have devastating consequences..."

In some regions, melting of glaciers can be serious from the standpoint of dry-season water supplies. For example, melts from glaciers in the Hindu Kush and the Himalayas now supply much of Asia, including China and India, with a dry-season water supply. Complete melting of these glacial systems would cause an exaggerated runoff for a few decades, after which there would be a drying out of some of the most densely populated regions of the world.

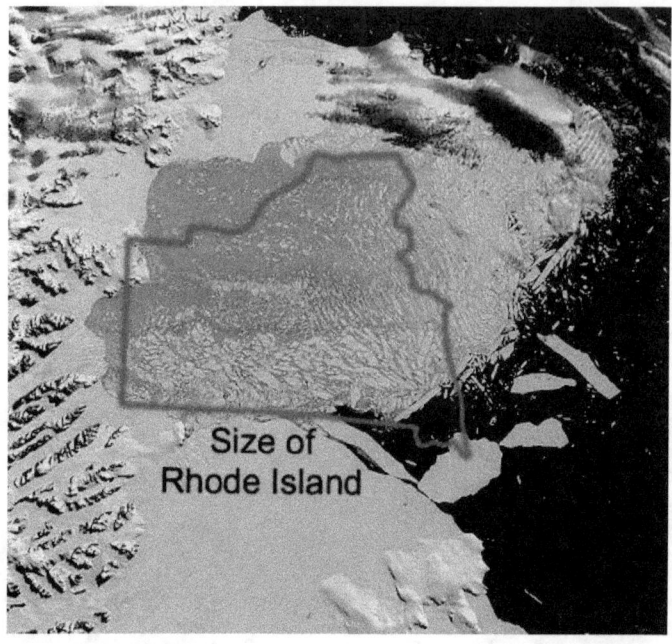

Fig. 2.4 The collapsing Larsen-B iceshelf in the Antarctic is similar in size to the US state of Rhode Island. (From Wikipedia.)

The threat of feed-back loops

The Discussion Paper presented by Stern on January 31, 2006, also notes that "Some of the potential risks could be irreversible and accelerate the process of global warming. Melting of permafrost in the Arctic could lead to the release of huge quantities of methane. Dieback of the Amazon forest could mean that the region starts to emit rather than absorb greenhouse gases. These feedbacks could lead to warming that is at least twice as fast as current high-emissions projections, leading to temperatures higher than seen in the past 50 million years. There are still uncertainties about how much warming would be needed to trigger these abrupt changes. Nevertheless, the consequences would be catastrophic if they do occur."

The much larger (700 page) Stern Report was made public on October 30, 2006. It explores not only the scientific basis for predictions of global warming but also the possible economic consequences. Unless we act promptly to prevent it, the Stern Report states, global warming could render swaths of the planet uninhabitable, and do economic damage equal to that inflicted by the two world wars.

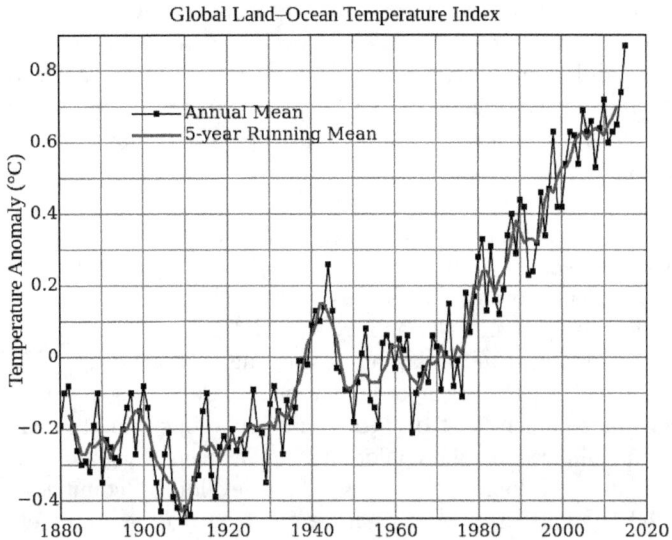

Fig. 2.5 Line plot of global mean land-ocean surface temperature index, 1880 to present, with the base period 1951-1980. The black line is the annual mean and the red line is the five-year running mean. 2016 is predicted to be the hottest year ever recorded, 1.3 °C above the 19th century baseline. (Public domain.)

A large United Nations Climate Conference (COP15) took place in Copenhagen from December 7 to December 18, 2009. In order to make the latest results of researchers available to the 15,000 expected participants, a preliminary meeting of scientists was held at the University of Copenhagen in March, 2009. 2,500 delegates from 80 countries attended the meeting. Among the conclusions of this international congress of scientists were the following:

- **Climatic trends**: "Recent observations confirm that, given the high rates of observed emissions, the worst-case IPCC scenario trajectories (or even worse) are being realized. For many key parameters, the climate system is already moving beyond the patterns of natural variability within which our society and economy developed and thrived. These parameters include global mean surface temperature, sea-level rise, ocean and ice sheet dynamics, ocean acidification and extreme climate events. There is a significant risk that many of the trends will accelerate, leading to an increasing risk of abrupt or irreversible climatic shifts."

- **Social disruption**: "Recent observations show that societies are highly vulnerable to even modest levels of climate change, with poor nations and communities particularly at risk. Temperature rises above 2°C will be very difficult for contemporary societies to cope with, and will increase the level of climate disruption through the rest of the century."

To avoid temperature increases of more than 2°C, the scientists said that it will be necessary for the world to reduce its CO_2 emissions by 90% by 2050. In other words if dangerous climate change is to be avoided, the fossil fuel era must essentially end by that date.

Despite these clear and unanimous warnings from the scientific community, the United Nations climate conference in Copenhagen failed to reach an agreement sufficiently strong to avoid dangerous climate change. The problem encountered by the conference was a deep disagreement between developed and developing countries. The developing countries correctly maintained that historically, they have not been to blame for emission of greenhouse gases. Meanwhile, the industrialized countries pointed to the future, saying (also correctly) that unless the developing countries accepted their future responsibilities, there would be no hope of avoiding disaster.

At the last moment, the United States, China, India, Brazil and South Africa hammered out a weak agreement, the *Copenhagen Accord*, which the other nations at the conference agreed to "take note of". The Copenhagen Accord recognizes the aim of limiting global warming to 2°C. However, it does not provide mechanisms sufficiently strong to reach that goal. Another UN climate conference will be held in Mexico in November, 2010, and it is to be hoped that during the intervening months, negotiators will be able to build on the very modest results of COP15 and put together an adequate and legally binding treaty.

Geological extinction events and runaway climate change

The melting of Arctic sea ice is taking place far more rapidly than was predicted by IPCC reports. David Wasdell, Director of the Apollo-Gaia Project, points out that the observed melting has been so rapid that within less than five years, the Arctic will be free of sea ice at the end of each summer. It will, of course continue to re-freeze during the winters, but the thickness and extent of the winter ice will diminish.

It has also been observed that both the Greenland ice cap and the Antarctic ice shelfs are melting much more rapidly than was predicted by the IPCC. Complete melting of both the Greenland ice cap and the Antarctic sea ice would raise ocean levels by 14 meters. It is hard to predict how fast this will take place, but certainly within 1-3 centuries.

Most worrying, however, is the threat that without an all-out effort by both developed and developing nations to immediately curb the release of greenhouse gases, climate change will reach a tipping point where feed-back loops will have taken over, and where it will then be beyond the power of human action to prevent exponentially accelerating warming.

By far the most dangerous of these feedback loops involves methane hydrates or clathrates. When organic matter is carried into the oceans by rivers, it decays to form methane. The methane then combines with water to form hydrate crystals, which are stable at the temperatures and pressures which currently exist on ocean floors. However, if the temperature rises, the crystals become unstable, and methane gas bubbles up to the surface. Methane is a greenhouse gas which is 70 times as potent as CO_2.

The worrying thing about the methane hydrate deposits on ocean floors is the enormous amount of carbon involved: roughly 10,000 gigatons. To put this huge amount into perspective, we can remember that the total amount of carbon in world CO_2 emissions since 1751 has only been 337 gigatons.

A runaway, exponentially increasing, feedback loop involving methane hydrates could lead to one of the great geological extinction events that have periodically wiped out most of the animals and plants then living. This must be avoided at all costs.[b]

The worst consequences of runaway climate change will not occur within our own lifetimes. However, we have a duty to all future human generations, and to the plants and animals with which we share our existence, to give them a future world in which they can survive.

Preventing a human-initiated 6th geological extinction event

Geologists studying the strata of rocks have observed 5 major extinction events. These are moments in geological time when most of the organisms then living suddenly became extinct. The largest of these was the Permian-Triasic extinction event, which occurred 252 million years ago. In this event, 96 percent of all marine species were wiped out, as well as 70 percent of all terrestrial vertebrates.

In 2012, the World Bank issued a report warning that without quick action to curb CO_2 emissions, global warming is likely to reach 4°C during the 21st century. This is dangerously close to the temperature which initiated the Permian-Triasic extinction event: 6°C above normal.[c]

---

[b] https://www.youtube.com/watch?v=MVwmi7HCmSI
https://www.youtube.com/watch?v=AjZaFjXfLec
https://www.youtube.com/watch?v=m6pFDu7lLV4

[c] http://www.worldbank.org/en/news/feature/2012/11/18/Climate-change-report-warns-dramatically-warmer-world-this-century

The Permian-Triasic thermal maximum seems to have been triggered by global warming and CO2 release from massive volcanic eruptions in a region of northern Russia known as the Siberian Traps. The amount of greenhouse gases produced by these eruptions is comparable to the amount emitted by human activities today.

Scientists believe that once the temperature passed 6°C above normal, a feedback loop was initiated in which methane hydrate crystals on the ocean floors melted, releasing methane, a potent greenhouse gas. The more methane released the more methane hydrate crystals were destabilized, raising the temperature still further, releasing more methane gas, and so on in a vicious circle. This feedback loop raised the global temperature to 15°C above normal, causing the Permian-Triasic mass extinction.[d]

No reputable doctor who diagnoses cancer would keep this knowledge from the patient. The reaction of the patient may be to reject the diagnosis and get another doctor, but no matter. It is very important that the threatened person should hear the diagnosis, because, with treatment, there is hope of a cure.

Similarly, the scientific community, when aware of a grave danger to our species and the biosphere, has a duty to bring this knowledge to the attention of as broad a public as possible, even at the risk of unpopularity. The size of the threatened catastrophe is so immense as to dwarf all other considerations. All possible efforts must be made to avoid it.

Consider what may be lost if a 6th mass extinction event occurs, caused by our own actions: It is possible that a few humans may survive in mountainous regions such as the Himalayas, but this will be a population of millions rather than billions. If an event comparable to the Permian-Triasic thermal maximum occurs, the family trees of virtually all of the people, animals and plants alive today will end in nothing.

The great and complex edifice of human civilization is a treasure whose value is almost above expression; and this may be lost unless we give up many of our present enjoyments. Each living organism, each animal or plant, is product of three billion years of evolution, and a miracle of harmony and complexity; and most of these will perish if we persist in our folly and greed.

Let us, for once, look beyond present pleasures, and acknowledge our duty to preserve a future world in which all forms of life can survive.

---

[d] https://www.youtube.com/watch?v=sRGVTK-AAvw
https://www.youtube.com/watch?v=MVwmi7HCmSI
https://www.youtube.com/watch?v=AjZaFjXfLec
https://www.youtube.com/watch?v=m6pFDu7lLV4

Loss of biodiversity

Agricultural monocultures

In modern agriculture it has become common to plant large regions with a single crop variety. For example, it is common to plant large regions with a single high-yield wheat variety. Monocultures of this kind offer farmers advantages of efficiency in the timing of planting and harvesting. With regard to pest and disease control, there may be short-term advantages, but these have to be weighed against the threat of long-term disasters. In the great Irish Potato Famine of 1845-1849, the potato monoculture which had sustained Ireland's growing population was suddenly devastated by Phytophthora infestans, commonly called "potato blight". The result was a catastrophic famine that resulted in the death or emigration of much of Ireland's population.

In general, monocultures are vulnerable to plant disease. Thus the replacement of traditional varieties with the high-yield crops developed by the "Green Revolution" carries serious risks. Adjustment to climate change also requires genetic diversity. In general, a genetically diverse population is far better to adjust to environmental changes than a genetically homogeneous population. This being so, it is vital to preserve civilization's heritage of genetically diverse crops.

Deforestation and loss of biodiversity

The earth's tropical rainforests are rapidly being destroyed for the sake of new agricultural land. Tropical rainforests are thought to be the habitat of more than half of the world's species of plants, animals and insects; and their destruction is accompanied by an alarming rate of extinction of species. The Harvard biologist, E.O. Wilson, estimates that the rate of extinction resulting from deforestation in the tropics may now exceed 4,000 species per year - 10,000 times the natural background rate (*Scientific American*, September, 1989).

The enormous biological diversity of tropical rainforests has resulted from their stability. Unlike northern forests, which have been affected by glacial epochs, tropical forests have existed undisturbed for millions of years. As a result, complex and fragile ecological systems have had a chance to develop. Professor Wilson expresses this in the following words:

"Fragile superstructures of species build up when the environment remains stable enough to support their evolution during long periods of time. Biologists now know that biotas, like houses of cards, can be brought tumbling down by relatively small perturbations in the physical environment. They are not robust at all."

The number of species which we have until now domesticated or used in medicine is very small compared with the number of potentially useful species still waiting in the world's tropical rainforests. When we destroy them, we damage our future. But we ought to regard the annual loss of thousands of species as a tragedy, not only because biological diversity is potential wealth for human society, but also because every form of life deserves our respect and protection.

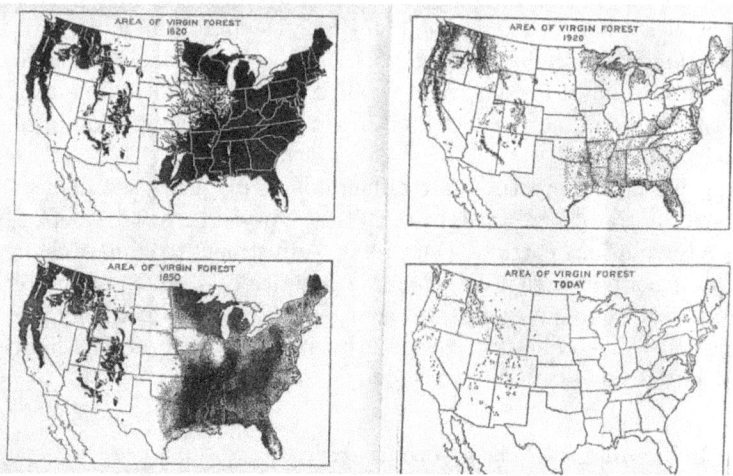

Fig. 2.6 **Deforestation in the United States between 1620 and the present.** (From William B. Greeley's The Relation of Geography to Timber Supply.)

Economics without growth

According to Adam Smith, the free market is the dynamo of economic growth. The true entrepreneur does not indulge in luxuries for himself and his family, but reinvests his profits, with the result that his business or factory grows larger, producing still more profits, which he again reinvests, and so on. This is indeed the formula for exponential economic growth.

Economists (with a few notable exceptions such as Aurelio Peccei and Herman Daly) have long behaved as though growth were synonymous with economic health. If the gross national product of a country increases steadily by 4% per year, most economists express approval and say that the economy is healthy. If the economy could be made to grow still faster (they maintain), it would be still more healthy. If the growth rate should fall, economic illness would be diagnosed. However, the basic idea of Malthus

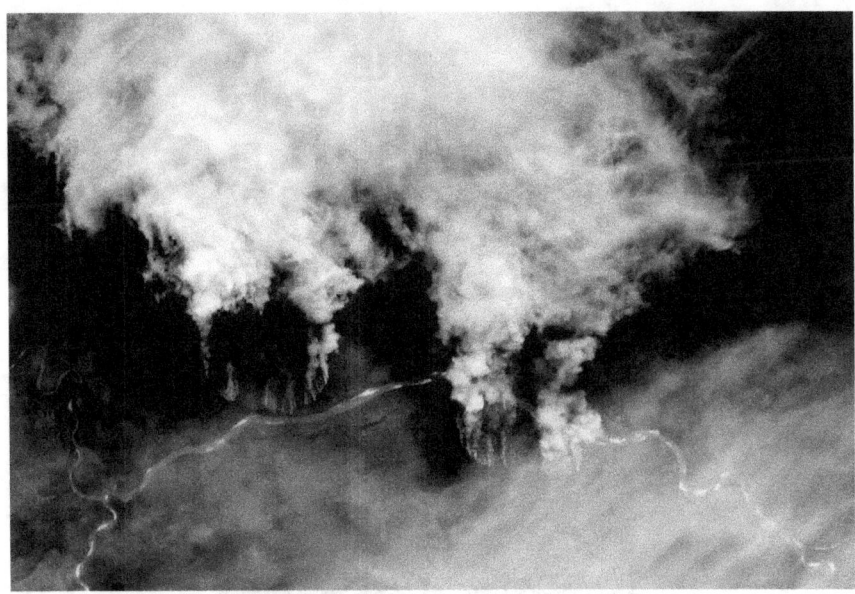

Fig. 2.7 This satellite photograph illustrates slash-and-burn forest clearing along the Rio Xingu (Xingu River) in the state of Mato Grosso, Brazil. (Public domain.)

is applicable to exponential increase of any kind. It is obvious that on a finite Earth, neither population growth nor resource-using and pollution-generating economic growth can continue indefinitely.

A "healthy" economic growth rate of 4% per year corresponds to an increase by a factor of 50 in a century. (The reader is invited to calculate the factor of increase in five centuries. The answer is $50^5 = 312,500,000$.) No one can maintain that this type of growth is sustainable except by refusing to look more than a short distance into the future. Sooner or later (perhaps surprisingly soon) an entirely new form of economics will be needed - not the empty-world economics of Adam Smith, but what might be called "full-world economics", or "steady-state economics".

Economic activity is usually divided into two categories, 1) production of goods and 2) provision of services. It is the rate of production of goods that will be limited by the carrying capacity of the global environment. Services that have no environmental impact will not be constrained in this way. Thus a smooth transition to a sustainable economy will involve a shift of a large fraction the work force from the production of goods to the provision of services.

Fig. 2.8 In 1968 Aurelio Pecci, Thorkil Kristensen and others founded the Club of Rome, an organization of economists and scientists devoted to studying the predicament of human society. One of the first acts of the organization was to commission an MIT study of future trends using computer models. The result was a book entitled "Limits to Growth", published in 1972. From the outset the book was controversial, but it became a best-seller. (From Great Change.)

In his recent popular book *The Rise of the Creative Class*, the economist Richard Florida points out that in a number of prosperous cities - for example Stockholm - a large fraction of the population is already engaged in what might be called creative work - a type of work that uses few resources, and produces few waste products - work which develops knowledge and culture rather than producing material goods. For example, producing computer software requires few resources and results in few waste products. Thus it is an activity with a very small ecological footprint. Similarly, education, research, music, literature and art are all activities that do not weigh heavily on the carrying capacity of the global environment. Furthermore, cultural activities lead in a natural way to global cooperation and internationalism. Florida sees this as a pattern for the future, and maintains that everyone is capable of creativity. He visualizes the transition to a sustainable future economy as one in which a large fraction of the work force moves from industrial jobs to information-related work. Meanwhile, as Florida acknowledges, industrial workers feel uneasy and threatened by such trends.

The present use of resources by the industrialized countries is extremely wasteful. A growing national economy must, at some point, exceed the real needs of the citizens. It has been the habit of the developed countries to create artificial needs by means of advertising, in order to allow economies to grow beyond the point where all real needs have been met; but this extra growth is wasteful, and in the future it will be important not to waste the earth's diminishing supply of non-renewable resources.

Thus, the times in which we live present a challenge: We need a revolution in economic thought. We must develop a new form of economics, taking into account the realities of the world's present situation - an economics based on real needs and on a sustainable equilibrium with the environment, not on the thoughtless assumption that growth can continue forever.

Adam Smith was perfectly correct in saying that the free market is the dynamo of economic growth; but rapid growth of human population and economic activity have brought us, in a surprisingly short time, from the empty-world situation in which he lived to a full-world situation. In today's world, we are pressing against the absolute limits of the earth's carrying capacity, and further growth carries with it the danger of future collapse. Full-world economics, the economics of the future, will no longer be able to rely on growth to give profits to stockbrokers or to solve problems of unemployment or to alleviate poverty. In the long run, growth of any kind is not sustainable (except perhaps growth of culture and knowledge); and we are now nearing the environmentally-imposed limits.

Transition to a sustainable economy

Like a speeding bus headed for a brick wall, the earth's rapidly-growing population of humans and its rapidly-growing resource-using and pollution-generating economic activity are headed for a collision with a very solid barrier - the carrying capacity of the global environment. As in the case of the bus and the wall, the correct response to the situation is to apply the brakes in time - but fear prevents us from doing this. What will happen if we slow down very suddenly? Will not many of the passengers be injured?

Undoubtedly. But what will happen if we hit the wall at full speed? Perhaps it would be wise, after all, to apply the brakes!

The memory of the great depression of 1929 makes us fear the consequences of an economic slowdown, especially since unemployment is already a serious problem in many parts of the world. Although the history of the 1929 depression is frightening, it may nevertheless be useful to look at the measures which were used then to bring the global economy back to its feet. A similar level of governmental responsibility may help us to avoid some of the more painful consequences of the necessary transition from the economics of growth to steady-state economics.

In the United States, President Franklin D. Roosevelt was faced with the difficult problems of the depression during his first few years in office. Roosevelt introduced a number of special governmental programs, such as the WPA, the Civilian Construction Corps and the Tennessee Valley Authority, which were designed to create new jobs on projects directed towards socially useful goals - building highways, airfields, auditoriums, harbors, housing projects, schools and dams. The English economist John Maynard Keynes, (1883-1946), provided an analysis of the factors that had caused the 1929 depression, and a theoretical justification of Roosevelt's policies.

The transition to a sustainable global society will require a similar level of governmental responsibility, although the measures needed are not the same as those which Roosevelt used to end the great depression. Despite the burst of faith in the free market which has followed the end of the Cold War, it seems unlikely that market mechanisms alone will be sufficient to solve problems of unemployment in the long-range future, or to achieve conservation of land, natural resources and environment.

The Worldwatch Institute, Washington D.C., lists the following steps as necessary for the transition to sustainability[e]:

(1) Stabilizing population
(2) Shifting to renewable energy
(3) Increasing energy efficiency
(4) Recycling resources
(5) Reforestation
(6) Soil conservation

All of these steps are labor-intensive; and thus, wholehearted governmental commitment to the transition to sustainability can help to solve the problem of unemployment.

---

[e]L.R. Brown and P. Shaw, 1982.

Fig. 2.9 Lester R. Brown, founder of the Worldwatch Institute, and for many years its President. He is now the leader of the Earth Policy Institute. His recent book, "Plan B", gives important information about the ecological crisis now facing the world. It may be downloaded free of charge from the website of the Earth Policy Institute. (From Famous Birthdays.)

In much the same spirit that Roosevelt (with Keynes' approval) used governmental powers to end the great depression, we must now urge our governments to use their powers to promote sustainability and to reduce the trauma of the transition to a steady-state economy. For example, an increase in the taxes on fossil fuels could make a number of renewable energy technologies economically competitive; and higher taxes on motor fuels would be especially useful in promoting the necessary transition from private automobiles to bicycles and public transportation. Tax changes could also be helpful in motivating smaller families.

The present economic recession offers us an opportunity to take steps towards the creation of a sustainable steady-state economic system.

Government measures to avoid unemployment could at the same time shift the work force to jobs that promote sustainability, i.e., jobs in the areas listed by the Worldwatch Institute.

Governments already recognize their responsibility for education. In the future, they must also recognize their responsibility for helping young people to make a smooth transition from education to secure jobs. If jobs are scarce, work must be shared, in a spirit of solidarity, among those seeking employment; hours of work (and if necessary, living standards) must be reduced to insure a fair distribution of jobs. Market forces alone cannot achieve this. The powers of government are needed.

Population and goods per capita

In the distant future, the finite carrying capacity of the global environment will impose limits on the amount of resource-using and waste-generating economic activity that it will be possible for the world to sustain. The consumption of goods per capita will be equal to this limited total economic activity divided by the number of people alive at that time. Thus, our descendants will have to choose whether they want to be very numerous and very poor, or less numerous and more comfortable, or very few and very rich. Perhaps the middle way will prove to be the best.

Given the fact that environmental carrying capacity will limit the sustainable level of resource-using economic activity to a fixed amount, average wealth in the distant future will be approximately inversely proportional to population over a certain range of population values. Obviously, if the number of people is reduced to such an extent that it approaches zero, the average wealth will not approach infinity, since a certain level of population is needed to maintain a modern economy. However, if the global population becomes extremely large, the average wealth will indeed approach zero.

In the 1970's the equation $I = P \times A \times T$ was introduced in the course of a debate between Barry Commoner, Paul R. Ehrlich and John P. Holdren. Here I represents environmental impact, P is population, while A represents goods per capita, and T is an adjustable factor that depends on the technology used to produce the goods. The assertion of the previous paragraph can be expressed by solving for A and setting I equal to a constant: $A = I/(P \times T)$. In the distant future, the environmental impact I will not be allowed to increase, and therefore for a given value of T, A will be inversely proportional to P.

If the environmental impact I is broken up into several components, a few of them have historically fallen with increasing values of $A \times P$ because of diminishing T (thus exhibiting the *environmental Kuznets curve*). However, most components of I, such as energy, land and resource use, have historically increased with increasing $A \times P$.

Paris, India and coal

The MIT Technology Review recently published an important article entitled "India's Energy Crisis".[f]

The article makes alarming reading in view of the world's urgent need to make a very rapid transition from fossil fuels to 100% renewable energy. We must make this change quickly in order to avoid a tipping point beyond which catastrophic climate change will be unavoidable.[g]

The MIT article states that "Since he took power in May, 2014, Prime Minister Narendra Modi has made universal access to electricity a key part of his administration's ambitions. At the same time, he has pledged to help lead international efforts to limit climate change. Among other plans, he has promised to increase India's total power generating capacity to 175 gigawatts, including 100 gigawatts of solar, by 2022. (That's about the total power generation of Germany.)"

However India plans to expand its industrial economy, and to do this, it is planning to very much increase its domestic production and use of coal. The MIT article continues, pointing out that:

"Such growth would easily swamp efforts elsewhere in the world to curtail carbon emissions, dooming any chance to head off the dire effects of global climate change. (Overall, the world will need to reduce its current annual emissions of 40 billion tons by 40 to 70 percent between now and 2050.) By 2050, India will have roughly 20 percent of the world's population. If those people rely heavily on fossil fuels such as coal to expand the economy and raise their living standards to the level people in the rich world have enjoyed for the last 50 years, the result will be a climate catastrophe regardless of anything the United States or even China does to decrease its emissions. Reversing these trends will require radical transformations in two main areas: how India produces electricity, and how it distributes it."

The Indian Minister of Power, Piyush Goyal, is an enthusiastic supporter of renewable energy expansion, but he also supports, with equal enthusiasm, the large-scale expansion of domestic coal production in India.

Meanwhile, the consequences of global warming are being felt by the people of India. For example, in May 2015, a heat wave killed over 1,400 people and melted asphalt streets.[h]

Have India's economic planners really thought about the long-term future? Have they considered the fact that drastic climate change could make

---

[f] http://www.technologyreview.com/featuredstory/542091/indias-energy-crisis/
[g] https://www.youtube.com/watch?v=2bRrg96UtMc
https://www.youtube.com/watch?v=MVwmi7HCmSI
https://www.youtube.com/watch?v=AjZaFjXfLec
https://www.youtube.com/watch?v=MVwmi7HCmSI
[h] https://www.rt.com/news/262641-india-heat-wave-killed/

Fig. 2.10 A logo for Coal India Limited. Although the government of India supports renewable energy, the country's use of coal is also increasing rapidly. (Wikipedia.)

India completely uninhabitable?

Jacob von Uexküll's speech to the World Future Council

In a recent speech to the World Future Council, the distinguished writer, philanthropist, activist and former politician Jacob von Uexküll outlined the future dangers facing our world with an accuracy and eloquence that has seldom been equaled. Here is a link to his speech:[i]

Jakob von Uexküll belongs to a brilliant family. His grandfather was a famous Baltic-German physiologist who founded the discipline of

---

[i] http://www.worldfuturecouncil.org/2016/03/15/world-future-forum-2016-opening-speech-jakob-von-uexkull/

Biosemiotics. Besides being a former Member of the European Parliament and a leader of the German Green Party, von Uexküll himself founded both the Right Livelihood Award (sometimes called the Alternative Nobel Prize) and also the World Future Council.[j]

Here are a few excerpts from his speech:

"Today we are heading for unprecedented dangers and conflicts, up to and including the end of a habitable planet in the foreseeable future, depriving all future generations of their right to life and the lives of preceding generations of meaning and purpose."

"This apocalyptic reality is the elephant in the room. Current policies threaten temperature increases triggering permafrost melting and the release of ocean methane hydrates which would make our earth unliveable, according to research presented by the British Government Met office at the Paris Climate Conference."

"The myth that climate change is conspiracy to reduce freedom is spread by a powerful and greedy elite which has largely captured governments to preserve their privileges in an increasingly unequal world. Long before that point, our prosperity, security, culture and identity will disintegrate. A Europe unable to cope with a few million war refugees will collapse under the weight of tens or even hundreds of millions of climate refugees."

Paris and the long-term future

We give our children loving care, but it makes no sense do so and at the same time to neglect to do all that is within our power to ensure that they and their descendants will inherit an earth in which they can survive. We also have a responsibility to all the other living organisms with which we share the gift of life.

Human emotional nature is such that we respond urgently to immediate temptations or dangers, while long-term considerations are pushed into the background. Thus the temptations of immediate profit or advantage motivate politicians and the executives of fossil fuel corporations; and the temptations of continued overconsumption and luxury blind the general public. Public fears of terrorism have been magnified by our perfidious mass media to such an extent that the equally perfidious French Government has been able to use this fear as an excuse to exclude democracy and proper care for the long-term future from the Paris Climate Conference.

However, our generation has an urgent duty to think of the distant future. The ultimate fate of human civilization and the biosphere is in

---

[j] http://www.rightlivelihood.org/
http://www.worldfuturecouncil.org/
http://www.worldfuturecouncil.org/gpact/

our hands. What we really have to fear, for the sake of our children and grandchildren and their descendants, is reaching a tipping point, beyond which uncontrollable feedback loops will make catastrophic climate change inevitable despite all human efforts to prevent it.

A feedback loop is a self-reinforcing cycle. The more it goes on, the stronger it becomes. An example of how such a feedback loop could drive climate change and make it uncontrollable is the albedo effect: When sunlight falls on sea ice in the Arctic or Antarctic, most of it is reflected by the white surface of the snow-covered ice. But when sunlight falls on dark sea water, it is almost totally absorbed. This cycle is self-reinforcing because warming the water reduces the ice cover. This is happening today, especially in the Arctic, and we have to stop it.

Another dangerous feedback loop involves the evaporation of sea water, which itself is a greenhouse gas. However, if we think of the long-term future, by far the most dangerous feedback loop is that which involves the melting of methane hydrate crystals, releasing the extremely powerful greenhouse gas methane into the atmosphere. Discussion of this highly dangerous feedback loop seems to be almost completely banned by our mass media.

When organic matter is carried into the oceans by rivers, it decays to form methane. The methane then combines with water to form hydrate crystals, which are stable at the temperatures and pressures which currently exist on ocean floors. However, if the temperature rises, the crystals become unstable, and methane gas bubbles up to the surface. Methane is a greenhouse gas which is much more potent than CO_2.

The worrying thing about the methane hydrate deposits on ocean floors is the enormous amount of carbon involved: roughly 10,000 gigatons. To put this huge amount into perspective, we can remember that the total amount of carbon in world CO_2 emissions since 1751 has only been 337 gigatons.[k]

A runaway, exponentially increasing feedback loop involving methane hydrates could lead to one of the great geological extinction events that have periodically wiped out most of the animals and plants then living. This must be avoided at all costs.

The worst consequences of runaway climate change will not occur within our own lifetimes. However, we have a duty to all future human generations, and to the plants and animals with which we share our existence, to give them a future world in which they can survive.

---

[k] https://www.youtube.com/watch?v=2bRrg96UtMc
https://www.youtube.com/watch?v=MVwmi7HCmSI
https://www.youtube.com/watch?v=AjZaFjXfLec
https://www.youtube.com/watch?v=MVwmi7HCmSI

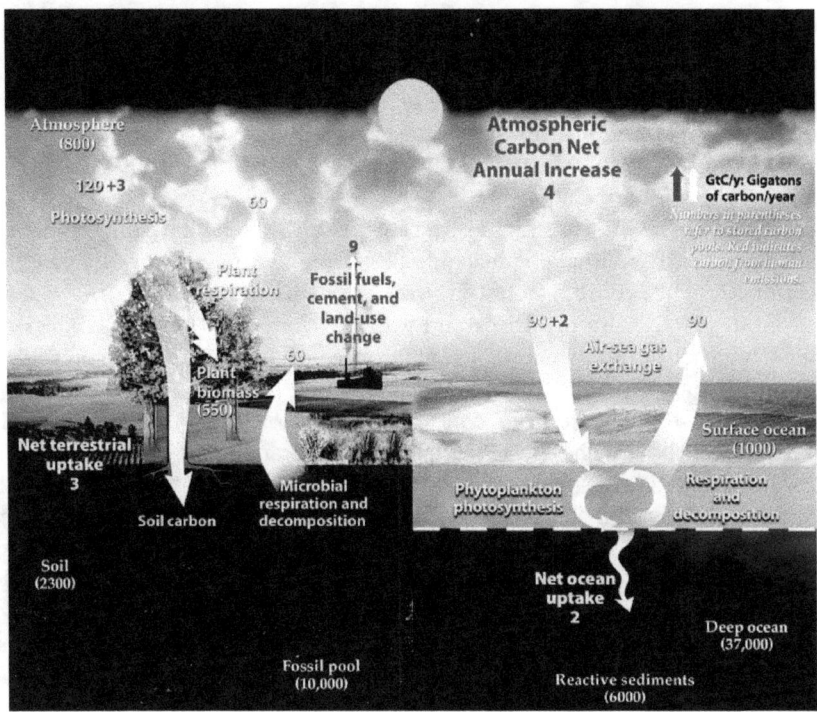

Fig. 2.11 This diagram shows a simplified representation of the contemporary global carbon cycle. Changes are measured in gigatons of carbon per year (GtC/y). (Public domain.)

We can also fear a catastrophic future famine, produced by a combination of climate change, population growth and the end of fossil-fuel-dependent high-yield modern agriculture.[1]

These very real and very large long-term disasters are looming on our horizon, but small short-term considerations blind us, so that we do not take the needed action. But what is at stake is the future of everyone's children and grandchildren and their progeny, your future family tree and mine, also the families of Francois Hollande and the executives of Exxon. They should think carefully about the consequences of making our beautiful world completely uninhabitable.

---

[1]http://human-wrongs-watch.net/2015/08/19/the-need-for-a-new-economic-system-part-vii-glreadoval-food-crisis/

OPEC oil and climate change

In an amazing display of collective schizophrenia, our media treat oil production and the global climate emergency as though they were totally disconnected. But the use of all fossil fuels, including oil, must stop almost immediately if the world is to have a chance of avoiding uncontrollable and catastrophic climate change.

The recent Doha summit meeting of the Oil Producing and Exporting Countries (OPEC) aimed at reaching an agreement on limiting the production of oil. This aim did not stem from the climate emergency but rather a from desire to raise oil prices and profits. However, the OPEC meeting failed to reach an agreement. Production continues to be extremely high and prices low.[m]

Our high-energy lifestyles continue. Our profligate use of fossil fuels continues as though the life-threatening climate emergency did not exist.

Meanwhile, early spring temperatures in 2016 have totally smashed all previous records, and this is especially pronounced in the Arctic and Antarctic regions. Polar ice caps are melting in an alarmingly rapid and non-linear way. The rate of melting of the icecaps is far greater than predicted by conventional modeling which does not include feedback loops. Many island nations and coastal cities are threatened, not in the very distant future, but by the middle of our present century. Here are a few links reporting what is happening:[n]

---

[m] http://www.cnbc.com/2016/04/17/doha-oil-producers-meeting-ends-without-an-agreement.html

[n] http://www.truth-out.org/news/item/35398-climate-disruption-in-overdrive-submerged-cities-and-melting-that-feeds-on-itself
http://www.commondreams.org/news/2016/03/14/nasa-drops-major-bomb-march-toward-ever-warmer-planet
http://www.theguardian.com/environment/2016/mar/15/record-breaking-temperatures-have-robbed-the-arctic-of-its-winter
http://www.truth-out.org/news/item/35283-arctic-sea-ice-volume-nears-record-low
http://dissidentvoice.org/2016/03/does-methane-threaten-life/
https://theconversation.com/meltdown-earth-the-shocking-reality-of-climate-change-kicks-in-but-who-is-listening-56255
http://www.countercurrents.org/bardi150316.htm
http://www.informationclearinghouse.info/article44427.htm
http://www.truth-out.org/news/item/35202-antarctica-on-the-brink-nasa-emeritus-scientist-warns-of-dramatic-loss-of-glaciers
http://ecowatch.com/2016/03/02/february-record-hot/
http://nsidc.org/arcticseaicenews/charctic-interactive-sea-ice-graph/
http://nsidc.org/arcticseaicenews/
http://thinkprogress.org/climate/2016/03/01/3754891/arctic-sea-ice-growth/
http://thinkprogress.org/climate/2016/02/16/3749815/carbon-pollution-hottest-12-months-january/
http://www.truth-out.org/news/item/35468-agriculture-on-the-brink

Fig. 2.12 Iranian oil and gas facilities. An Iran-Iraq-Syria pipeline has been proposed. Catastrophic climate change threatens to destroy human civilization and much of the biosphere. The production and consumption of oil must stop almost immediately; but pipeline wars continue just as though these threats did not exist. (Public domain.)

http://www.commondreams.org/news/2016/03/21/after-unprecedented-year-warming-un-warns-we-must-curb-emissions-now
http://www.commondreams.org/news/2016/04/15/hottest-march-record-earth-keeps-hurtling-past-temperature-milestones
http://www.commondreams.org/news/2016/04/04/amid-climate-fueled-food-crisis-filipino-forces-open-fire-starving-farmers
http://ecowatch.com/2016/04/08/mckibben-break-free/
http://www.motherjones.com/environment/2016/04/water-scarcity-wikileaks
http://www.commondreams.org/news/2016/04/15/towards-common-good-mr-sanders-goes-vatican
http://ecowatch.com/2016/04/22/dicaprio-paris-climate-change-agreement/

In the long-term future, catastrophic anthropogenic climate change threatens to destroy human civilization and to drive the majority of plant and animal species into extinction. To prevent this from happening, we need to stop subsidizing and accepting fossil fuel production. We need to vigorously support the transition to a sustainable economy based on renewable energy.

Our duty to future generations

Many traditional agricultural societies have an ethical code that requires them to preserve the fertility of the land for future generations. This recognition of a duty towards the distant future is in strong contrast to the shortsightedness of modern economists. For example, John Maynard Keynes has been quoted as saying "In the long run, we will all be dead", meaning that we need not look that far ahead. By contrast, members of traditional societies recognize that their duties extend far into the distant future, since their descendants will still be alive.

Here is an ethical principle of the Native Americans: "Treat the earth well. It was not given to you by your parents. It was loaned to you by your children." They also say: "We must protect the forests for our children, grandchildren, and children yet to be born. We must protect the forests for those who cannot speak for themselves, such as the birds, animals, fish and trees."

In his book, "The Land of the Spotted Eagle", the Lakota chief Luther Standing Bear (ca. 1834-1908) wrote: "The Lakota was a true lover of Nature. He loved the earth and all things of the earth... From Waken Tanka (the Great Spirit) there came a great unifying life force that flowered in and through all things: the flowers of the plains, blowing winds, rocks, trees, birds, animals, and was the same force that had been breathed into the first man. Thus all things were kindred and were brought together by the same Great Mystery."

In some parts of Africa, a man who plans to cut down a tree offers a prayer of apology, telling the tree why necessity has forced him to harm it. This preindustrial attitude is something from which industrialized countries could learn. In industrial societies, land "belongs" to someone, and the owner has the "right" to ruin the land or to kill the communities of creatures

http://www.informationclearinghouse.info/article44510.htm
http://www.informationclearinghouse.info/article44519.htm
http://www.truth-out.org/news/item/35796-temperatures-in-2016-are-already-near-cop21-limit
https://www.newscientist.com/article/2084835-unprecedented-global-warming-as-2016-approaches-1-5-c-mark/

Fig. 2.13 Chief Luther Standing Bear (ca. 1873-1908), author of "Land of the Spotted Eagle". (Public domain.)

living on it, if this happens to give some economic advantage, in much the same way that a Roman slave-owner was thought to have the "right" to kill his slaves. Preindustrial societies have a much less rapacious and much more custodial attitude towards the land and towards its non-human inhabitants.

On April 22, 2010, the World People's Conference on Climate Change and the Rights of Mother Earth in Cochabamba, Bolivia, adopted a Universal Declaration of the Rights of Mother Earth. Here is a link: http://therightsofnature.org/universal-declaration/
Contrast this expression of the deep ethical convictions of the world's people with the cynical, money-centered results of various intergovernmental conferences on climate change!

Our economic system is built on the premise that individuals act out of self-interest, and as things are today, they do so with a vengeance.There is no place in the system for thoughts about the environment and the long-term future. All that matters is the bottom line. The machine moves on relentlessly, exhausting non-renewable resources, turning fertile land into deserts, driving animal species into extinction, felling the last of the world's tropical rainforests, pumping greenhouse gases into the atmosphere, and

Fig. 2.14 Mother Nature image, 17th century alchemical text, Atalanta Fugiens. (Public domain.)

sponsoring TV programs that deny the reality of climate change, or other programs that extol the concept of never-ending industrial growth. But the economists, bankers, bribed politicians and corporation chiefs who destroy the earth today, are destroying the future for their own children, grandchildren and great-grandchildren. Does it make sense for them to saw off the branch on which they, like all of us, are sitting?

Recently an extremely grave danger to the long-term future of human civilization and the biosphere has become clear. The latest observations show that Arctic sea ice is melting far faster than was predicted by the IPCC. It now seems likely that the September Arctic sea ice will vanish by as early as 2016 or 2017. It will, of course, re-freeze in the winters, but its average total mass will continue to rapidly decrease.

The rapid and non-linear vanishing of Arctic sea ice is due to a feedback loop involving albedo, i.e the high reflectivity of white ice compared with dark sea water which absorbs most of the radiation that falls onto it. As Arctic sea ice disappears more radiation is absorbed, the Arctic temperature rises still further, still more ice melts, and so on in a vicious circle.

At present Arctic temperatures are roughly 4°C higher than preindustrial levels, and this has led to increasingly rapid melting of the Greenland ice cap. It is now observed that during the summers, lakes of melted water form on the surface of Greenland's inland ice. These lakes feed rivers

Fig. 2.15 Aion, Gaia and four children. (Public domain.)

Fig. 2.16 Demonstrators declaring the rights of Mother Earth. (Earth Peoples.)

that run for some distance along the surface of the ice cap, but which ultimately fall through fissures to the bottom of the sheet, where they lubricate its flow. Through this mechanism, the Greenland ice cap is flowing more quickly and calving into massive icebergs much more rapidly than climate scientists expected.

Fig. 2.17 We are sawing off the branch on which we are sitting. (Commentsmeme.com.)

Complete melting of the Greenland ice cap would raise ocean levels by 7 meters. Antarctic sea ice is also breaking up much more rapidly than expected. When it is totally gone, the disappearance of Antarctic sea ice would add another 7 meters to ocean levels, making a total of 14 meters. It is hard to predict how soon this will happen, but certainly within 1-3 centuries.

However, by far the most worrying threat to our long-term future comes from the danger of an out-of-control and exponentially accelerating feedback loop involving methane hydrates. When rivers carry organic matter into the ocean, it decays, forming methane, a powerful greenhouse gas. At the temperatures and pressures currently prevailing on ocean floors, the methane combines with water molecules to form stable crystals called methane hydrates. The amount of carbon stored in methane hydrates is immense: roughly 10,000 gigatons. By comparison, the amount of carbon emitted by human activities since preindustrial times is only 337 gigatons.

Geologists have observed that life on earth has experienced 5 major extinction events, the largest of which was the Permian-Triassic event, when 96 percent of all marine species and 70 percent of all terrestrial vertebrates disappeared from the fossil record. Predictions based on current CO_2 emission rates predict that early in the 22nd century, global temperature increases will have reached 6°C, the temperature that is thought to have initiated the Permian-Triassic extinction event. These dangers are eloquently discussed in a short, important and clear video prepared by Thom Hartmann and his coworkers. It is available on www.lasthours.org

Must there be a human-initiated 6th geological extinction event? Is it inevitable that the long-term future will witness the disappearance of human civilization and most of the plants and animals that are alive today? No! Absolutely not! It is only inevitable if we persist in our greed and folly. It is only inevitable if we continue to value money more than nature. It is only inevitable if we are afraid to question the authority of corrupt politicians. It is only inevitable if we fail to cooperate globally, and if we fail to develop a new economic system with both a social conscience and an ecological conscience.

We are living today in a time of acute crisis. We need to act with a sense of urgency never before experienced. We need to have great courage to meet an unprecedented challenge. We need to fulfil our duty to future generations.

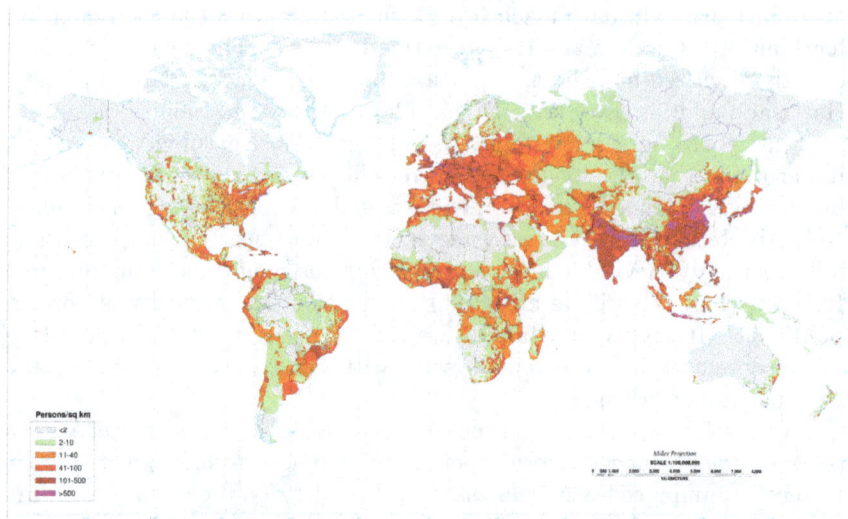

Fig. 2.18 Population density (people per km^2) map of the world in 1994. It is vitally important that all countries should pass quickly through the demographic transition from high birth and death rates to low birth and death rates. (Public domain.)

In the world as it is, the population is increasing so fast that it doubles every thirty-nine years. Most of this increase is in the developing countries, and in many of these, the doubling time is less than twenty-five years. Famine is already present, and it threatens to become more severe and widespread in the future.

In the world as it could be, population would be stabilized at a level that could be sustained comfortably by the world's food and energy resources. Each country would be responsible for stabilizing its own population.

Fig. 2.19 There is a danger that a tipping point may be reached, where drying of tropical forests leads to their destruction by fires. (World Visits.)

In the world as it is, large areas of tropical rain forest are being destroyed by excessive timber cutting. The cleared land is generally unsuitable for farming.

In the world as it could be, it would be recognized that the conversion of carbon dioxide into oxygen by tropical forests is necessary for the earth's climatic stability. Tropical forests would also be highly valued because of their enormous diversity of plant and animal life, and large remaining areas of forest would be protected.

Fig. 2.20 Air pollution from a fossil fuel power station. Among the pollutants released into the atmosphere, greenhouse gases are the most dangerous. To avoid catastrophic climate change, we must very rapidly make the transition from fossil fuels to renewable energy. (Public domain.)

In the world as it is, pollutants are dumped into our rivers, oceans and atmosphere. Some progress has been made in controlling pollution, but far from enough.

In the world as it could be, a stabilized and perhaps reduced population would put less pressure on the environment. Strict international laws would prohibit the dumping of pollutants into our common rivers, oceans and atmosphere. The production of greenhouse gases would also be limited by international laws.

Suggestions for further reading

(1) A. Gore, *An Inconvenient Truth: The Planetary Emergency of Global Warming and What We Can Do About It*, Rodale Books, New York, (2006).
(2) A. Gore, *Earth in the Balance: Forging a New Common Purpose*, Earthscan, (1992).
(3) A.H. Ehrlich and P.R. Ehrlich, *Earth*, Thames and Methuen, (1987).pro Simon and Schuster, (1990).
(4) P.R. Ehrlich and A.H. Ehrlich, *Healing the Planet: Strategies for Resolving the Environmental Crisis*, Addison-Wesley, (1991).
(5) P.R. Ehrlich and A.H. Ehrlich, *Betrayal of Science and Reason: How Anti-Environmental Rhetoric Threatens our Future*, Island Press, (1998).
(6) P.R. Ehrlich and A.H. Ehrlich, *One With Nineveh: Politics, Consumption and the Human Future*, Island Press, (2004).
(7) A.H. Ehrlich and U. Lele, *Humankind at the Crossroads: Building a Sustainable Food System*, in *Draft Report of the Pugwash Study Group: The World at the Crossroads*, Berlin, (1992).
(8) P.R. Ehrlich, *The Population Bomb*, Sierra/Ballentine, New York, (1972).
(9) P.R. Ehrlich, A.H. Ehrlich and J. Holdren, *Human Ecology*, W.H. Freeman, San Francisco, (1972).
(10) P.R. Ehrlich, A.H. Ehrlich and J. Holdren, *Ecoscience: Population, Resources, Environment*, W.H. Freeman, San Francisco, (1977)
(11) P.R. Ehrlich and A.H. Ehrlich, *Extinction*, Victor Gollancz, London, (1982).
(12) D.H. Meadows, D.L. Meadows, J. Randers, and W.W. Behrens III, *The Limits to Growth: A Report for the Club of Rome's Project on the Predicament of Mankind*, Universe Books, New York, (1972).
(13) D.H. Meadows et al., *Beyond the Limits. Confronting Global Collapse and Envisioning a Sustainable Future*, Chelsea Green Publishing, Post Mills, Vermont, (1992).
(14) D.H. Meadows, J. Randers and D.L. Meadows, *Limits to Growth: the 30-Year Update*, Chelsea Green Publishing, White River Jct., VT 05001, (2004).
(15) A. Peccei and D. Ikeda, *Before it is Too Late*, Kodansha International, Tokyo, (1984).
(16) A. Peccei, *The Human Quality*, Pergamon Press, Oxford, (1977).
(17) A. Peccei, *One Hundred Pages for the Future*, Pergamon Press, New York, (1977).
(18) V.K. Smith, ed., *Scarcity and Growth Reconsidered*, Johns Hopkins

University Press, Baltimore, (1979).
(19) R. Costannza, ed., *Ecological Economics: The Science and Management of Sustainability*, Colombia University Press, New York, (1991).
(20) IPCC, Intergovernmental Panel on Climate Change, *Climate Change 2001: The Scientific Basis*, (1001).
(21) N. Stern et al., *The Stern Review*, www.sternreview.org.uk, (2006).
(22) T.M. Swanson, ed., *The Economics and Ecology of Biodiversity Decline: The Forces Driving Global Change*, Cambridge University Press, (1995).
(23) P.M. Vitousek, H.A. Mooney, J. Lubchenco and J.M. Melillo, *Human Domination of Earth's Ecosystems*, Science, **277**, 494-499, (1997).
(24) P.M. Vitousek, P.R. Ehrlich, A.H. Ehrlich and P.A. Matson, *Human Appropriation of the Products of Photosynthesis*, Bioscience, *34*, 368-373, (1986).
(25) D. King, *Climate Change Science: Adapt, Mitigate or Ignore*, Science, **303** (5655), pp. 176-177, (2004).
(26) S. Connor, *Global Warming Past Point of No Return*, The Independent, (116 September, 2005).
(27) D. Rind, *Drying Out the Tropics*, New Scientist (6 May, 1995).
(28) J. Patz et al., *Impact of Regional Climate Change on Human Health*, Nature, (17 November, 2005).
(29) M. McCarthy, *China Crisis: Threat to the Global Environment*, The Independent, (19 October, 2005).
(30) L.R. Brown, *The Twenty-Ninth Day*, W.W. Norton, New York, (1978).
(31) N. Myers, *The Sinking Ark*, Pergamon, New York, (1972).
(32) N. Myers, *Conservation of Tropical Moist Forests*, National Academy of Sciences, Washington D.C., (1980).
(33) National Academy of Sciences, *Energy and Climate*, NAS, Washington D.C., (1977).
(34) W. Ophuls, *Ecology and the Politics of Scarcity*, W.H. Freeman, San Francisco, (1977).
(35) E. Eckholm, *Losing Ground: Environmental Stress and World Food Prospects*, W.W. Norton, New York, (1975).
(36) E. Eckholm, *The Picture of Health: Environmental Sources of Disease*, New York, (1976).
(37) Economic Commission for Europe, *Air Pollution Across Boundaries*, United Nations, New York, (1985).
(38) G. Hagman and others, *Prevention is Better Than Cure*, Report on Human Environmental Disasters in the Third World, Swedish Red Cross, Stockholm, Stockholm, (1986).
(39) G. Hardin, "The Tragedy of the Commons", *Science*, December 13, (1968).
(40) K. Newland, *Infant Mortality and the Health of Societies*, Worldwatch

Paper 47, Worldwatch Institute, Washington D.C., (1981).
(41) D.W. Orr, *Ecological Literacy*, State University of New York Press, Albany, (1992).
(42) E. Pestel, *Beyond the Limits to Growth*, Universe Books, New York, (1989).
(43) D.C. Pirages and P.R. Ehrlich, *Ark II: Social Responses to Environmental Imperatives*, W.H. Freeman, San Francisco, (1974).
(44) Population Reference Bureau, *World Population Data Sheet*, PRM, 777 Fourteenth Street NW, Washington D.C. 20007, (published annually).
(45) R. Pressat, *Population*, Penguin Books Ltd., (1970).
(46) M. Rechcigl (ed.), *Man/Food Equation*, Academic Press, New York, (1975).
(47) J.C. Ryan, *Life Support: Conserving Biological Diversity*, Worldwatch Paper 108, Worldwatch Institute, Washington D.C., (1992).
(48) J. Shepard, *The Politics of Starvation*, Carnegie Endowment for International Peace, Washington D.C., (1975).
(49) B. Stokes, *Local Responses to Global Problems: A Key to Meeting Basic Human Needs*, Worldwatch Paper 17, Worldwatch Institute, Washington D.C., (1978).
(50) L. Timberlake, *Only One Earth: Living for the Future*, BBC/ Earthscan, London, (1987).
(51) UNEP, *Environmental Data Report*, Blackwell, Oxford, (published annually).
(52) UNESCO, *International Coordinating Council of Man and the Biosphere*, MAB Report Series No. 58, Paris, (1985).
(53) United Nations Fund for Population Activities, *A Bibliography of United Nations Publications on Population*, United Nations, New York, (1977).
(54) United Nations Fund for Population Activities, *The State of World Population*, UNPF, 220 East 42nd Street, New York, 10017, (published annually).
(55) United Nations Secretariat, *World Population Prospects Beyond the Year 2000*, U.N., New York, (1973).
(56) J. van Klinken, *Het Dierde Punte*, Uitgiversmaatschappij J.H. Kok-Kampen, Netherlands (1989).
(57) B. Ward and R. Dubos, *Only One Earth*, Penguin Books Ltd., (1973).
(58) WHO/UNFPA/UNICEF, *The Reproductive Health of Adolescents: A Strategy for Action*, World Health Organization, Geneva, (1989).
(59) E.O. Wilson, *Sociobiology*, Harvard University Press, (1975).
(60) E.O. Wilson (ed.), *Biodiversity*, National Academy Press, Washington D.C., (1988).
(61) E.O. Wilson, *The Diversity of Life*, Allen Lane, The Penguin Press,

London, (1992).
(62) G. Woodwell (ed.), *The Earth in Transition: Patterns and Processes of Biotic Impoverishment*, Cambridge University Press, (1990).
(63) World Resources Institute (WRI), *Global Biodiversity Strategy*, The World Conservation Union (IUCN), United Nations Environment Programme (UNEP), (1992).
(64) World Resources Institute, *World Resources 200-2001: People and Ecosystems: The Fraying Web of Life*, WRI, Washington D.C., (2000).
(65) D.W. Pearce and R.K. Turner, *Economics of Natural Resources and the Environment*, Johns Hopkins University Press, Baltimore, (1990).
(66) P. Bartelmus, *Environment, Growth and Development: The Concepts and Strategies of Sustainability*, Routledge, New York, (1994).
(67) H.E. Daly and K.N. Townsend, (editors), *Valuing the Earth. Economics, Ecology, Ethics*, MIT Press, Cambridge, Massachusetts, (1993)
(68) C. Flavin, *Slowing Global Warming: A Worldwide Strategy*, Worldwatch Paper 91, Worldwatch Institute, Washington D.C., (1989).
(69) S.H. Schneider, *The Genesis Strategy: Climate and Global Survival*, Plenum Press, (1976).
(70) WHO/UNFPA/UNICEF, *The Reproductive Health of Adolescents: A Strategy for Action*, World Health Organization, Geneva, (1989).
(71) World Commission on Environment and Development, *Our Common Future*, Oxford University Press, (1987).
(72) W. Jackson, *Man and the Environment*, Wm. C. Brown, Dubuque, Iowa, (1971).
(73) T. Berry, *The Dream of the Earth*, Sierra Club Books, San Francisco, (1988).
(74) T.M. Swanson, ed., *The Economics and Ecology of Biodiversity Decline: The Forces Driving Global Change*, Cambridge University Press, (1995).

Chapter 3

Growing Population, Vanishing Resources

"Let us try to translate pollution and ruthless exploitation of the environment into economic language: Both of these mean that we are spending our capital, i.e., we are spending the earth's riches of coal, oil and raw materials, as well as our inheritance of clean air, clean water, and places where one can be free from noise pollution. It is clear that economic growth, as we experience it today, means that we are spending more and more of humankind's natural wealth. This cannot continue indefinitely."

Professor Thorkil Kristensen, former Secretary General of the OECD

Introduction

Is a transition to 100% renewable energy possible? One answer to this question is that the transition must take place within a century or so because coal, oil and natural gas will become too rare and expensive to be used as fuels. But the vital point which we must remember is that, to avoid the threat of catastrophic climate change, the transition must take place very rapidly, within a few decades. This will require lifestyle changes in the industrialized countries, which at present use energy at a rate too high to be supported by the renewable energy that is likely to be available in the near future.

To avoid widespread famine, the less industrialized countries will also need to change their lifestyles. The impact of the end of the fossil fuel era, as well as the unavoidable early effects of climate change, will make food very expensive. It is therefore vital that countries with rapidly-growing populations should make information and materials for birth control available to all their citizens.

Fossil fuels: a long-term view

In Chapter 2 we saw that in order to avoid dangerous climate change, the world will have to reduce its CO_2 emissions by 90% by 2050. Thus the fossil fuel era will have to end by the middle of the 21st century in order to avoid disastrous climate change. But even if it were not for these considerations, the fossil fuel era would end within a century because of vanishing resources.

As oil becomes scarce, it is likely that coal will be converted to liquid fuels, as was done in Germany during World War II, and in South Africa during the oil embargo. In this process, coal is gasified to form syngas, which is a mixture of CO and H_2. These two gases are then converted to light hydrocarbons by means of Fischer-Tropsch catalysts. Both gasoline and diesel fuel can be made in this way.

If coal is converted to liquid fuels on a large scale, the rate of use of coal will increase. Thus the projected date for the exhaustion of coal reserves based on the present consumption of coal is unrealistic. It is more accurate to lump all fossil fuels together and to predict a future date for their exhaustion based on the lumped consumption of coal, natural gas and oil. Doing so gives a figure of 95 years; but the true figure is likely to be less because of increased rates of consumption. We must remember also that the conversion of coal to liquid fuels requires energy. Of course, neither coal, nor oil, nor natural gas will disappear entirely, but they will become so expensive that their use as fuels will seem inappropriate, and they will be reserved as starting materials for synthesis.

The date at which the possibility for nuclear energy will end is more controversial and difficult to predict. However, it seems likely that if nuclear reactors are used as an energy source despite their great dangers, finite reserves of uranium and thorium will be exhausted by the end of the 21st century.

Optimists point to the possibility of using fusion of light elements, such as hydrogen, to generate power. However, although this can be done on a very small scale (and at great expense) in laboratory experiments, the practical generation of energy by means of thermonuclear reactions remains a mirage rather than a realistic prospect on which planners can rely. The reason for this is the enormous temperature required to produce thermonuclear reactions. This temperature is comparable to that existing in the interior of the sun, and it is sufficient to melt any ordinary container. Elaborate "magnetic bottles" have been constructed to contain thermonuclear reactions, and these have been used in successful very small scale experiments. However, despite 50 years of heavily-financed research, there has been absolutely no success in producing thermonuclear energy on a large scale, or at anything remotely approaching commercially competitive prices.

Thus, after the end of the fossil fuel era, our industrial civilization will probably have to rely on renewable sources to supply our energy needs. These sources include hydropower, wind and tidal power, biomass, geothermal energy and solar energy. Let us try to survey how much energy these sources can be expected to produce.

Before the start of the industrial era, human society relied exclusively on renewable energy sources - but can we do so again, with our greatly increased population and greatly increased demands? Will we ultimately be forced to reduce the global population or our per capita use of energy, or both? Let us now try to examine these questions.

Global energy resources

The total ultimately recoverable resources of fossil fuels amount to roughly 1260 terawatt-years of energy (1 terawatt-year $\equiv 10^{12}$ Watt-years $\equiv$ 1 TWy is equivalent to 5 billion barrels of oil or 1 billion tons of coal). Of this total amount, 760 TWy is coal, while oil and natural gas each constitute roughly 250 TWy.[a] In 1890, the rate of global consumption of energy was 1 terawatt, but by 1990 this figure had grown to 13.2 TW, distributed as follows: oil, 4.6; coal, 3.2; natural gas, 2.4; hydropower, 0.8; nuclear, 0.7; fuelwood, 0.9; crop wastes, 0.4; and dung, 0.2. By 2005, the rate of oil, natural gas and coal consumption had risen to 6.0 TW, 3.7 TW and 3.5 TW respectively. Thus, if we continue to use oil at the 2005 rate, it will last for 42 years, while

---

[a]British Petroleum, "B.P. Statistical Review of World Energy", London, 1991.

natural gas will last for 68 years. The reserves of coal are much larger; and used at the 2005 rate, coal would last for 217 years. However, it seems likely that as oil and natural gas become depleted, coal will be converted to liquid and gaseous fuels, and its rate of use will increase. Also, the total global energy consumption is likely to increase because of increasing population and rising standards of living in the developing countries.

The industrialized countries use much more than their fair share of global resources. For example, with only a quarter of world's population they use more than two thirds of its energy; and in the U.S.A. and Canada the average per capita energy consumption is 12 kilowatts, compared with 0.2 kilowatts in Bangladesh. If we are to avoid severe damage to the global environment, the industrialized countries must rethink some of their economic ideas, especially the assumption that growth can continue forever.

Hubbert peaks for oil and gas

One can predict that as the reserves of oil become exhausted, the price will rise to such an extent that production and consumption will diminish. Thus oil experts do not visualize a special date in the future after which oil will totally disappear, but rather a date at which the production and consumption of oil will reach a maximum and afterward diminish because of scarcity of the resource and increase in price. Such a peak in the production of any nonrenewable resource is called a *Hubbert peak*, after Dr. M. King Hubbert, who applied the idea to oil reserves.

Most experts agree that the Hubbert peak for oil will occur within a decade or two. Thus the era of cheap petroleum is rapidly approaching its end, and we must be prepared for the serious economic and political impacts of rising oil prices, as well as great changes in lifestyle in the industrialized countries. Halfway through the present century, petroleum will become too expensive and rare to be used as a fuel. It will be reserved almost exclusively for lubrication and as a starting material for the manufacture of plastics, paint, fertilizers and pharmaceuticals.

The United States uses petroleum at the rate of more than 7 billion barrels (7 Gb) per year, while that country's estimated reserves and undiscovered resources are respectively 50.7 Gb and 49.0 Gb. Thus if the United States were to rely only on its own resources for petroleum, then, at the 2001 rate of use, these would be exhausted within 14 years. In fact, the United States already imports more than half of its oil. According to the "National Energy Policy" report (sometimes called the "Cheney Report" after its chief author) US domestic oil production will decline from 3.1 Gb/y in 2002 to 2.6 Gb/y in 2020, while US consumption will rise from 7.2 Gb/y to 9.3 Gb/y. Thus the United States today imports 57% of its oil, but the report predicts that by 2020 this will rise to 72%. The predicted increment in US imports of oil between 2002 and 2020 is greater than the present oil consumption of China.

It is clear from these figures that if the United States wishes to maintain its enormous rate of petroleum use, it will have to rely on imported oil, much of it coming from regions of the world that are politically unstable, or else unfriendly to America, or both.

As the per-capita oil consumption of India and China increases, global production will fail to meet demand. For example, if the consumption in these two countries were to increase to 12 barrels per person per year (half the North American level), it would amount to 27 billion barrels per year - roughly the same amount of oil that the whole world uses today. Even a smaller increase in petroleum use by China and India may soon produce an energy crisis. One can anticipate that many voices will then be raised favoring widespread use of nuclear energy. However there would be great dangers associated with such a development.

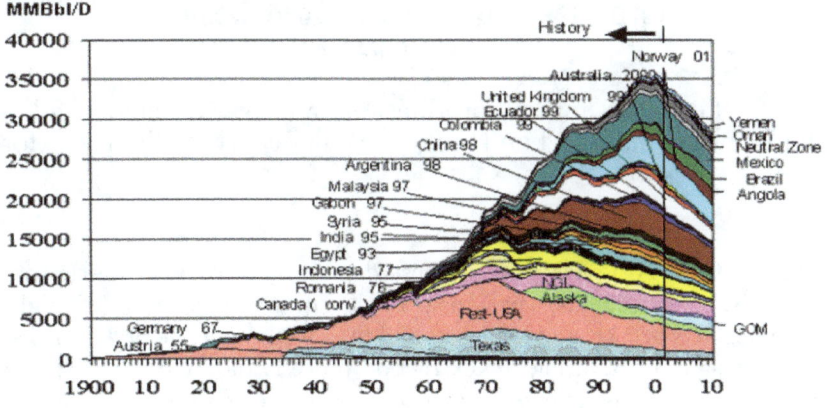

Fig. 3.1 The Hubbert Peak for non-OPEC non-FSU oil. (Wikipedia.)

Oilsands, tarsands and heavy oil

When the Hubbert peak for conventional oil has been passed, the price of oil will steadily increase, and this will make the extraction of oil from unconventional sources more economically feasible. For example, very large deposits of oilsands and tarsands exist in northern Alberta, Canada, a few

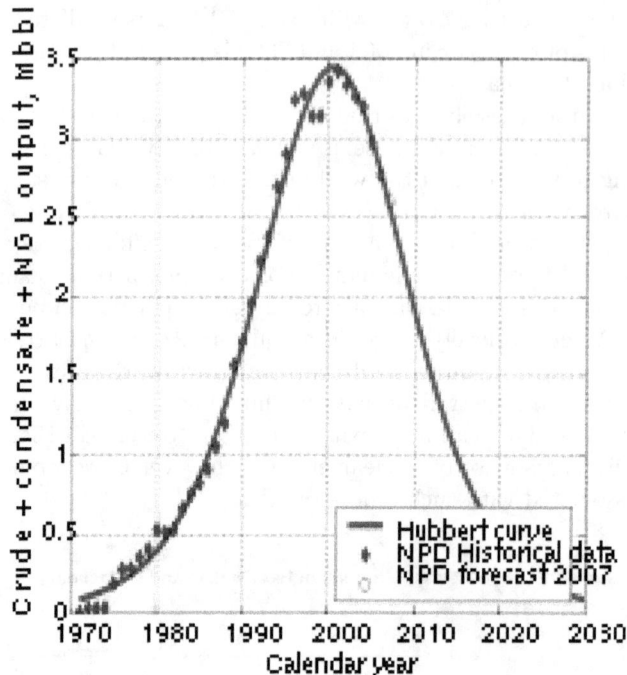

Fig. 3.2 The data for oil production by Norway closely follow the Hubbert model. The Hubbert Peak occurred slightly before 2000. (Wikipedia.)

miles north of Fort McMurray. These deposits, known as the Athabasca oil sands, consist of sand layers near to the surface. Each grain of sand in these deposits is surrounded by a thin film of water, outside of which there is a coating of oil. During the extraction process, the sand is transported to tanks where oil is stripped away from the grains by a hot water flotation process. The oil recovered in this way is too viscous to be pumped, but it can be upgraded to a pumpable fluid by the addition of naphtha. Besides the Athabasca deposit, whose area is twice the size of Lake Ontario, Alberta also has three other smaller oilsand deposits.

The energy inputs for extraction of oil from oilsands are high. It has been estimated that three barrels of oil in the sands can produce only one net barrel of output oil, because the other two barrels are needed to supply energy for the extraction process.

The world's largest deposit of superheavy oil is the "cinturon de la brea" (belt of tar) in Venezuela. This semi-solid material can be made more fluid by the addition of hydrogen. Alternatively it can be emulsified, and the emulsion can be burned in power plants.

The extremely large deposits of unconventional oil in Canada and Venezuela will to some extent cushion the economic shocks produced by scarcity of conventional oil. Nevertheless, because of the high extraction costs of unconventional oil, we must still anticipate that the price of oil will rise steadily after the Hubbert peak has been reached.

Coal

The remaining reserves of coal in the world amount to about 1 exagram, i.e. 10^{18} grams or 10^{12} metric tons. The average energy density of coal is 760 Watt-years/ton, and therefore the world's coal reserves correspond to 760 TWy. If coal continues to be consumed at the present rate of 3.5 TW, the global reserves will last a little more than two centuries. However, it seems likely that as petroleum becomes prohibitively expensive, coal will be converted into liquid fuels, so that the rate of use of coal will increase. Therefore it is more realistic to lump all fossil fuels together and to divide the total supply (1260 TWy) by the the total rate of use (13.2 TW). The result is a prediction that the era of inexpensive fossil fuels will end in less than a century, as is shown in Table 7.4.

67% of the world recoverable reserves of coal are located in four countries:

(1) United States, 27%
(2) Russia, 17%
(3) China, 13%
(4) India, 10%

The present rate of use of coal by China and India is 1.5 billion metric tons per year which is equal to 1.1 TW. However, the rate of coal use by China and India is expected to double by 2030.

Renewable energy

Solar energy

Biomass, wind energy, hydropower and wave power derive their energy indirectly from the sun, but in addition, various methods are available for utilizing the power of sunlight directly. These include photovoltaic panels, solar designs in architecture, solar systems for heating water and cooking, concentrating photovoltaic systems, and solar thermal power plants.

Photovoltaic cells and concentrating photovoltaic systems

Solar photovoltaic cells are thin coated wafers of a semiconducting material (usually silicon). The coatings on the two sides are respectively charge donors and charge acceptors. Cells of this type are capable of trapping solar energy and converting it into direct-current electricity. The electricity generated in this way can be used directly (as it is, for example, in pocket calculators) or it can be fed into a general power grid. Alternatively it can be used to split water into hydrogen and oxygen. The gases can then be compressed and stored, or exported for later use in fuel cells. In the future, we may see solar photovoltaic arrays in sun-rich desert areas producing hydrogen as an export product. As their petroleum reserves become exhausted, the countries of the Middle East and Africa may be able to shift to this new technology and still remain energy exporters.

The cost of manufacturing photovoltaic cells is currently falling at the rate of 3%-5% per year. The cost in 2006 was $4.50 per peak Watt. Usually photovoltaic panels are warranted for a life of 20 years, but they are commonly still operational after 30 years or more. The cost of photovoltaic electricity is today 2-5 times the cost of electricity generated from fossil fuels, but photovoltaic costs are falling rapidly, while the costs of fossil fuels are rising equally rapidly.

Concentrating photovoltaic systems are able to lower costs still further by combining silicon solar cells with reflectors that concentrate the sun's rays. The most inexpensive type of concentrating reflector consists of a flat piece of aluminum-covered plastic material bent into a curved shape along one of its dimensions, forming a trough-shaped surface. (Something like this shape results when we hold a piece of paper at the top and bottom with our two hands, allowing the center to sag.) The axis of the reflector can be oriented so that it points towards the North Star. A photovoltaic array placed along the focal line will then receive concentrated sunlight throughout the day.

Photovoltaic efficiency is defined as the ratio of the electrical power produced by a cell to the solar power striking its surface. For commercially available cells today, this ratio is between 9% and 14%. If we assume 5 hours of bright sunlight per day, this means that a photocell in a desert area near to the equator (where 1 kW/m^2 of peak solar power reaches the earth's surface) can produce electrical energy at the average rate of 20-30 W$_e$/m^2, the average being taken over an entire day and night. (The subscript e means "in the form of electricity". Energy in the form of heat is denoted by the subscript t, meaning "thermal".) Thus the potential power per unit area for photovoltaic systems is far greater than for biomass. However, the mix of renewable energy sources most suitable for a particular country depends on many factors. We saw above that biomass is a promising future source of energy for Sweden, because of Sweden's low population density and high rainfall. By contrast, despite the high initial investment required, photovoltaics are undoubtedly a more promising future energy source for southerly countries with clear skies.

In comparing photovoltaics with biomass, we should be aware of the difference between electrical energy and energy contained in a the chemical bonds of a primary fuel such as wood or rapeseed oil. If Sweden (for example) were to supply all its energy needs from biomass, part of the biomass would have to be burned to generate electricity. The efficiency of energy conversion in electricity generation from fuel is 20%-35%. Of course, in dual use power plants, part of the left-over heat from electrical power generation can be used to heat homes or greenhouses. However, hydropower, wind power and photovoltaics have an advantage in generating electrical power, since they do so directly and without loss, whereas generation of electricity from biomass involves a loss from the inefficiency of the conversion from fuel energy to electrical energy. Thus a rational renewable energy program for Sweden should involve a mixture of biomass for heating and direct fuel use, with hydropower and wind power for generation of electricity. Perhaps photovoltaics will also play a role in Sweden's future electricity generation, despite the country's northerly location and frequently cloudy skies.

The global market for photovoltaics is expanding at the rate of 30% per year. This development is driven by rising energy prices, subsidies to photovoltaics by governments, and the realization of the risks associated with global warming and consequent international commitments to reduce carbon emissions. The rapidly expanding markets have resulted in lowered photovoltaic production costs, and hence further expansion, still lower costs, etc. - a virtuous feedback loop.

Solar thermal power plants

Solar Parabolic Troughs can be used to heat a fluid, typically oil, in a pipe running along the focal axis. The heated fluid can then be used to generate electrical power. The liquid that is heated in this way need not be oil. In a solar thermal power plant in California, reflectors move in a manner that follows the sun's position and they concentrate solar energy onto a tower, where molten salt is heated to a temperature of 1050 degrees F (566°C). The molten salt stores the heat, so that electricity can be generated even when the sun is not shining. The California plant, now in a three-year operating and testing phase, generates 10 MW_e.

Fig. 3.3 Part of the 354 MW SEGS solar complex in northern San Bernardino County, California. (Public domain.)

Solar designs in architecture

At present, the average global rate of use of primary energy is roughly 2 kW$_t$ per person. In North America, the rate is 12 kW$_t$ per capita, while in Europe, the figure is 6 kW$_t$. In Bangladesh, it is only 0.2 kW$_t$. This wide variation implies that considerable energy savings are possible, through changes in lifestyle, and through energy efficiency.

Wind energy

Wind parks in favorable locations, using modern wind turbines, are able to generate 10 MW$_e$/km^2 or 10 W$_e$/m^2. Often wind farms are placed in offshore locations. When they are on land, the area between the turbines can be utilized for other purposes, for example for pasturage. For a country like Denmark, with good wind potential but cloudy skies, wind turbines can be expected to play a more important future role than photovoltaics. Denmark is already a world leader both in manufacturing and in using wind turbines. Today, 23% of all electricity used in Denmark is generated by wind power, and the export of wind turbines makes a major contribution to the Danish economy.

Fig. 3.4 The 11 megawatt Serpa photovoltaic installation in Portugal. (Wikipedia.)

Globally, only 1.5% of all electricity generated comes from wind power. This corresponds to 121 GW_e or 0.121 TW_e. However, the use of wind power is currently growing at the rate of 38% per year. In the United States, it is the fastest-growing form of electricity generation.

The location of wind parks is important, since the energy obtainable from wind is proportional to the cube of the wind velocity. We can understand this cubic relationship by remembering that the kinetic energy of a moving object is proportional to the square of its velocity multiplied by the mass. Since the mass of air moving past a wind turbine is proportional to the wind velocity, the result is the cubic relationship just mentioned.

Before the decision is made to locate a wind park in a particular place, the wind velocity is usually carefully measured and recorded over an entire year. For locations on land, mountain passes are often very favorable locations, since wind velocities increase with altitude, and since the wind is concentrated in the passes by the mountain barrier. Other favorable locations include shorelines and offshore locations on sand bars. This is because onshore winds result when warm air rising from land heated by the sun is replaced by cool marine air. Depending on the season, the situation may

Fig. 3.5 Students perform an experiment using a solar cooker built out of an umbrella. (Public domain.)

be reversed at night, and an offshore wind may be produced if the water is warmer than the land.

The cost of wind-generated electrical power is currently about 5 US cents per kilowatt hour, i.e., lower than the cost of electricity generated by burning fossil fuels.

The "energy payback ratio" of a power installation is defined as the ratio of the energy produced by the installation over its lifetime, divided by the energy required to manufacture, construct, operate and decommission the installation. For wind turbines, this ratio is 17-39, compared with 11 for coal-burning plants. The construction energy of a wind turbine is usually paid back within three months.

Besides the propeller-like design for wind turbines there are also designs where the rotors turn about a vertical shaft. One such design was patented in 1927 by the French aeronautical engineer Georges Jean Marie Darrieus. The blades of a Darrieus wind turbine are airfoils similar to the wings of an aircraft. As the rotor turns in the wind, the stream of air striking the airfoils produces a force similar to the "lift" of an airplane wing. This force pushes the rotor in the direction that it is already moving. The Darrieus design has some advantages over conventional wind turbine design, since

Fig. 3.6 A geothermal power plant at Nesjavellir in Iceland. (Wikipedia.)

the generator can be placed at the bottom of the vertical shaft, where it may be more easily serviced. Furthermore, the vertical shaft can be lighter than the shaft needed to support a conventional wind turbine.

One problem with wind power is that it comes intermittently, and demand for electrical power does not necessarily come at times when the wind is blowing most strongly. To deal with the problem of intermittency, wind power can be combined with other electrical power sources in a grid. Alternatively, the energy generated can be stored, for example by pumped hydroelectric storage or by using hydrogen technology, as will be discussed below.

Bird lovers complain that birds are sometimes killed by rotor blades. This is true, but the number killed is small. For example, in the United States, about 70,000 birds per year are killed by turbines, but this must be compared with 57 million birds killed by automobiles and 97.5 million killed by collisions with plate glass.

The aesthetic aspects of wind turbines also come into the debate. Perhaps in the future, as wind power becomes more and more a necessity and less a matter of choice, this will be seen as a "luxury argument".

Fig. 3.7 Erection of an Enercon E70-4 in Germany. (Wikipedia.)

The case of Samsø

The Danish island of Samsø is only 112 square kilometers in size, and its population numbers only 4,300. Nevertheless, it has a unique distinction. Samsø was the first closed land area to declare its intention of relying entirely on renewable energy, and it has now achieved this aim, provided that one stretches the definitions slightly.

In 1997, the Danish Ministry of Environment and Energy decided to sponsor a renewable-energy contest. In order to enter, communities had to submit plans for how they could make a transition from fossil fuels to renewable energy. An engineer (who didn't live there) thought he knew how the island could do this, and together with the island's mayor he submitted a plan which won the contest. As a result, the islanders became interested in renewable energy. They switched from furnaces to heat pumps, and formed cooperatives for the construction of windmill parks in the sea near to the island. By 2005, Samsø was producing, from renewable sources, more energy than it was using. The islanders still had gasoline-driven automobiles, but they exported from their windmill parks an amount of electrical energy that balanced the fossil fuel energy that they imported. This is a story that can give us hope for the future, although a farming community like Samsø cannot serve as a model for the world.

Biomass

Biomass is defined as any energy source based on biological materials produced by photosynthesis - for example wood, sugar beets, rapeseed oil,[b] crop wastes, dung, urban organic wastes, processed sewage, etc. Using biomass for energy does not result in the net emission of CO_2, since the CO_2 released by burning the material had previously been absorbed from the atmosphere during photosynthesis. If the biological material had decayed instead of being burned, it would have released the same amount of CO_2 as in the burning process.

The solar constant has the value 1.4 kilowatts/m^2. It represents the amount of solar energy per unit area[c] that reaches the earth, before the sunlight has entered the atmosphere. Because the atmosphere reflects 6% and absorbs 16%, the peak power at sea level is reduced to 1.0 kW/m^2. Clouds also absorb and reflect sunlight. Average cloud cover reduces the energy of sunlight a further 36%. Also, we must take into account the fact that the sun's rays do not fall perpendicularly onto the earth's surface. The angle that they make with the surface depends on the time of day, the season and the latitude.

In Sweden, which lies at a northerly latitude, the solar energy per unit of horizontal area is less than for countries nearer the equator. Nevertheless, Göran Persson, the Prime Minister of Sweden, recently announced that his government intends to make the country independent of imported oil by 2020 through a program that includes energy from biomass.

In his thesis, *Biomass in a Sustainable Energy System*, the Swedish researcher Pål Börjesson states that of various crops grown as biomass, the largest energy yields come from short-rotation forests (*Salix viminalis*, a species of willow) and sugar beet plantations. These have an energy yield of from 160 to 170 GJ_t per hectare-year. One can calculate that this is equivalent to about 0.5 MW_t/km^2, or 0.5 W_t/m^2. Thus, although 1.0 kW/m^2 of solar energy reaches the earth at noon at the equator, the trees growing in northerly Sweden can harvest a day-and-night and seasonal average of only 0.5 Watts of thermal energy per horizontal square meter.[d]

---

[b]Canola, a cultivar of *Brassica napus* or *Brassica rapa*.

[c]The area is assumed to be perpendicular to the sun's rays.

[d]In tropical regions, the rate of biomass production can be more than double this amount.

Since Sweden's present primary energy use is approximately 0.04 TW_t, it follows that if no other sources of energy were used, a square area of *Salix* forest 290 kilometers on each side would supply Sweden's present energy needs. This corresponds to an area of 84,000 km^2, about 19% of Sweden's total area.[e] Of course, Sweden's renewable energy program will not rely exclusively on energy crops, but on a mixture of sources, including biomass from municipal and agricultural wastes, hydropower, wind energy and solar energy.

At present, both Sweden and Finland derive about 30% of their electricity from biomass, which is largely in the form of waste from the forestry and paper industries of these two countries.

Despite their northerly location, the countries of Scandinavia have good potentialities for developing biomass as an energy source, since they have small population densities and adequate rainfall. In Denmark, biodiesel oil derived from rapeseed (canola) has been used as fuel for experimental buses. Rapeseed fields produce oil at the rate of between 1,000 and 1,300 liters per hectare-crop. The energy yield is 3.2 units of fuel product energy for every unit of fuel energy used to plant the rapeseed, and to harvest and process the oil. After the oil has been pressed from rapeseed, two-thirds of the seed remains as a protein-rich residue which can be fed to cattle.

Miscanthus is a grassy plant found in Asia and Africa. Some forms will also grow in Northern Europe, and it is being considered as an energy crop in the United Kingdom. *Miscanthus* can produce up to 18 dry tonnes per hectare-year, and it has the great advantage that it can be cultivated using ordinary farm machinery. The woody stems are very suitable for burning, since their water content is low (20-30%).

Jatropha is a fast-growing woody shrub about 4 feet in height, whose seeds can be used to produce diesel oil at the cost of about $43 per barrel. The advantage of *Jatropha* is that is a hardy plant, requiring very little fertilizer and water. It has a life of roughly 50 years, and can grow on wasteland that is unsuitable for other crops. The Indian State Railway has planted 7.5 million *Jatropha* shrubs beside its right of way. The oil harvested from these plants is used to fuel the trains.

For some southerly countries, honge oil, derived from the plant *Pongamia pinnata* may prove to be a promising source of biomass energy. Studies conducted by Dr. Udishi Shrinivasa at the Indian Institute of Sciences in Bangalore indicate that honge oil can be produced at the cost of $150 per ton. This price is quite competitive when compared with other potential fuel oils.

Recent studies have also focused on a species of algae that has an oil

---

[e] Additional land area would be needed to supply the energy required for planting, harvesting, transportation and utilization of the wood.

content of up to 50%. Algae can be grown in desert areas, where cloud cover is minimal. Farm waste and excess CO_2 from factories can be used to speed the growth of the algae.

It is possible that in the future, scientists will be able to create new species of algae that use the sun's energy to generate hydrogen gas. If this proves to be possible, the hydrogen gas may then be used to generate electricity in fuel cells, as will be discussed below in the section on hydrogen technology. Promising research along this line is already in progress at the University of California, Berkeley.

Biogas is defined as the mixture of gases produced by the anaerobic digestion of organic matter. This gas, which is rich in methane (CH_4), is produced in swamps and landfills, and in the treatment of organic wastes from farms and cities. The use of biogas as a fuel is important not only because it is a valuable energy source, but also because methane is a potent greenhouse gas, which should not be allowed to reach the atmosphere. Biogas produced from farm wastes can be used locally on the farm, for cooking and heating, etc. When biogas has been sufficiently cleaned so that it can be distributed in a pipeline, it is known as "renewable natural gas". It may then be distributed in the natural gas grid, or it can be compressed and used in internal combustion engines. Renewable natural gas can also be used in fuel cells, as will be discussed below in the section on Hydrogen Technology.

Biofuels are often classified according to their generation. Those that can be used alternatively as food are called first-generation biofuels. By contrast, biofuels of the second generation are those that make use of crop residues or other cellulose-rich materials. Cellulose molecules are long chains of sugars, and by breaking the inter-sugar bonds in the chain using enzymes or other methods, the sugars can be freed for use in fermentation. In this way lignocellulosic ethanol is produced. The oil-producing and hydrogen-producing algae mentioned above are examples of third-generation biofuels.

We should notice that growing biofuels locally (even first-generation ones) may be of great benefit to smallholders in developing countries, since they can achieve local energy self-reliance in this way.

Competition between food and biofuels

Although it is largely a future problem rather than a present one, we can see today the start of a competition between food production and first-generation biofuels. In 2007, 27% of the US corn (maize) crop was used to produce ethanol for motor fuel, an increase by a factor of more than 5 from 1996. In 1996, none of the soybean oil produced in the US was used for biodiesel, while in 2007, more than 17% of the crop was used for fuel. South American soybean oil is also used for biodiesel. In Europe, biodiesel comes mainly from rapeseed (canola) oil, but this can also be used for food. Brazilian ethanol comes from sugar cane produced on land that alternatively could be used for food production. Biodiesel from South East Asia is mainly edible palm oil. Thus, many (but not all!) of the major biofuels are produced from feedstocks that could be used to produce food, or are grown on land that could be used for food production.

There is a danger that a food-versus-fuel competition will develop between the world's 860 million automobiles and its 2 billion poorest people. As Lester Brown puts it in his book *Plan B*, "Suddenly the world is faced by a moral and political issue for which there is no precedent: Should we use grain to fuel cars or to feed people? The average income of the world's automobile owners is roughly $30,000 a year; the 2 billion poorest people earn on average less than $3,000 a year. The market says, 'Let's fuel the cars'." It is up to the world's collective conscience to overrule the market on this point.

Future food output will also be decreased because, as petroleum prices become prohibitively high, synthetic fibers based on petroleum feedstocks will be less and less used. The additional land area needed to produce wool, cotton or linen for clothing will have to be subtracted from the area available for growing food. Finally, as petroleum disappears, draft animals may again be used in farming, and pasturage for them will have to be subtracted from the land available for agriculture. These factors will contribute to the predicted global food crisis discussed in Chapter 4.

Hydroelectric power

At present 20% of the world's electricity comes from hydroelectric power. In the developed countries, the potential for increasing this percentage is small, because most of the suitable sites for dams are already in use. Mountainous regions of course have the greatest potential for hydroelectric power, and this correlates well with the fact that virtually all of the electricity generated in Norway comes from hydro, while in Iceland and Austria the figures are respectively 83% and 67%. Among the large hydroelectric power stations now in use are the La Grande complex in Canada (16 GW_e) and the Itapú station on the border between Brazil and Paraguay (14 GW_e). The Three Gorges Dam under construction in China is planned to produce 18.2 GW_e by 2009.

Even in regions where the percentage of hydro in electricity generation is not so high, it plays an important role because hydropower can be used selectively at moments of peak demand. Pumping of water into reservoirs can also be used to store energy.

The creation of lakes behind new dams in developing countries often involves problems, for example relocation of people living on land that will be covered by water, and loss of the land for other purposes.[f] However the energy gain per unit area of lake can be very large - over 100 W_e/m^2. Fish ladders can be used to enable fish to reach their spawning grounds above dams. In addition to generating electrical power, dams often play useful roles in flood control and irrigation.

At present, hydroelectric power is used in energy-intensive industrial processes, such as the production of aluminum. However, as the global energy crisis becomes more severe, we can expect that metals derived from electrolysis, such as aluminum and magnesium, will be very largely replaced by other materials, because the world will no longer be able to afford the energy needed to produce them.

Fig. 3.8 Construction of the Three Gorges Dam (2006). (Wikimedia Commons.)

---

[f]Over a million people were displaced by the construction of the Three Gorges Dam in China, and many sites of cultural value were lost.

Geothermal energy

The ultimate source of geothermal energy is the decay of radioactive nuclei in the interior of the earth. Because of the heat produced by this radioactive decay, the temperature of the earth's core is 4300°C. The inner core is composed of solid iron, while the outer core consists of molten iron and sulfur compounds. Above the core is the mantle, which consists of a viscous liquid containing compounds of magnesium, iron, aluminum, silicon and oxygen. The temperature of the mantle gradually decreases from 3700°C near the core to 1000°C near the crust. The crust of the earth consists of relatively light solid rocks and it varies in thickness from 5 to 70 km.

The outward flow of heat from radioactive decay produces convection currents in the interior of the earth. These convection currents, interacting with the earth's rotation, produce patterns of flow similar to the trade winds of the atmosphere. One result of the currents of molten conducting material in the interior of the earth is the earth's magnetic field. The crust is divided into large sections called "tectonic plates", and the currents of molten material in the interior of the earth also drag the plates into collision with each other. At the boundaries, where the plates collide or split apart, volcanic activity occurs. Volcanic regions near the tectonic plate boundaries are the best sites for collection of geothermal energy.

The entire Pacific Ocean is ringed by regions of volcanic and earthquake activity, the so-called Ring of Fire. This ring extends from Tierra del Fuego at the southernmost tip of South America, northward along the western coasts of both South America and North America to Alaska. The ring then crosses the Pacific at the line formed by the Aleutian Islands, and it reaches the Kamchatka Peninsula in Russia. From there it extends southward along the Kuril Island chain and across Japan to the Philippine Islands, Indonesia and New Zealand. Many of the islands of the Pacific are volcanic in nature. Another important region of volcanic activity extends northward along the Rift Valley of Africa to Turkey, Greece and Italy. In the Central Atlantic region, two tectonic plates are splitting apart, thus producing the volcanic activity of Iceland. All of these regions are very favorable for the collection of geothermal power.

The average rate at which the energy created by radioactive decay in the interior of the earth is transported to the surface is 0.06 W_t/m^2. However, in volcanic regions near the boundaries of tectonic plates, the rate at which the energy is conducted to the surface is much higher - typically 0.3 W_t/m^2. If we insert these figures into the thermal conductivity law

$$q = K_T \frac{\Delta T}{z}$$

we can obtain an understanding of the types of geothermal resources available throughout the world. In the thermal conductivity equation, q is the power conducted per unit area, while K_T is the thermal conductivity of the material through the energy is passing. For sandstones, limestones and most crystalline rocks, thermal conductivities are in the range 2.5-3.5 $W_t/(m\ ^oC)$. Inserting these values into the thermal conductivity equation, we find that in regions near tectonic plate boundaries we can reach temperatures of 200 ^oC by drilling only 2 kilometers into rocks of the types named above. If the strata at that depth contain water, it will be in the form of highly-compressed steam. Such a geothermal resource is called a *high-enthalpy* resource.[g]

In addition to high-enthalpy geothermal resources there are *low-enthalpy* resources in nonvolcanic regions of the world, especially in basins covered by sedimentary rocks. Clays and shales have a low thermal conductivity, typically 1-2 $W_t/(m\ ^oC)$. When we combine these figures with the global average geothermal power transmission, $q = 0.06\ W_t/m^2$, the thermal conduction equation tells us that $\Delta T/z = 0.04\ ^oC/m$. In such a region the geothermal resources may not be suitable for the generation of electrical power, but nevertheless adequate for heating buildings. The Creil district heating scheme north of Paris is an example of a project where geothermal energy from a low enthalpy resource is used for heating buildings.

The total quantity of geothermal electrical power produced in the world today is 8 GW_e, with an additional 16 GW_t used for heating houses and buildings. In the United States alone, 2.7 GW_e are derived from geothermal sources. In some countries, for example Iceland and Canada, geothermal energy is used both for electrical power generation and for heating houses.

There are three methods for obtaining geothermal power in common use today: Deep wells may yield dry steam, which can be used directly to drive turbines. Alternatively water so hot that it boils when brought to the surface may be pumped from deep wells in volcanic regions. The steam is then used to drive turbines. Finally, if the water from geothermal wells is less hot, it may be used in binary plants, where its heat is exchanged with an organic fluid which then boils. In this last method, the organic vapor drives the turbines. In all three methods, water is pumped back into the wells to be reheated. The largest dry steam field in the world is The Geysers, 145 kilometers north of San Francisco, which produces 1,000 MW_e.

There is a fourth method of obtaining geothermal energy, in which water is pumped down from the surface and is heated by hot dry rocks. In order

---

[g]Enthalpy $\equiv H \equiv U + PV$ is a thermodynamic quantity that takes into account not only the internal energy U of a gas, but also energy PV that may be obtained by allowing it to expand.

to obtain a sufficiently large area for heat exchange the fissure systems in the rocks must be augmented, for example by pumping water down at high pressures several hundred meters away from the collection well. The European Union has established an experimental station at Soultz-sous-Forets in the Upper Rhine to explore this technique. The experiments performed at Soultz will determine whether the "hot dry rock" method can be made economically viable. If so, it can potentially offer the world a very important source of renewable energy.

The molten lava of volcanoes also offers a potential source of geothermal energy that may become available in the future, but at present, no technology has been developed that is capable of using it.

Hydrogen technologies

When water containing a little acid is placed in a container with two electrodes and subjected to an external direct current voltage greater than 1.23 Volts, bubbles of hydrogen gas form at one electrode (the cathode), while bubbles of oxygen gas form at the other electrode (the anode). At the cathode, the half-reaction

$$2H_2O(l) \to O_2(g) + 4H^+(aq) + 4e^- \qquad E^0 = -1.23\ Volts$$

takes place, while at the anode, the half-reaction

$$4H^+(aq) + 4e^- \to 2H_2(g) \qquad E^0 = 0$$

occurs.

Half-reactions differ from ordinary chemical reactions in containing electrons either as reactants or as products. In electrochemical reactions, such as the electrolysis of water, these electrons are either supplied or removed by the external circuit. When the two half-reactions are added together, we obtain the total reaction:

$$2H_2O(l) \to O_2(g) + 2H_2(g) \qquad E^0 = -1.23\ Volts$$

Notice that $4H^+$ and $4e^-$ cancel out when the two half-reactions are added. The total reaction does not occur spontaneously (as is discussed in Appendix A), but it can be driven by an external potential E, provided that the magnitude of E is greater than 1.23 volts.

When this experiment is performed in the laboratory, platinum is often used for the electrodes, but electrolysis of water can also be performed using electrodes made of graphite.

Electrolysis of water to produce hydrogen gas has been proposed as a method for energy storage in a future renewable energy system. For example, it might be used to store energy generated by photovoltaics in desert areas of the world. Compressed hydrogen gas could then be transported to other regions and used in fuel cells. Electrolysis of water and storage of hydrogen could also be used to solve the problem of intermittency associated with wind energy or solar energy.

Hydrogen fuel cells

Fuel cells allow us to convert the energy of chemical reactions directly into electrical power. In hydrogen fuel cells, for example, the exact reverse of the electrolysis of water takes place. Hydrogen reacts with oxygen, and produces electricity and water, the reaction being

$$O_2(g) + 2H_2(g) \to 2H_2O(l) \qquad E^0 = 1.23\ Volts$$

The arrangement of the a hydrogen fuel cell is such that the hydrogen cannot react directly with the oxygen, releasing heat. Instead, two half reactions take place, one at each electrode, as was just mentioned in connection with the electrolysis of water. In a hydrogen fuel cell, hydrogen gas produces electrons and hydrogen H^+ ions at one of the electrodes.

$$2H_2(g) \to 4H^+(aq) + 4e^- \qquad E^0 = 0$$

The electrons flow through the external circuit to the oxygen electrode, while the hydrogen ions complete the circuit by flowing through the interior of the cell (from which the hydrogen and oxygen molecules are excluded by semipermeable membranes) to the oxygen electrode. Here the electrons react with oxygen molecules and H^+ ions to form water.

$$O_2(g) + 4H^+(aq) + 4e^- \to 2H_2O(l) \qquad E^0 = 1.23\ Volts$$

In this process, a large part of the chemical energy of the reaction becomes available as electrical power.

We can recall that the theoretical maximum efficiency of a heat engine operating between a cold reservoir at temperature T_C and a hot reservoir at T_H is 1-T_C/T_H, where the temperatures are expressed on the Kelvin scale. Since fuel cells are not heat engines, their theoretical maximum efficiency is not limited in this way. Thus it can be much more efficient to generate electricity by reacting hydrogen and oxygen in a fuel cell than it would be to burn the hydrogen in a heat engine and then use the power of the engine to drive a generator.

Hydrogen technologies are still at an experimental stage. Furthermore, they do not offer us a source of renewable energy, but only means for storage, transportation and utilization of energy derived from other sources.

Nevertheless, it seems likely that hydrogen technologies will have great importance in the future.

Germany's ban on internal combustion energy vehicles

The Parliament of Germany recently voted to ban the sale of internal combustion engine motor vehicles after 2030.[h]

Germany's Bundesrat, its upper house of parliament, passed a bipartisan resolution calling for a ban on sales of new vehicles powered by internal combustion engines, which includes both gasoline and diesel.

In a statement reported in the newsmagazine Der Spiegel, the Green Party lawmaker Oliver Krischer said:"If the Paris agreement to curb climate-warming emissions is to be taken seriously, no new combustion engine cars should be allowed on roads after 2030".

This remarkably farsighted action by the German Parliament will have enormously important consequences. More than a million electric vehicles are already on the world's roads.[i] They need a network of stations at which their batteries can be recharged, but once a critical number is passed, such recharging stations will become very widespread. Germany's farsighted action gives advance warning to the automotive industry, so that it can start today to re-tool for the coming revolution in transportation.

Some concluding remarks on energy

It can be seen from our discussion of renewable energy technologies that they can potentially offer a partial replacement for the fossil fuels on which the world is now dependent. All forms of renewable energy should be developed simultaneously, since all will be needed. Energy conservation and changes of lifestyle will also be necessary. Much of the limited amount of energy that will be available in the future will be needed for agriculture, and therefore less energy will be available for transportation and industry.

It seems likely that photovoltaics, solar thermal power, wind power, biomass and wave power will become the major energy sources of the future. In addition, hydropower is extremely helpful in overcoming the problem of intermittency, while other forms of renewable energy may have great advantages in certain locations.

The transition to renewable energy will require wholehearted governmental commitment, tax changes, and a considerable investment in research. At present nuclear energy, nuclear research and the oil industry all

---

[h] http://www.ecowatch.com/germany-bans-combustion-engine-cars-2037788435.html

[i] http://www.ecowatch.com/1-million-electric-cars-are-now-on-the-worlds-roads-1891162177.html

receive enormous governmental support. It is vital that this support should go instead to renewable energy technologies.

The time factor is also important. The Hubbert peak for oil will occur in a decade, and the peak for natural gas in two decades. After that, the outlook for the future is that petroleum and natural gas will become more and more expensive - finally so expensive that they will not be used as fuels. To minimize the shock of these events, and to avoid dangerous climate change, serious work on substitutes must begin immediately, and on a large scale. At present the development of renewable energy is proceeding so slowly that if the trend is not corrected, we can anticipate a period of great energy scarcity and economic trauma.

The transition to renewable energy will involve rededication of much land from agriculture to energy generation. This will be easiest in countries where the population density is low, and difficult in countries that already have problems in feeding their people.

Metals

W. David Menzie (Chief of the Minerals Information Team of the U.S. Geological Survey) testified to a committee of the U.S. House of Representatives in 2006 that global reserves of copper are approximately 470 million tons. He also stated that world consumption of copper in 2000 was 14.9 million tons per year, but that it is increasing at 3.1% per year and is expected to reach 27 Mt/y by 2020. Menzie predicted that most of this increase will be in the developing countries. For example, China's use of copper is expected to increase from 2 Mt/y in 2000 to 5.6 Mt/y in 2020, while for India, the increase will be from 0.4 Mt/y to 1.6 Mt/y.

At the 2000 rate of use, global copper reserves will be exhausted in 31 years, while if used at a higher rate, the reserves will last for a shorter time. It is predicted that a Hubbert peak will occur for copper, analogous to the Hubbert peaks for petroleum and natural gas. Thus, copper will not disappear entirely, but there will be a date when the production of copper will reach a maximum and afterward decline because of rising prices.

Reserve indices

The reserve index of a metal is defined as the size of its reserves divided by the current annual rate of production. Today, many metals have reserve indices between 10 years and 100 years. These include indium, tantalum, gold, bismuth, silver, cadmium, cobalt, arsenic, tungsten, molybdenum, tin, nickel, lead, zinc, and copper, while magnesium and iron have reserve indices of approximately 100 years.[j]

Recycling metals

Future exploration may increase the size of known reserves of metals; and future advances in technology may also make it possible to use lower grade ores. However, we must remember that the extraction of metals from their ores requires much energy. In the long-term future, energy will probably not be available for the production of (for example) iron, steel, and aluminum on the scale that we know today. Thus, recycling will assume great importance.

Substitutes for metals

It seems likely that composite materials, such as carbon-fiber-reinforced plastic and glass-reinforced plastic (fiberglass), will become important in the future as substitutes for metals. Carbon fiber consists of threads of carbon as thin as 6 microns (0.006 mm). The carbon atoms in such a fiber are bonded together in crystals, aligned along the axis of the fiber, and in this configuration they have incredible strength in relation to their weight. In the composite material, the carbon fibers are protected by a resin. The result is a material that has both toughness and an extremely high strength-to-weight ratio.

Suggestions for further reading

(1) P.R. Ehrlich and A.H. Ehrlich, *One With Nineveh: Politics, Consumption and the Human Future*, Island Press, (2004).
(2) D.H. Meadows, D.L. Meadows, J. Randers, and W.W. Behrens III, *The Limits to Growth: A Report for the Club of Rome's Project on the Predicament of Mankind*, Universe Books, New York, (1972).
(3) D.H. Meadows et al., *Beyond the Limits. Confronting Global Collapse and Envisioning a Sustainable Future*, Chelsea Green Publishing, Post Mills, Vermont, (1992).
(4) D.H. Meadows, J. Randers and D.L. Meadows, *Limits to Growth: the 30-Year Update*, Chelsea Green Publishing, White River Jct., VT 05001, (2004).
(5) A. Peccei and D. Ikeda, *Before it is Too Late*, Kodansha International, Tokyo, (1984).

---

[j]Craig, J.R., Vaugn, D.J. and Skinner, B.J., *Resources of the Earth: Origin, Use and Environmental Impact*, Third Edition, page 64.

(6) V.K. Smith, ed., *Scarcity and Growth Reconsidered*, Johns Hopkins University Press, Baltimore, (1979).
(7) British Petroleum, *BP Statistical Review of World Energy*, (published yearly).
(8) R. Costannza, ed., *Ecological Economics: The Science and Management of Sustainability*, Colombia University Press, New York, (1991).
(9) J. Darmstadter, *A Global Energy Perspective*, Sustainable Development Issue Backgrounder, Resources for the Future, (2002).
(10) D.C. Hall and J.V. Hall, *Concepts and Measures of Natural Resource Scarcity*, Journal of Environmental Economics and Management, **11**, 363-379, (1984).
(11) M.K. Hubbert, *Energy Resources*, in *Resources and Man: A Study and Recommendations*, Committee on Resources and Man, National Academy of Sciences, National Research Council, W.H. Freeman, San Francisco, (1969).
(12) J.A. Krautkraemer, *Nonrenewable Resource Scarcity*, Journal of Economic Literature, bf 36, 2065-2107, (1998).
(13) C.J. Cleveland, *Physical and Economic Aspects of Natural Resource Scarcity: The Cost of Oil Supply in the Lower 48 United States 1936-1987*, Resources and Energy **13**, 163-188, (1991).
(14) C.J. Cleveland, *Yield Per Effort for Additions to Crude Oil Reserves in the Lower 48 States, 1946-1989*, American Association of Petroleum Geologists Bulletin, **76**, 948-958, (1992).
(15) M.K. Hubbert, *Technique of Prediction as Applied to the Production of Oil and Gas*, in *NBS Special Publication 631*, US Department of Commerce, National Bureau of Standards, (1982).
(16) L.F. Ivanhoe, *Oil Discovery Indices and Projected Discoveries*, Oil and Gas Journal, **11**, 19, (1984).
(17) L.F. Ivanhoe, *Future Crude Oil Supplies and Prices*, Oil and Gas Journal, July 25, 111-112, (1988).
(18) L.F. Ivanhoe, *Updated Hubbert Curves Analyze World Oil Supply*, World Oil, November, 91-94, (1996).
(19) L.F. Ivanhoe, *Get Ready for Another Oil Shock!*, The Futurist, January-February, 20-23, (1997).
(20) Energy Information Administration, *International Energy Outlook, 2001*, US Department of Energy, (2001).
(21) Energy Information Administration, *Caspian Sea Region*, US Department of Energy, (2001).
(22) National Energy Policy Development Group, *National Energy Policy*, The White House, (2004). (http://www.whitehouse.gov/energy/)
(23) IEA, *CO_2 from Fuel Combustion Fact-Sheet*, International Energy Agency, (2005).

(24) H. Youguo, *China's Coal Demand Outlook for 2020 and Analysis of Coal Supply Capacity*, International Energy Agency, (2003).
(25) R.H. Williams, *Advanced Energy Supply Technologies*, in *World Energy Assessment: Energy and the Challenge of Sustainability*, UNDP, (2000).
(26) H. Lehmann, *Energy Rich Japan*, Institute for Sustainable Solutions and Innovations, Achen, (2003).
(27) W.V. Chandler, *Materials Recycling: The Virtue of Necessity*, Worldwatch Paper 56, Worldwatch Institute, Washington D.C, (1983).
(28) W.C. Clark and others, *Managing Planet Earth*, Special Issue, *Scientific American*, September, (1989).
(29) B. Commoner, *The Closing Circle: Nature, Man and Technology*, Bantam Books, New York, (1972).
(30) J.R. Frisch, *Energy 2000-2020: World Prospects and Regional Stresses*, World Energy Conference, Graham and Trotman, (1983).
(31) J. Holdren and P. Herrera, *Energy*, Sierra Club Books, New York, (1971).
(32) National Academy of Sciences, *Energy and Climate*, NAS, Washington D.C., (1977).
(33) W. Ophuls, *Ecology and the Politics of Scarcity*, W.H. Freeman, San Francisco, (1977).
(34) C. Pollock, *Mining Urban Wastes: The Potential for Recycling*, Worldwatch Paper 76, Worldwatch Institute, Washington D.C., (1987).
(35) World Resources Institute, *World Resources*, Oxford University Press, New York, (published annually).
(36) World Resources Institute, *World Resources 2000-2001: People and Ecosystems: The Fraying Web of Life*, WRI, Washington D.C., (2000).
(37) J.E. Young, John E., *Mining the Earth*, Worldwatch Paper 109, Worldwatch Institute, Washington D.C., (1992).
(38) J.R. Craig, D.J. Vaughan and B.J. Skinner, *Resources of the Earth: Origin, Use and Environmental Impact*, Third Edition, Prentice Hall, (2001).
(39) W. Youngquist, *Geodestinies: The Inevitable Control of Earth Resources Over Nations and Individuals*, National Book Company, Portland Oregon, (1997).
(40) M. Tanzer, *The Race for Resources. Continuing Struggles Over Minerals and Fuels*, Monthly Review Press, New York, (1980).
(41) C.B. Reed, *Fuels, Minerals and Human Survival*, Ann Arbor Science Publishers Inc., Ann Arbor Michigan, (1975).
(42) M.K. Hubbert, *Energy Resources*, in *Resources and Man: A Study and Recommendations*, Committee on Resources and Man, National Academy of Sciences, National Research Council, W.H. Freeman, San

Francisco, (1969).
(43) J.A. Krautkraemer, *Nonrenewable Resource Scarcity, Journal of Economic Literature*, bf 36, 2065-2107, (1998).
(44) C.J. Cleveland, *Physical and Economic Aspects of Natural Resource Scarcity: The Cost of Oil Supply in the Lower 48 United States 1936-1987*, Resources and Energy **13**, 163-188, (1991).
(45) C.J. Cleveland, *Yield Per Effort for Additions to Crude Oil Reserves in the Lower 48 States, 1946-1989*, American Association of Petroleum Geologists Bulletin, **76**, 948-958, (1992).
(46) M.K. Hubbert, *Technique of Prediction as Applied to the Production of Oil and Gas*, in *NBS Special Publication 631*, US Department of Commerce, National Bureau of Standards, (1982).
(47) Energy Information Administration, *International Energy Outlook, 2001*, US Department of Energy, (2001).
(48) Energy Information Administration, *Caspian Sea Region*, US Department of Energy, (2001).
(49) National Energy Policy Development Group, *National Energy Policy*, The White House, (2004). (http://www.whitehouse.gov/energy/)
(50) M. Klare, *Bush-Cheney Energy Strategy: Procuring the Rest of the World's Oil*, Foreign Policy in Focus, (Interhemispheric Resource Center/Institute for Policy Studies/SEEN), Washington DC and Silver City NM, January, (2004).
(51) IEA, *CO2 from Fuel Combustion Fact-Sheet*, International Energy Agency, (2005).
(52) H. Youguo, *China's Coal Demand Outlook for 2020 and Analysis of Coal Supply Capacity*, International Energy Agency, (2003).
(53) R.H. Williams, *Advanced Energy Supply Technologies*, in *World Energy Assessment: Energy and the Challenge of Sustainability*, UNDP, (2000).
(54) H. Lehmann, *Energy Rich Japan*, Institute for Sustainable Solutions and Innovations, Achen, (2003).
(55) W.V. Chandler, *Materials Recycling: The Virtue of Necessity*, Worldwatch Paper 56, Worldwatch Institute, Washington D.C, (1983).
(56) J.R. Frisch, *Energy 2000-2020: World Prospects and Regional Stresses*, World Energy Conference, Graham and Trotman, (1983).
(57) J. Gever, R. Kaufmann, D. Skole and C. Vorosmarty, *Beyond Oil: The Threat to Food and Fuel in the Coming Decades*, Ballinger, Cambridge MA, (1986).
(58) J. Holdren and P. Herrera, *Energy*, Sierra Club Books, New York, (1971).
(59) National Academy of Sciences, *Energy and Climate*, NAS, Washington D.C., (1977).

(60) W. Ophuls, *Ecology and the Politics of Scarcity*, W.H. Freeman, San Francisco, (1977).
(61) P.B. Smith, J.D. Schilling and A.P. Haines, *Introduction and Summary*, in *Draft Report of the Pugwash Study Group: The World at the Crossroads*, Berlin, (1992).
(62) World Resources Institute, *World Resources*, Oxford University Press, New York, (published annually).
(63) J.R. Craig, D.J. Vaughan and B.J. Skinner, *Resources of the Earth: Origin, Use and Environmental Impact*, Third Edition, Prentice Hall, (2001).
(64) W. Youngquist, *Geodestinies: The Inevitable Control of Earth Resources Over Nations and Individuals*, National Book Company, Portland Oregon, (1997).
(65) M. Tanzer, *The Race for Resources. Continuing Struggles Over Minerals and Fuels*, Monthly Review Press, New York, (1980).
(66) C.B. Reed, *Fuels, Minerals and Human Survival*, Ann Arbor Science Publishers Inc., Ann Arbor Michigan, (1975).
(67) A.A. Bartlett, *Forgotten Fundamentals of the Energy Crisis*, American Journal of Physics, **46**, 876-888, (1978).
(68) N. Gall, *We are Living Off Our Capital*, Forbes, September, (1986).

Chapter 4

The Global Food and Refugee Crisis

"Unless progress with agricultural yields remains very strong, the next century will experience human misery that, on a sheer numerical scale, will exceed everything that has come before."

Nobel Laureate Norman Borlaug speaking of a global food crisis in the 21st century

Introduction

As glaciers melt in the Himalayas, depriving India and China of summer water supplies; as sea levels rise, drowning the fertile rice fields of Vietnam and Bangladesh; as drought threatens the productivity of grain-producing regions of North America; and as the end of the fossil fuel era impacts modern high-yield agriculture, there is a threat of wide-spread famine. There is a danger that the 1.5 billion people who are undernourished today will not survive an even more food-scarce future.

People threatened with famine will become refugees, desperately seeking entry into countries where food shortages are less acute. Wars, such as those currently waged in the Middle East, will add to the problem.

What can we do to avoid this crisis, or at least to reduce its severity? We must urgently address the problem of climate change; and we must shift money from military expenditure to the support of birth control programs and agricultural research. We must also replace the institution of war by a system of effective global governance and enforcible international laws.

Optimum population in the distant future

What is the optimum population of the world? It is certainly not the maximum number that can be squeezed onto the globe by eradicating every species of plant and animal that cannot be eaten. The optimum global population is one that can be supported in comfort, equality and dignity - and with respect for the environment.

In 1848 (when there were just over one billion people in the world), John Stuart Mill described the optimal global population in the following words:

"The density of population necessary to enable mankind to obtain, in the greatest degree, all the advantages of cooperation and social intercourse, has, in the most populous countries, been attained. A population may be too crowded, although all be amply supplied with food and raiment."

"... Nor is there much satisfaction in contemplating the world with nothing left to the spontaneous activity of nature; with every rood of land brought into cultivation, which is capable of growing food for human beings; every flowery waste or natural pasture plowed up, all quadrupeds or birds which are not domesticated for man's use exterminated as his rivals for food, every hedgerow or superfluous tree rooted out, and scarcely a place left where a wild shrub or flower could grow without being eradicated as a weed in the name of improved agriculture. If the earth must lose that great portion of its pleasantness which it owes to things that the unlimited increase of wealth and population would extirpate from it, for the mere

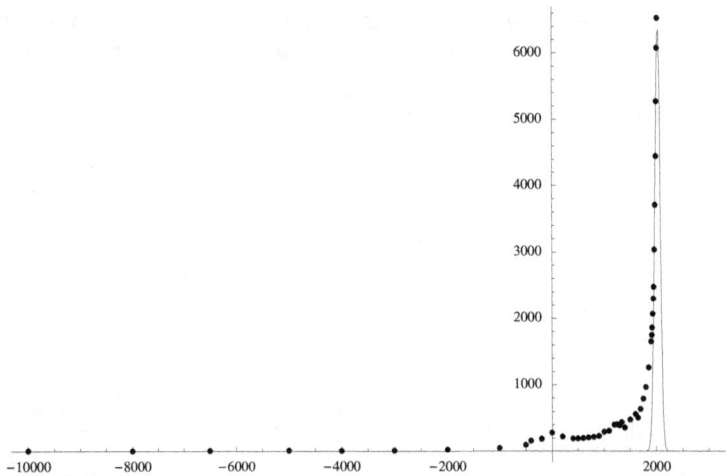

Fig. 4.1 Population growth and fossil fuel use, seen on a time-scale of several thousand years. The dots are population estimates in millions from the US Census Bureau. Fossil fuel use appears as a spike-like curve, rising from almost nothing to a high value, and then falling again to almost nothing in the space of a few centuries. When the two curves are plotted together, the explosive rise of global population is seen to be simultaneous with, and perhaps partially driven by, the rise of fossil fuel use. This raises the question of whether the world's population is headed for a crash when the fossil fuel era has ended. (Author's own graph.)

purpose of enabling it to support a larger, but not better or happier population, I sincerely hope, for the sake of posterity, that they will be content to be stationary, long before necessity compels them to it."[a]

Has the number of humans in the world already exceeded the earth's sustainable limits? Will the global population of humans crash catastrophically after having exceeded the carrying capacity of the environment? There is certainly a danger that this will happen - a danger that the 21st century will bring very large scale famines to vulnerable parts of the world, because modern energy-intensive agriculture will be dealt a severe blow by prohibitively high petroleum prices, and because climate change will reduce the world's agricultural output. When the major glaciers in the Himalayas have melted, they will no longer be able to give India and China summer water supplies; rising oceans will drown much agricultural land; and

---

[a]John Stuart Mill, *Principles of Political Economy, With Some of Their Applications to Social Philosophy*, (1848).

aridity will reduce the output of many regions that now produce much of the world's grain. Falling water tables in overdrawn aquifers, and loss of topsoil will add to the problem. We should be aware of the threat of a serious global food crisis in the 21st century if we are to have a chance of avoiding it.

The term *ecological footprint* was introduced by William Rees and Mathis Wackernagel in the early 1990's to compare demands on the environment with the earth's capacity to regenerate. In 2005, humanity used environmental resources at such a rate that it would take 1.3 earths to renew them. In other words, we have already exceeded the earth's carrying capacity. Since eliminating the poverty that characterizes much of the world today will require more resources per capita, rather than less. it seems likely that in the era beyond fossil fuels, the optimum global population will be considerably less than the present population of the world.

Population growth and the Green Revolution

Limitations on cropland

In 1944 the Norwegian-American plant geneticist Norman Borlaug was sent to Mexico by the Rockefeller Foundation to try to produce new wheat varieties that might increase Mexico's agricultural output. Borlaug's dedicated work on this project was spectacularly successful. He remained with the project for 16 years, and his group made 6,000 individual crossings of wheat varieties to produce high-yield disease-resistant strains.

In 1963, Borlaug visited India, bringing with him 100 kg. of seeds from each of his most promising wheat strains. After testing these strains in Asia, he imported 450 tons of the Lerma Rojo and Sonora 64 varieties - 250 tons for Pakistan and 200 for India. By 1968, the success of these varieties was so great that school buildings had to be commandeered to store the output. Borlaug's work began to be called a "Green Revolution". In India, the research on high-yield crops was continued and expanded by Prof. M.S. Swaminathan and his coworkers. The work of Green Revolution scientists, such Norman Borlaug and M.S. Swaminathan, has been credited with saving the lives of as many as a billion people.

Despite these successes, Borlaug believes that the problem of population growth is still a serious one. "Africa and the former Soviet republics", Borlaug states, "and the Cerrado,[b] are the last frontiers. After they are in use, the world will have no additional sizable blocks of arable land left to put into production, unless you are willing to level whole forests, which you

---

[b]The Cerrado is a large savanna region of Brazil.

Fig. 4.2 Professor M.S. Swaminathan, father of the Green Revolution in India. (Open and Shut7.)

should not do. So, future food-production increases will have to come from higher yields. And though I have no doubt that yields will keep going up, whether they can go up enough to feed the population monster is another matter. Unless progress with agricultural yields remains very strong, the next century will experience human misery that, on a sheer numerical scale, will exceed the worst of everything that has come before."

With regard to the prospect of increasing the area of cropland, a report by the United Nations Food and Agricultural Organization (*Provisional*

Fig. 4.3 Norman Borlaug and agronomist George Harrer in 1943. (Human Wrongs Watch.)

Indicative World Plan for Agricultural Development, FAO, Rome, 1970) states that "In Southern Asia,... in some countries of Eastern Asia, in the Near East and North Africa... there is almost no scope for expanding agricultural area... In the drier regions, it will even be necessary to return to permanent pasture the land that is marginal and submarginal for cultivation. In most of Latin America and Africa south of the Sahara, there are still considerable possibilities for expanding cultivated areas; but the costs of development are high, and it will often be more economical to intensify the utilization of areas already settled." Thus there is a possibility of increasing the area of cropland in Africa south of the Sahara and in Latin America, but only at the cost of heavy investment and at the additional cost of destruction of tropical rainforests.

Rather than an increase in the global area of cropland, we may encounter a future loss of cropland through soil erosion, salination, desertification, loss of topsoil, depletion of minerals in topsoil, urbanization and failure of water supplies. In China and in the southwestern part of the United States, water tables are falling at an alarming rate. The Ogallala aquifer (which supplies water to many of the plains states in the central and southern parts of the United States) has a yearly overdraft of 160%.

In the 1950's, both the U.S.S.R and Turkey attempted to convert arid grasslands into wheat farms. In both cases, the attempts were defeated by drought and wind erosion, just as the wheat farms of Oklahoma were overcome by drought and dust in the 1930's.

If irrigation of arid lands is not performed with care, salt may be deposited, so that the land is ruined for agriculture. This type of desertification can be seen, for example, in some parts of Pakistan. Another type of desertification can be seen in the Sahel region of Africa, south of the

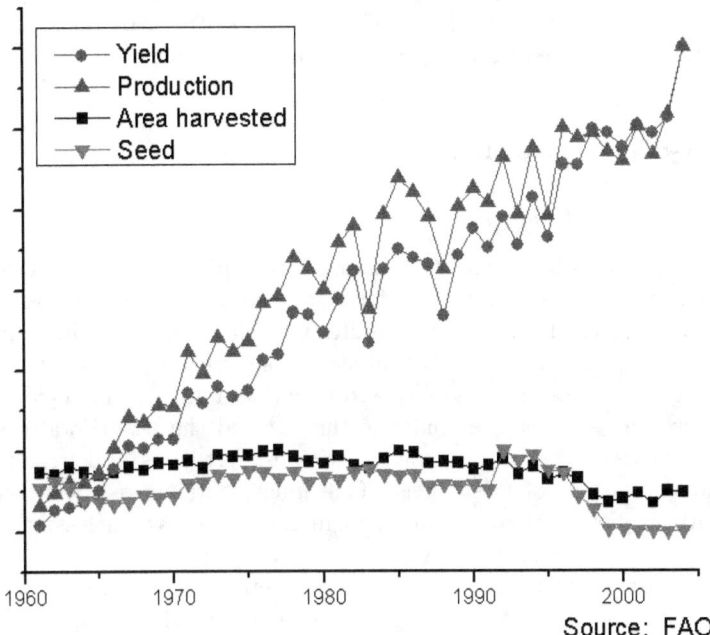

Fig. 4.4 This graph shows the total world production of coarse grain between 1960 and 2004. Because of high-yield varieties, the yield of grain increased greatly. Notice, however, that the land under cultivation remained almost constant. High-yield agriculture depends on large inputs of fossil fuel energy and irrigation, and may be difficult to maintain in the future. (FAO.)

Sahara. Rapid population growth in the Sahel has led to overgrazing, destruction of trees, and wind erosion, so that the land has become unable to support even its original population.

Especially worrying is a prediction of the International Panel on Climate Change concerning the effect of global warming on the availability of water: According to Model A1 of the IPCC, global warming may, by the 2050's, have reduced by as much as 30% the water available in large areas of world that now a large producers of grain.[c]

Added to the agricultural and environmental problems, are problems of finance and distribution. Famines can occur even when grain is available somewhere in the world, because those who are threatened with starvation

---
[c]See the discussion of the Stern Report in Chapter 7.

may not be able to pay for the grain, or for its transportation. The economic laws of supply and demand are not able to solve this type of problem. One says that there is no "demand" for the food (meaning demand in the economic sense), even though people are in fact starving.

Energy-dependence of modern agriculture

Food prices and energy prices

A very serious problem with Green Revolution plant varieties is that they require heavy inputs of pesticides, fertilizers and irrigation. Because of this, the use of high-yield varieties contributes to social inequality, since only rich farmers can afford the necessary inputs. Monocultures, such as the Green Revolution varieties may also prove to be vulnerable to future epidemics of plant diseases, such as the epidemic that caused the Irish Potato Famine in 1845. Even more importantly, pesticides, fertilizers and irrigation all depend on the use of fossil fuels. One must therefore ask whether high agricultural yields can be maintained in the future, when fossil fuels are expected to become prohibitively scarce and expensive.

Modern agriculture has become highly dependent on fossil fuels, especially on petroleum and natural gas. This is especially true of production of the high-yield grain varieties introduced in the Green Revolution, since these require especially large inputs of fertilizers, pesticides and irrigation. Today, fertilizers are produced using oil and natural gas, while pesticides are synthesized from petroleum feedstocks, and irrigation is driven by fossil fuel energy. Thus agriculture in the developed countries has become a process where inputs of fossil fuel energy are converted into food calories. If one focuses only on the farming operations, the fossil fuel energy inputs are distributed as follows:

(1) Manufacture of inorganic fertilizer, 31%
(2) Operation of field machinery, 19%
(3) Transportation, 16%
(4) Irrigation, 13%
(5) Raising livestock (not including livestock feed), 8%
(6) Crop drying, 5%
(7) Pesticide production, 5%
(8) Miscellaneous, 8%

The ratio of the fossil fuel energy inputs to the food calorie outputs depends on how many energy-using elements of food production are included in the accounting. David Pimental and Mario Giampietro of Cornell

University estimated in 1994 that U.S. agriculture required 0.7 kcal of fossil fuel energy inputs to produce 1.0 kcal of food energy. However, this figure was based on U.N. statistics that did not include fertilizer feedstocks, pesticide feedstocks, energy and machinery for drying crops, or electricity, construction and maintenance of farm buildings. A more accurate calculation, including these inputs, gives an input/output ratio of approximately 1.0. Finally, if the energy expended on transportation, packaging and retailing of food is included, Pimental and Giampietro found that the input/output ratio for the U.S. food system was approximately 10, and this figure did not include energy used for cooking.

The Brundtland Report[d] estimate of the global potential for food production assumes "that the area under food production can be around 1.5 billion hectares (3.7 billion acres - close to the present level), and that the average yields could go up to 5 tons of grain equivalent per hectare (as against the present average of 2 tons of grain equivalent)." In other words, the Brundtland Report assumes an increase in yields by a factor of 2.5. This would perhaps be possible if traditional agriculture could everywhere be replaced by energy-intensive modern agriculture using Green Revolution plant varieties. However, Pimental and Giampietro's studies show that modern energy-intensive agricultural techniques cannot be maintained after fossil fuels have been exhausted.

At the time when the Brundtland Report was written (1987), the global average of 2 tons of grain equivalent per hectare included much higher yields from the sector using modern agricultural methods. Since energy-intensive petroleum-based agriculture cannot be continued in the post-fossil-fuel era, future average crop yields will probably be much less than 2 tons of grain equivalent per hectare.

The 1987 global population was approximately 5 billion. This population was supported by 3 billion tons of grain equivalent per year. After fossil fuels have been exhausted, the total world agricultural output is likely to be considerably less than that, and therefore the population that it will be possible to support will probably be considerably less than 5 billion, assuming that our average daily per capita use of food calories remains the same, and assuming that the amount of cropland and pasturage remains the same (1.5 billion hectares cropland, 3.0 billion hectares pasturage).

The Brundtland Report points out that "The present (1987) global average consumption of plant energy for food, seed and animal feed amounts to 6,000 calories daily, with a range among countries of 3,000-15,000 calories, depending on the level of meat consumption." Thus there is a certain

---

[d] World Commission on Environment and Development, *Our Common Future*, Oxford University Press, (1987). This book is often called "The Brundtland Report" after Gro Harlem Brundtland, the head of WCED, who was then Prime Minister of Norway.

flexibility in the global population that can survive on a given total agricultural output. If the rich countries were willing to eat less meat, more people could be supported.

Effects of climate change on agriculture

Effects of temperature increase on crops

There is a danger that when climate change causes both temperature increases and increased aridity in regions like the US grain belt, yields will be very much lowered. Of the three main grain types (corn, wheat and rice) corn is the most vulnerable to the direct effect of increases in temperature. One reason for this is the mechanism of pollination of corn: A pollen grain lands on one end of a corn-silk strand, and the germ cell must travel the length of the strand in order to fertilize the kernel. At high temperatures, the corn silk becomes dried out and withered, and is unable to fulfill its biological function. Furthermore, heat can cause the pores on the underside of the corn leaf to close, so that photosynthesis stops.

According to a study made by Mohan Wali and coworkers at Ohio State University, the photosynthetic activity of corn increases until the temperature reaches 20°C. It then remains constant until the temperature reaches 35°C, after which it declines. At 40°C and above, photosynthesis stops altogether.

Scientists in the Philippines report that the pollination of rice fails entirely at 40°C, leading to crop failures. Wheat yields are also markedly reduced by temperatures in this range.

Predicted effects on rainfall

According to the Stern Report, some of the major grain-producing areas of the world might loose up to 30% of their rainfall by 2050. These regions include much of the United States, Brazil, the Mediterranean region, Eastern Russia and Belarus, the Middle East, Southern Africa and Australia. Of course possibilities for agriculture may simultaneously increase in other regions, but the net effect of climate change on the world's food supply is predicted to be markedly negative.

Unsustainable use of groundwater

It may seem surprising that fresh water can be regarded as a non-renewable resource. However, groundwater in deep aquifers is often renewed very slowly. Sometimes renewal requires several thousand years. When the rate of withdrawal of groundwater exceeds the rate of renewal, the carrying capacity of the resource has been exceeded, and withdrawal of water becomes analogous to mining a mineral. However, it is more serious than ordinary mining because water is such a necessary support for life.

In many regions of the world today, groundwater is being withdrawn faster than it can be replenished, and important aquifers are being depleted. In China, for example, groundwater levels are falling at an alarming rate. Considerations of water supply in relation to population form the background for China's stringent population policy.

At a recent lecture, Lester Brown of the Worldwatch Institute was asked by a member of the audience to name the resource for which shortages would most quickly become acute. Most of the audience expected him to name oil, but instead he replied "water". Lester Brown then cited China's falling water table. He predicted that within decades, China would be unable to feed itself. He said that this would not cause hunger in China itself: Because of the strength of China's economy, the country would be able to purchase grain on the world market. However Chinese purchases of grain would raise the price, and put world grain out of reach of poor countries in Africa. Thus water shortages in China will produce famine in parts of Africa, Brown predicted.

Under many desert areas of the world are deeply buried water tables formed during glacial periods when the climate of these regions was wetter. These regions include the Middle East and large parts of Africa. Water can be withdrawn from such ancient reservoirs by deep wells and pumping, but only for a limited amount of time.

In oil-rich Saudi Arabia, petroenergy is used to drill wells for ancient water and to bring it to the surface. Much of this water is used to irrigate wheat fields, and this is done to such an extent that Saudi Arabia exports wheat. The country is, in effect, exporting its ancient heritage of water, a policy that it may, in time, regret. A similarly short-sighted project is Muammar Qaddafi's enormous pipeline, which will bring water from ancient sub-desert reservoirs to coastal cities of Libya.

In the United States, the great Ogallala aquifer is being overdrawn. This aquifer is an enormous stratum of water-saturated sand and gravel underlying parts of northern Texas, Oklahoma, New Mexico, Kansas, Colorado, Nebraska, Wyoming and South Dakota. The average thickness of the aquifer is about 70 meters. The rate of water withdrawal from the aquifer exceeds the rate of recharge by a factor of eight.

Thus we can see that in many regions, the earth's present population is living on its inheritance of water, rather than its income. This fact, coupled with rapidly increasing populations and climate change, may contribute to a food crisis partway through the 21st century.

Glacial melting and summer water supplies

Fig. 4.5 Whitechuck Glacier in the North Cascades National Park in 1973. (Nicholas College.)

Fig. 4.6 The same glacier in 2006 (Nicholas College.)

The summer water supplies of both China and India are threatened by the melting of glaciers. The Gangotri glacier, which is the principle glacier feeding India's great Ganges River, is reported to be melting at an accelerating rate, and it could disappear within a few decades. If this happens,the Ganges could become seasonal, flowing only during the monsoon season.

Chinese agriculture is also threatened by disappearing Himalayan glaciers, in this case those on the Tibet-Quinghai Plateau. The respected Chinese glaciologist Yao Tandong estimates that the glaciers feeding the Yangtze and Yellow Rivers are disappearing at the rate of 7% per year.

The Indus and Mekong Rivers will be similarly affected by the melting of glaciers. Lack of water during the summer season could have a serious impact on the irrigation of rice and wheat fields.

Forest loss and climate change

Mature forests contain vast amounts of sequestered carbon, not only in their trees, but also in the carbon-rich soil of the forest floor. When a forest is logged or burned to make way for agriculture, this carbon is released into the atmosphere. One fifth of the global carbon emissions are at present due to destruction of forests. This amount is greater than the CO_2 emissions for the world's transportation systems.

An intact forest pumps water back into the atmosphere, increasing inland rainfall and benefiting agriculture. By contrast, deforestation, for example in the Amazonian rainforest, accelerates the flow of water back into the ocean, thus reducing inland rainfall. There is a danger that the Amazonian rainforest may be destroyed to such an extent that the region will become much more dry. If this happens, the forest may become vulnerable to fires produced by lightning strikes. This is one of the feedback loops against which the Stern Report warns - the drying and burning of the Amazonian rainforest may become irreversible, greatly accelerating climate change, if destruction of the forest proceeds beyond a certain point.

Erosion of topsoil

Besides depending on an adequate supply of water, food production also depends on the condition of the thin layer of topsoil that covers the world's croplands. This topsoil is being degraded and eroded at an alarming rate: According to the World Resources Institute and the United Nations Environment Programme, "It is estimated that since World War II, 1.2 billion hectares... has suffered at least moderate degradation as a result of human activity. This is a vast area, roughly the size of China and India combined." This area is 27% of the total area currently devoted to agriculture.[e] The

---

[e]The total area devoted to agriculture throughout the world is 1.5 billion hectares of cropland and 3.0 billion hectares of pasturage.

report goes on to say that the degradation is greatest in Africa.

The risk of topsoil erosion is greatest when marginal land is brought into cultivation, since marginal land is usually on steep hillsides which are vulnerable to water erosion when wild vegetation is removed.

David Pimental and his associates at Cornell University pointed out in 1995 that "Because of erosion-associated loss of productivity and population growth, the per capita food supply has been reduced over the past 10 years and continues to fall. The Food and Agricultural Organization reports that the per capita production of grains which make up 80% of the world's food supply, has been declining since 1984."

Pimental et al. add that "Not only is the availability of cropland per capita decreasing as the world population grows, but arable land is being lost due to excessive pressure on the environment. For instance, during the past 40 years nearly one-third of the world's cropland (1.5 billion hectares) has been abandoned because of soil erosion and degradation. Most of the replacement has come from marginal land made available by removing forests. Agriculture accounts for 80% of the annual deforestation."

Topsoil can also be degraded by the accumulation of salt when irrigation water evaporates. The worldwide area of irrigated land has increased from 8 million hectares in 1800 to more than 100 million hectares today. This land is especially important to the world food supply because it is carefully tended and yields are large in proportion to the area. To protect this land from salination, it should be irrigated in such a way that evaporation is minimized.

Finally cropland with valuable topsoil is being be lost to urban growth and highway development, a problem that is made more severe by growing populations and by economic growth.

Laterization

Every year, more than 100,000 square kilometers of rain forest are cleared and burned, an area which corresponds to that of Switzerland and the Netherlands combined. Almost half of the world's tropical forests have already been destroyed. Ironically, the land thus cleared often becomes unsuitable for agriculture within a few years.

Tropical soils may seem to be fertile when covered with luxuriant vegetation, but they are usually very poor in nutrients because of leeching by heavy rains. The nutrients which remain are contained in the vegetation itself; and when the forest cover is cut and burned, the nutrients are rapidly lost.

Often the remaining soil is rich in aluminum oxide and iron oxide. When such soils are exposed to oxygen and sun-baking, a rocklike substance called

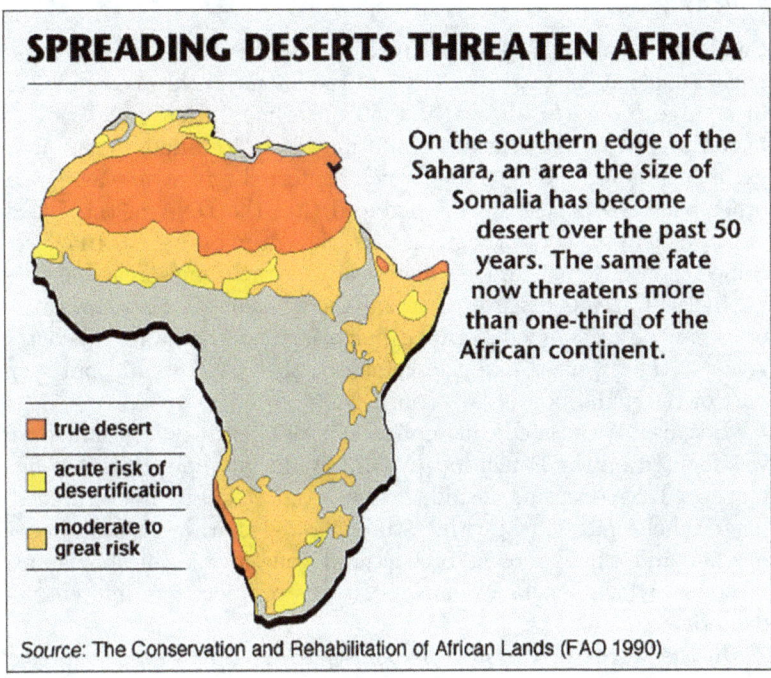

Fig. 4.7 Desert regions of the Africa that are in danger of spreading. (FAO.)

Laterite is formed. The temples of Angkor Wat in Cambodia are built of Laterite; and it is thought that laterization of the soil contributed to the disappearance of the Khmer civilization, which built these temples.

Harmful effects of industrialized farming

A major global public health crisis may soon be produced by the wholesale use of antibiotics in the food of healthy farm animals. The resistance factors produced by shovelling antibiotics into animal food produces resistance factors (plasmids) which can easily be transferred to human pathogens. A related problem is the excessive use of pesticides and artificial fossil-fuel-derived fertilizers in agriculture. Farming is not a joke. It is a serious threat.[f]

---

[f] http://ecowatch.com/2014/03/06/misuse-antibiotics-fatal-superbug-crisis/
http://ecowatch.com/2013/12/06/8-scary-facts-about-antibiotic-resistance/

Plasmids

Bacteria belong to a class of organisms (prokaryotes) whose cells do not have a nucleus. Instead, the DNA of the bacterial chromosome is arranged in a large loop. In the early 1950's, Joshua Lederberg discovered that bacteria can exchange genetic information. He found that a frequently-exchanged gene, the F-factor (which conferred fertility), was not linked to other bacterial genes; and he deduced that the DNA of the F-factor was not physically a part of the main bacterial chromosome. In 1952, Lederberg coined the word "plasmid" to denote any extrachromosomal genetic system.

In 1959, it was discovered in Japan that genes for resistance to antibiotics can be exchanged between bacteria; and the name "R-factors" was given to these genes. Like the F-factors, the R-factors did not seem to be part of the main loop of bacterial DNA.

Because of the medical implications of this discovery, much attention was focused on the R-factors. It was found that they were plasmids, small loops of DNA existing inside the bacterial cell, but not attached to the bacterial chromosome. Further study showed that, in general, between one percent and three percent of bacterial genetic information is carried by plasmids, which can be exchanged freely even between different species of bacteria.

In the words of the microbiologist, Richard Novick, "Appreciation of the role of plasmids has produced a rather dramatic shift in biologists' thinking about genetics. The traditional view was that the genetic makeup of a species was about the same from one cell to another, and was constant over long periods of time. Now a significant proportion of genetic traits are known to be variable (present in some individual cells or strains, absent in others), labile (subject to frequent loss or gain) and mobile, all because those traits are associated with plasmids or other atypical genetic systems."

Because of the ease with which plasmids conferring resistance to antibiotics can be transferred from animal bacteria to the bacteria carrying human disease, the practice of feeding antibiotics to healthy farm animals is becoming a major human health hazard. The World Health Organization has warned that if we lose effective antibiotics through this mechanism, "Many common infections will no longer have a cure, and could kill unabated". The US Center for Disease Control has pointed to the emergence

---

http://ecowatch.com/2015/03/27/obama-fight-superbug-crisis/
http://ecowatch.com/2014/03/12/fda-regulation-antibiotics-factory-farms/
http://www.bbc.com/news/health-35153795
http://www.bbc.com/news/health-21702647
http://www.bbc.com/news/health-34857015
http://sustainableagriculture.net/about-us/
https://pwccc.wordpress.com/programa/

of "nightmare bacteria", and the chief medical officer for England Prof Dame Sally Davies has evoked parallels with the "apocalypse".

Pesticides, artificial fertilizers and topsoil

A closely analogous danger results from the overuse of pesticides and petroleum-derived fertilizers in agriculture. A very serious problem with Green Revolution plant varieties is that they require heavy inputs of pesticides, fertilizers and irrigation. Because of this, the use of high-yield varieties contributes to social inequality, since only rich farmers can afford the necessary inputs. Monocultures, such as the Green Revolution varieties may also prove to be vulnerable to future plant diseases, such as the epidemic that caused the Irish Potato Famine in 1845. Even more importantly, pesticides, fertilizers and irrigation all depend on the use of fossil fuels. One must ask, therefore, whether high-yield agriculture can be maintained in the post-fossil-fuel era.

Topsoil is degraded by excessive use of pesticides and artificial fertilizers. Natural topsoil is rich in organic material, which contains sequestered carbon that would otherwise be present in our atmosphere in the form of greenhouse gases. In addition, natural topsoil contains an extraordinarily rich diversity of bacteria and worms that act to convert agricultural wastes from one year's harvest into nutrients for the growth of next year's crop. Pesticides kill these vital organisms, and make the use of artificial fertilizers necessary.

Finally, many small individual farmers, whose methods are sustainable, are being eliminated by secret land-grabs or put out of business because they cannot compete with unsustainable high-yield agriculture. Traditional agriculture contains a wealth of knowledge and biodiversity, which it would be wise for the world to preserve.

The demographic transition

The phrase "developing countries" is more than a euphemism; it expresses the hope that with the help of a transfer of technology from the industrialized nations, all parts of the world can achieve prosperity. Some of the forces that block this hope have just been mentioned. Another factor that prevents the achievement of worldwide prosperity is population growth.

In the words of Dr. Halfdan Mahler, former Director General of the World Health Organization, "Country after country has seen painfully achieved increases in total output, food production, health and educational facilities and employment opportunities reduced or nullified by excessive population growth."

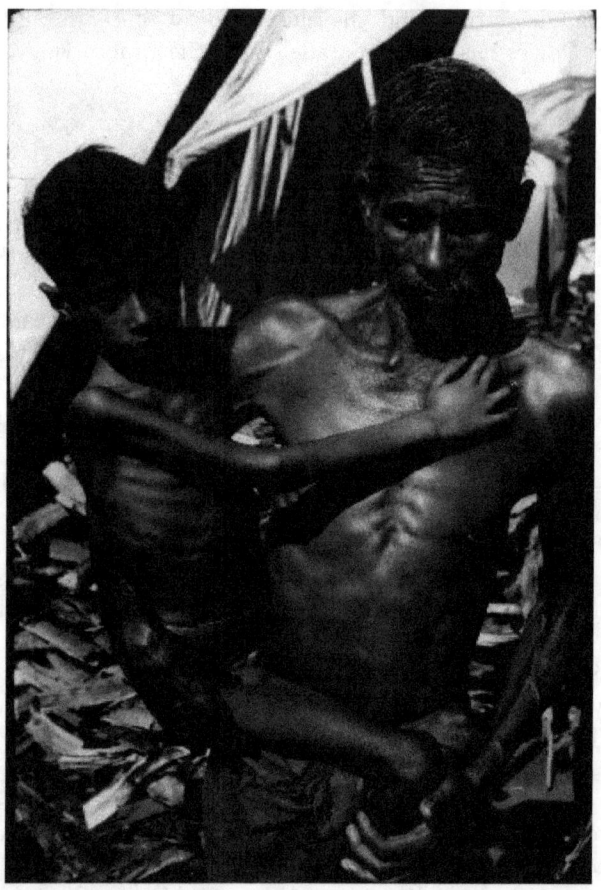

Fig. 4.8 Child suffering with the deficiency disease Marasmus in India. (Public domain.)

The growth of population is linked to excessive urbanization, infrastructure failures and unemployment. In rural districts in the developing countries, family farms are often divided among a growing number of heirs until they can no longer be subdivided. Those family members who are no longer needed on the land have no alternative except migration to overcrowded cities, where the infrastructure is unable to cope so many new arrivals. Often the new migrants are forced to live in excrement-filled makeshift slums, where dysentery, hepatitis and typhoid are endemic, and where the conditions for human life sink to the lowest imaginable level. In Brazil, such shanty towns are called "favelas".

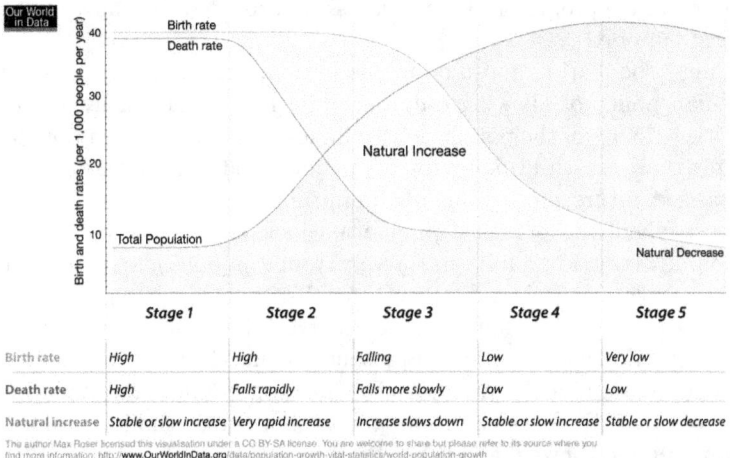

Fig. 4.9 A schematic graph showing the demographic transition from an old equilibrium with a high birth rate and high death rate to a new equilibrium where both the birth rate and the death rate are low. (Wikimedia.)

If modern farming methods are introduced in rural areas while population growth continues, the exodus to cities is aggravated, since modern techniques are less labor-intensive and favor large farms. In cities, the development of adequate infrastructure requires time, and it becomes a hopeless task if populations are growing rapidly. Thus, population stabilization is a necessary first step for development.

It can be observed that birth rates fall as countries develop. However, development is sometimes blocked by the same high birth rates that economic progress might have prevented. In this situation (known as the "demographic trap"), economic gains disappear immediately because of the demands of an exploding population.

For countries caught in the demographic trap, government birth control programs are especially important, because one cannot rely on improved social conditions to slow birth rates. Since health and lowered birth rates should be linked, it is appropriate that family-planning should be an important part of programs for public health and economic development.

A recent study conducted by Robert F. Lapham of Demographic Health Surveys and W. Parker Maudlin of the Rockefeller Foundation has shown that the use of birth control is correlated both with socio-economic setting and with the existence of strong family-planning programs. The implication of this study is that even in the absence of increased living standards, family-planning programs can be successful, provided they have strong government support.

China, the world's most populous nation, has adopted the somewhat draconian policy of allowing only one child for families in living in towns and cities (35.9% of the population). Chinese leaders obtained popular support for their one-child policy by means of an educational program which emphasized future projections of diminishing water resources and diminishing cropland per person if population increased unchecked. Like other developing countries, China has a very young population, which will continue to grow even when fertility has fallen below the replacement level because so many of its members are contributing to the birth rate rather than to the death rate. China's present population is 1.3 billion. Its projected population for the year 2025 is 1.5 billion. China's one-child policy is supported by 75% of the country's people, but the methods of enforcement are sometimes criticized, and it has led to a M/F sex ratio of 1.17/1.00. The natural baseline for the sex ratio ranges between 1.03/1.00 and 1.07/1.00.

Education of women and higher status for women are vitally important measures, not only for their own sake, but also because in many countries these social reforms have proved to be the key to lower birth rates. Religious leaders who oppose programs for the education of women and for family planning on "ethical" grounds should think carefully about the scope and consequences of the catastrophic global famine which will undoubtedly occur within the next 50 years if population is allowed to increase unchecked. Do these leaders really wish to be responsible for the suffering and death from starvation of hundreds of millions of people?

At the United Nations Conference on Population and Development, held in Cairo in September, 1994, a theme which emerged very clearly was that one of the most important keys to controlling the global population explosion is giving women better education and equal rights. These goals are desirable for the sake of increased human happiness, and for the sake of the uniquely life-oriented point of view which women can give us; but in addition, education and improved status for women have shown themselves to be closely connected with lowered birth rates. When women lack education and independent careers outside the home, they can be forced into the role of baby-producing machines by men who do not share in the drudgery of cooking, washing and cleaning; but when women have educational, legal, economic, social and political equality with men, experience has shown that they choose to limit their families to a moderate size.

Sir Partha Dasgupta of Cambridge University has pointed out that the changes needed to break the cycle of overpopulation and poverty are all desirable in themselves. Besides education and higher status for women, they include state-provided social security for old people, provision of water

Fig. 4.10 Education of women and higher status for women are vitally important measures, not only for their own sake, but also because these social reforms have proved to be the key to lower birth rates. (Kundan Srivastava.)

supplies near to dwellings, provision of health services to all, abolition of child labor and general economic development.

The UN summit on addressing large movements of refugees and migrants

On September 19, 2016, the United Nations General Assembly held a 1-day summit meeting to address the pressing problem of refugees. It is a problem that has been made acute by armed conflicts in the Middle East and Africa, and by climate change.

One of the outcomes of the summit was the a Declaration for Refugees and Migrants. Here is a statement of the severity of the problem from paragraph 3 of the Declaration:

"We are witnessing in today's world an unprecedented level of human mobility. More people than ever before live in a country other than the one in which they were born. Migrants are present in all countries of the world. Most of them move without incident. In 2015, their number surpassed 244 million, growing at a rate faster than the world's population. However, there are 65 million forcibly displaced persons, including over 21 million refugees, 3 million asylum seekers and over 40 million internally displaced persons."

Sadly, the world's response to the tragic plight of refugees fleeing from zones of armed conflict has been less than generous. Men, women and many children, trying to escape from almost certain death in the war-torn Middle East, have been met, not with sympathy and kindness, but with barbed wire and tear gas.

Germany's Chancellor, Angela Merkel, courageously made arrangements for her country to accept a large number of refugees, but as a consequence her party has suffered political setbacks. On the whole, European governments have moved to the right, as anti-refugee parties gained strength. The United States, Canada Australia and Russia, countries that could potentially save the lives of many refugees, have accepted almost none. In contrast, tiny Lebanon, despite all its problems, has become the home of so many refugees that they are a very large fraction of the country's total population.

As the effects of climate change become more pronounced, we can expect the suffering and hopelessness of refugees to become even more severe. This is a challenge which the world must meet with humanity and solidarity.

The World Cities Report, 2016

According to the World Cities Report,[g] by 2030, two thirds of the world's population will be living in cities. As the urban population increases, the land area occupied by cities is increasing at a higher rate. It is projected that by 2030, the urban population of developing countries will double, while the area covered by cites could triple.

Commenting on this, the UN-Habitat Executive Director, Joan Clos, said: "In the twenty years since the Habitat II conference, the world has seen a gathering of its population in urban areas. This has been accompanied by socioeconomic growth in many instances. But the urban landscape is changing and with it, the pressing need for a cohesive and realistic approach to urbanization".

"Such urban expansion is wasteful in terms of land and energy consumption and increases greenhouse gas emissions. The urban centre of gravity, at least for megacities, has shifted to the developing regions."

One can foresee that in the future, as fossil fuels become increasingly scarce, the problem of feeding urban populations will become acute.

---

[g] http://wcr.unhabitat.org/

Suggestions for further reading

(1) P. Dasgupta, *Population, Resources and Poverty*, Ambio, **21**, 95-101, (1992).
(2) L.R. Brown, *Who Will Feed China?*, W.W. Norton, New York, (1995).
(3) L.R. Brown, et al., *Saving the Planet. How to Shape and Environmentally Sustainable Global Economy*, W.W. Norton, New York, (1991).
(4) L.R. Brown, *Postmodern Malthus: Are There Too Many of Us to Survive?*, The Washington Post, July 18, (1993).
(5) L.R. Brown and H. Kane, *Full House. Reassessing the Earth's Population Carrying Capacity*, W.W. Norton, New York, (1991).
(6) L.R. Brown, *Seeds of Change*, Praeger Publishers, New York, (1970).
(7) L.R. Brown, *The Worldwide Loss of Cropland*, Worldwatch Paper 24, Worldwatch Institute, Washington, D.C., (1978).
(8) L.R. Brown, and J.L. Jacobson, *Our Demographically Divided World*, Worldwatch Paper 74, Worldwatch Institute, Washington D.C., (1986).
(9) L.R. Brown, and J.L. Jacobson, *The Future of Urbanization: Facing the Ecological and Economic Constraints*, Worldwatch Paper 77, Worldwatch Institute, Washington D.C., (1987).
(10) L.R. Brown, and others, *State of the World*, W.W. Norton, New York, (published annually).
(11) H. Brown, *The Human Future Revisited. The World Predicament and Possible Solutions*, W.W. Norton, New York, (1978).
(12) H. Hanson, N.E. Borlaug and N.E. Anderson, *Wheat in the Third World*, Westview Press, Boulder, Colorado, (1982).
(13) A. Dil, ed., *Norman Borlaug and World Hunger*, Bookservice International, San Diego/Islamabad/Lahore, (1997).
(14) N.E. Borlaug, *The Green Revolution Revisitied and the Road Ahead*, Norwegian Nobel Institute, Oslo, Norway, (2000).
(15) N.E. Borlaug, *Ending World Hunger. The Promise of Biotechnology and the Threat of Antiscience Zealotry*, Plant Physiology, **124**, 487-490, (2000).
(16) M. Giampietro and D. Pimental, *The Tightening Conflict: Population, Energy Use and the Ecology of Agriculture*, in Negative Population Forum, L. Grant ed., Negative Population Growth, Inc., Teaneck, N.J., (1993).
(17) H.W. Kendall and D. Pimental, *Constraints on the Expansion of the Global Food Supply*, Ambio, **23**, 198-2005, (1994).
(18) D. Pimental et al., *Natural Resources and Optimum Human Population*, Population and Environment, **15**, 347-369, (1994).
(19) D. Pimental et al., *Environmental and Economic Costs of Soil Erosion and Conservation Benefits*, Science, **267**, 1117-1123, (1995).

(20) D. Pimental et al., *Natural Resources and Optimum Human Population*, Population and Environment, **15**, 347-369, (1994).

(21) D. Pimental and M. Pimental, *Food Energy and Society*, University Press of Colorado, Niwot, Colorado, (1996).

(22) D. Pimental et al., *Environmental and Economic Costs of Soil Erosion and Conservation Benefits*, Science, **267**, 1117-1123, (1995).

(23) RS and NAS, *The Royal Society and the National Academy of Sciences on Population Growth and Sustainability*, Population and Development Review, **18**, 375-378, (1992).

(24) A.M. Altieri, *Agroecology: The Science of Sustainable Agriculture*, Westview Press, Boulder, Colorado, (1995).

(25) G. Conway, *The Doubly Green Revolution*, Cornell University Press, (1997).

(26) J. Dreze and A. Sen, *Hunger and Public Action*, Oxford University Press, (1991).

(27) G. Bridger, and M. de Soissons, *Famine in Retreat?*, Dent, London, (1970).

(28) W. Brandt, *World Armament and World Hunger: A Call for Action*, Victor Gollanz Ltd., London, (1982).

(29) A.K.M.A. Chowdhury and L.C. Chen, *The Dynamics of Contemporary Famine*, Ford Foundation, Dacca, Pakistan, (1977)

(30) J. Shepard, *The Politics of Starvation*, Carnegie Endowment for International Peace, Washington D.C., (1975).

(31) M.E. Clark, *Ariadne's Thread: The Search for New Modes of Thinking*, St. Martin's Press, New York, (1989).

(32) J.-C. Chesnais, *The Demographic Transition*, Oxford, (1992).

(33) C.M. Cipola, *The Economic History of World Population*, Penguin Books Ltd., (1974).

(34) E. Draper, *Birth Control in the Modern World*, Penguin Books, Ltd., (1972).

(35) Draper Fund Report No. 15, *Towards Smaller Families: The Crucial Role of the Private Sector*, Population Crisis Committee, 1120 Nineteenth Street, N.W., Washington D.C. 20036, (1986).

(36) E. Eckholm, *Losing Ground: Environmental Stress and World Food Prospects*, W.W. Norton, New York, (1975).

(37) E. Havemann, *Birth Control*, Time-Life Books, (1967).

(38) J. Jacobsen, *Promoting Population Stabilization: Incentives for Small Families*, Worldwatch Paper 54, Worldwatch Institute, Washington D.C., (1983).

(39) N. Keyfitz, *Applied Mathematical Demography*, Wiley, New York, (1977).

(40) W. Latz (ed.), *Future Demographic Trends*, Academic Press, New York,

(1979).
(41) World Bank, *Poverty and Hunger: Issues and Options for Food Security in Developing Countries*, Washington D.C., (1986).
(42) J.E. Cohen, *How Many People Can the Earth Support?*, W.W. Norton, New York, (1995).
(43) J. Amos, *Climate Food Crisis to Deepen*, BBC News (5 September, 2005).
(44) J. Vidal and T. Ratford, *One in Six Countries Facing Food Shortage*, The Guardian, (30 June, 2005).
(45) J. Mann, *Biting the Environment that Feeds Us*, The Washington Post, July 29, 1994.
(46) G.R. Lucas, Jr., and T.W. Ogletree, (editors), *Lifeboat Ethics. The Moral Dilemmas of World Hunger*, Harper and Row, New York.
(47) J.L. Jacobson, *Gender Bias: Roadblock to Sustainable Development*, Worldwatch Paper 110, Worldwatch Institute, Washington D.C., (1992).
(48) J. Gever, R. Kaufmann, D. Skole and C. Vorosmarty, *Beyond Oil: The Threat to Food and Fuel in the Coming Decades*, Ballinger, Cambridge MA, (1986).
(49) M. ul Haq, *The Poverty Curtain: Choices for the Third World*, Columbia University Pres, New York, (1976).
(50) H. Le Bras, *La Planète au Village*, Datar, Paris, (1993).
(51) E. Mayr, *Population, Species and Evolution*, Harvard University Press, Cambridge, (1970).

Chapter 5

Intolerable Economic Inequality

"Every Night & every Morn,
Some to Misery are Born.
Every Night & every Morn,
Some are Born to sweet delight.
Some are Born to sweet delight,
Some are Born to Endless Night."

William Blake

"Whatever happens, we have got
The Maxim gun, and they have not."

Hilaire Beloc

Introduction

The excessive inequality that we can see today, both within countries and between countries, has many harmful effects, and these are experienced by both poor and rich. For example, crime, drug use, and mental illness are much more common in very unequal societies.

On a global scale, the vast chasm of economic inequality between countries blocks efforts to make the United Nations more effective, since rich countries fear that a more effective UN will rob them of their privileged position.

We must also remember that inequality between nations is often maintained by means of military force, regime-change, and interference by powerful nations in the internal affairs of weaker ones.

Oxfam's report on inequality

A recent report by Oxfam[a] has revealed that the wealth of the poorest half of the world's population has fallen by a trillion dollars since 2010, a drop of 38%. Meanwhile, the wealth of the richest 62 people in the world has increased to 1.76 trillion dollars. In fact, the wealthiest 62 individuals now own more than the poorest half of the world's population. Enormous contrasts exist today, not only between nations, but also within nations.

Winnie Byanyima, Oxfam's International Executive Director stated that "It is simply unacceptable that the poorest half of the world's population owns no more than a few dozen super-rich people who could fit onto one bus. World leaders' concern about the escalating inequality has so far not translated into concrete action; the world has become a much more unequal place, and the trend is accelerating. We cannot continue to allow hundreds of millions of people to go hungry while resources that could be used to help them are sucked up by those at the top."

Speaking at the Davos Forum in Switzerland, she continued: "I challenge the governments and elites at Davos to play their part in in ending the era of tax havens, which is fueling economic inequality and preventing hundreds of millions of people from lifting themselves out of poverty. Multinational companies and wealthy elites are playing by different rules than everyone else, refusing to pay the taxes that society needs to function. The fact that 188 of 201 leading companies have a presence in at least one tax haven shows that it it time to act."

Oxfam estimates that globally, 7.6 trillion dollars of individual's wealth sits offshore, and this includes as much as 38% of African financial wealth.

---

[a] https://www.oxfam.org/en/research/economy-1

Persistent effects of colonialism

Part of the extreme economic inequality that exists in today's world is due to colonial and neocolonial wars.

The Industrial Revolution opened up an enormous gap in military strength between the industrialized nations and the rest of the world. Taking advantage of their superior weaponry, Europe, the United States and Japan rapidly carved up the remainder of the world into colonies, which acted as sources of raw materials and food, and as markets for manufactured goods. Between 1800 and 1914, the percentage of the earth under the domination of colonial powers increased to 85 percent, if former colonies are included.

The English economist and Fabian, John Atkinson Hobson (1858-1940), offered a famous explanation of the colonial era in his book "Imperialism: A Study" (1902). According to Hobson, the basic problem that led to colonial expansion was an excessively unequal distribution of incomes in the industrialized countries. The result of this unequal distribution was that neither the rich nor the poor could buy back the total output of their society. The incomes of the poor were insufficient, and rich were too few in number. The rich had finite needs, and tended to reinvest their money. As Hobson pointed out, reinvestment in new factories only made the situation worse by increasing output.

Hobson had been sent as a reporter by the Manchester Guardian to cover the Second Boer War. His experiences had convinced him that colonial wars have an economic motive. Such wars are fought, he believed, to facilitate investment of the excess money of the rich in African or Asian plantations and mines, and to make possible the overseas sale of excess manufactured goods. Hobson believed imperialism to be immoral, since it entails suffering both among colonial peoples and among the poor of the industrial nations. The cure that he recommended was a more equal distribution of incomes in the manufacturing countries.

Neocolonialism?

In his book, *Neocolonialism, The Last Stage of Imperialism* (Thomas Nielsen, London, 1965), Kwami Nkrumah defined neocolonialism with the following words: "The essence of neocolonialism is that the State which is subject to it is, in theory independent, and has all the outward trappings of international sovereignty. In reality its economic system and thus its political policy is directed from the outside. The methods and form of this direction can take various shapes. For example, in an extreme case, the

Fig. 5.1 A late 19th century French cartoon showing England, Germany, Russia, France and Japan slicing up the pie of China. (Public domain.)

troops of the imperial power may garrison the territory of the neocolonial State and control the government of it. More often, however, neocolonial control is exercised through monetary means... The struggle against neo-colonialism is not aimed at excluding the capital of the developed world from operating in less developed countries. It is aimed at preventing the financial power of the developed countries from being used in such a way as to impoverish the less developed."

Fig. 5.2 A cartoon showing Cecil Rhodes' colonial ambitions for Africa. The thread in his hands represents a proposed Cape-Town-to-Cairo telegraph line. He wanted to "paint the map British red", and declared, "If I could, I would annex other planets." (Public domain.)

The resource curse

The way in which the industrialized countries maintain their control over less developed nations can be illustrated by the "resource curse", i.e. the fact that resource-rich developing countries are no better off economically than those that lack resources, but are cursed with corrupt and undemocratic governments. This is because foreign corporations extracting local resources under unfair agreements exist in a symbiotic relationship with corrupt local officials.

One might think that taxation of foreign resource-extracting firms would provide developing countries with large incomes. However, there is at present no international law governing multinational tax arrangements. These are usually agreed to on a bilateral basis, and the industrialized countries have stronger bargaining powers in arranging the bilateral agreements.

Manufacture and export of small arms

Another important poverty-generating factor in the developing countries is war - often civil war. The five permanent members of the U.N. Security Council are, ironically, the five largest exporters of small arms. Small arms have a long life. The weapons poured into Africa by both sides during the Cold War are still there, and they contribute to political chaos and civil wars that block development and cause enormous human suffering.

The United Nations website on Peace and Security through Disarmament states that "Small arms and light weapons destabilize regions; spark, fuel and prolong conflicts; obstruct relief programmes; undermine peace initiatives; exacerbate human rights abuses; hamper development; and foster a 'culture of violence'."

An estimated 639 million small arms and light weapons are in circulation worldwide, one for every ten people. Approximately 300,000 people are killed every year by these weapons, many of them women and children.

Examples of endemic conflict

In several regions of Africa, long-lasting conflicts have prevented development and caused enormous human misery. These regions include Ethiopia, Eritiria, Somalia (Darfur), Chad, Zimbabwe and the Democratic Republic of Congo. In the Congo, the death toll reached 5.4 million in 2008, with most of the victims dying of disease and starvation, but with war as the root cause. In view of these statistics, the international community can be seen to have a strong responsibility to stop supplying small arms and ammunition to regions of conflict. There is absolutely no excuse for the large-scale manufacture and international sale of small arms that exists today.

The plight of indigenous peoples

Readers of Charles Darwin's book describing *The Voyage of the Beagle* will remember his horrifying account of General Rosas' genocidal war against

Fig. 5.3 Children sleeping in Mulberry Street, New York City, 1890 (Jacob Riis photo) (Public domain.)

the Amerind population of Argentina. Similar genocidal violence has been experienced by indigenous peoples throughout South and Central America, and indeed throughout the world. In general, the cultures of indigenous peoples require much land, and greed for this land is the motive for violence against them. However, the genetic and cultural heritage of indigenous peoples can potentially be of enormous value to humanity, and great efforts should be made to protect them.

The resurgence of infectious disease

Tropical diseases

Endemic disease is strongly linked to poverty. Great improvements in reducing the effects of diseases like HIV/AIDS, malaria, schistosomiasis, trichoniosis, and river blindness could be made if pharmaceutical companies could be induced to do more research on tropical diseases and to provide

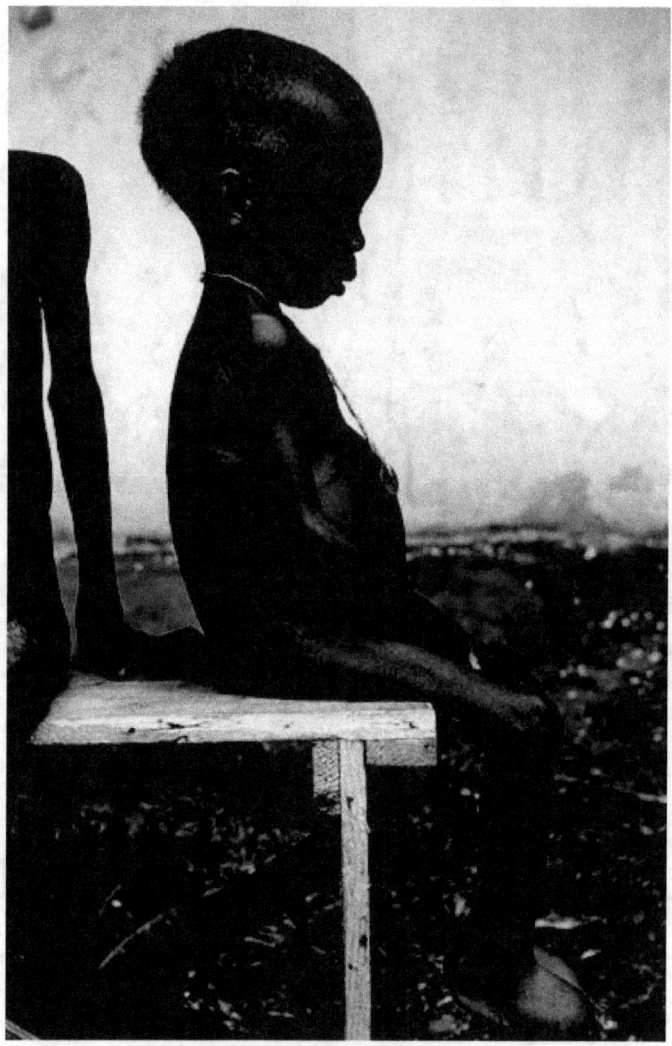

Fig. 5.4 20,000 children die each day from starvation. (Public domain.)

drugs to developing countries at affordable prices. Other important measures would be universal vaccination programs, and the provision of safe water to all. It is in the interests of developed countries to promote health in the developing world, because air travel can quickly spread epidemics from one region to another.

Fig. 5.5 Men from Bathurst Island in Australia's Northern Territories. Indigenous people everywhere in the world are under great pressure from those who desire their land. Indigenous cultures and languages are in danger of being lost. (Wikipedia.)

HIV/AIDS

In 2004, there were approximately 39.4 million people living with HIV, 4.9 million new HIV infections, and 3.1 million deaths due to AIDS. It is estimated that in five populous countries, Nigeria, Ethiopia, Russia, India and China, the number of people infected with HIV/AIDS will grow from 14-23 million currently to 50-75 million by 2010. 95% of those living with HIV/AIDS do not know that they are infected with the disease.

Tuberculosis

Approximately 2 billion people (one third of the world's population!) are infected with TB, often in a latent form. 90% of the burden of TB falls on the developing countries; on India alone, 30%. Roughly 2 million people die from TB each year. It is the number one killer of women of childbearing age.

Malaria

Every year there are 300 million cases of malaria, and it causes about one million deaths. There are roughly 10 new cases of malaria every second, 90% of which are in Africa. A quarter of all childhood deaths in Africa are due to malaria.

Slavery

Debt slavery

At the moment, the issue of debt slavery is very topical because of the case of Greece; but it is an issue that has a far more general significance.

Usury, the charging of interest on loans, has a history of being forbidden by several major religions, including not only the three Abrahamic religions, Judaism, Christianity and Islam, but also the ancient Vedic Scriptures of India.

Perhaps the reason for these religious traditions can be found in the remarkable properties of exponential growth. If any quantity, for example indebtedness, is growing at the rate of 3% per year, it will double in 23.1 years; if it is growing at the rate of 4% per year, the doubling time is 17.3 years. For a 5% growth rate, the doubling time is 13.9 years, if the growth rate is 7%, the doubling time is only 9.9 years. It follows that if a debt remains unpaid for a few years, most of the repayments will go for interest, rather than for reducing the amount of the debt.

In the case of the debts of third world countries to private banks in the industrialized parts of the world and to the IMF, many of the debts were incurred in the 1970's for purposes which were of no benefit to local populations, for example purchase of military hardware. Today the debts remain, although the amount paid over the years by the developing countries is very many times the amount originally borrowed. Third world debt can be regarded as a means by which the industrialized nations extract raw materials from developing countries without any repayment whatever. In fact, besides extracting raw materials, they extract money.

Child labor and child slavery

The reform movement's efforts, especially those of the Fabians, overcame the worst horrors of early 19th century industrialism, but today their hard-won achievements are being undermined and lost because of uncritical and unregulated globalization. Today, a factory owner or CEO, anxious to avoid high labor costs, and anxious to violate environmental regulations merely moves his factory to a country where laws against child labor and rape of the environment do not exist or are poorly enforced. In fact, he must do so or be fired, since the only thing that matters to the stockholders is the bottom line. One might say (as someone has done), that Adam Smith's invisible hand is at the throat of the world's peoples and at the throat of the global environment.

The movement of a factory from Europe or North America to a country with poorly enforced laws against environmental destruction, child labor and slavery puts workers into unfair competition. Unless they are willing to accept revival of the unspeakable conditions of the early Industrial Revolution, they are unable to compete.

Today, child labor accounts for 22% of the workforce in Asia, 32% in Africa, and 17% in Latin America. Large-scale slavery also exists today, although there are formal laws against it in every country. There are more slaves now than ever before. Their number is estimated to be between 12 million and 27 million. Besides outright slaves, who are bought and sold for as little as 100 dollars, there many millions of workers whose lack of options and dreadful working conditions must be described as slavelike.

Enforcement, in all countries, of laws against child labor would help to stabilize the world's rapidly growing population. When children are regarded as a source of income, or are sold into slavery or prostitution, parents aim for very large families. Thus, slavery or slavelike exploitation of children is a factor behind the global population explosion.

Political and geopolitical consequences of inequality

The intolerable economic inequality of today's world is closely linked with the problem of war:

- Military force is used to maintain neocolonialism and unfair trade relationships between countries.
- Billionaires and corporations use their enormous wealth to dominate governments and media. When this happens, democracy is replaced by oligarchy, and motives of profit take the place of social and ecological goals. The military-industrial complex also gains control of governmental budgets.
- The enormous amounts of money used for war could have been used for education, infrastructure, public health (including information and materials for family planning), sanitary drinking water, and social services.

- An effective system of international law is needed for the abolition of war. But at present, economic inequality between countries is so great that rich countries fear effective global governance because they fear taxation.

Benefits of equality

Interestingly, TED Talks was recently under fire from many progressive groups for censoring a short talk by the venture capitalist, Nick Hanauer, entitled "Income Inequality". In this talk, Hanauer says exactly the same thing as John Hobson, but he applies the ideas, not to colonialism, but to current unemployment in the United States. Hanauer says that the rich are unable to consume the products of society because they are too few in number. To make an economy work, demand must be increased, and for this to happen, the distribution of incomes must become much more equal than it is today in the United States.

TED has now posted Hanauer's talk, and the interested reader can find another wonderful TED talk dealing with the same issues from the standpoint of health and social problems. In a splendid lecture entitled "How economic inequality harms societies", Richard Wilkinson demonstrates that there is almost no correlation between gross national product and a number of indicators of the quality of life, such as physical health, mental health, drug abuse, education, imprisonment, obesity, social mobility, trust, violence, teenage pregnancies and child well-being. On the other hand he offers comprehensive statistical evidence that these indicators are strongly correlated with the degree of inequality within countries, the outcomes being uniformly much better in nations where income is more equally distributed.

Warren Buffet famously remarked, "There's class warfare, all right. But it's my class, the rich class, that's making war, and we're winning." However, the evidence presented by Hobson, Hanauer and Wilkinson shows conclusively that no one wins in a society where inequality is too great, and everyone wins when incomes are more evenly distributed.

We must decrease economic inequality

In his Apostolic Exhortation, "Evangelii Gaudium", Pope Francis said:
"In our time humanity is experiencing a turning-point in its history, as we can see from the advances being made in so many fields. We can only praise the steps being taken to improve peoples welfare in areas such as

health care, education and communications. At the same time we have to remember that the majority of our contemporaries are barely living from day to day, with dire consequences. A number of diseases are spreading. The hearts of many people are gripped by fear and desperation, even in the so-called rich countries. The joy of living frequently fades, lack of respect for others and violence are on the rise, and inequality is increasingly evident. It is a struggle to live and, often, to live with precious little dignity."

"This epochal change has been set in motion by the enormous qualitative, quantitative, rapid and cumulative advances occurring in the sciences and in technology, and by their instant application in different areas of nature and of life. We are in an age of knowledge and information, which has led to new and often anonymous kinds of power."

"Just as the commandment 'Thou shalt not kill' sets a clear limit in order to safeguard the value of human life, today we also have to say 'thou shalt not' to an economy of exclusion and inequality. Such an economy kills. How can it be that it is not a news item when an elderly homeless person dies of exposure, but it is news when the stock market loses two points? This is a case of exclusion. Can we continue to stand by when food is thrown away while people are starving? This is a case of inequality. Today everything comes under the laws of competition and the survival of the fittest, where the powerful feed upon the powerless. As a consequence, masses of people find themselves excluded and marginalized: without work, without possibilities, without any means of escape."

"In this context, some people continue to defend trickle-down theories which assume that economic growth, encouraged by a free market, will inevitably succeed in bringing about greater justice and inclusiveness in the world. This opinion, which has never been confirmed by the facts, expresses a crude and naive trust in the goodness of those wielding economic power and in the sacralized workings of the prevailing economic system. Meanwhile, the excluded are still waiting."

In a recent speech, Senator Bernie Sanders quoted Pope Francis extensively and added: "We have a situation today, Mr. President, incredible as it may sound, where the wealthiest 85 people in the world own more wealth than the bottom half of the world's population."[b]

The social epidemiologist Prof. Richard Wilkinson, has documented the ways in which societies with less economic inequality do better than more unequal societies in a number of areas, including increased rates of life expectancy, mathematical performance, literacy, trust, social mobility,

---

[b] https://www.youtube.com/watch?v=9_LJpN893Vg
https://www.oxfam.org/en/tags/inequality
https://www.oxfam.org/sites/www.oxfam.org/files/file_attachments/cr-even-it-up-extreme-inequality-291014-en.pdf

together with decreased rates of infant mortality, homicides, imprisonment, teenage births, obesity and mental illness, including drug and alcohol addiction.[c] We must also remember that according to the economist John A. Hobson, the basic problem that led to imperialism was an excessively unequal distribution of incomes in the industrialized countries. The result of this unequal distribution was that neither the rich nor the poor could buy back the total output of their society. The incomes of the poor were insufficient, and rich were too few in number.

---

[c] https://www.youtube.com/watch?v=cZ7LzE3u7Bw
https://en.wikipedia.org/wiki/Richard_G._Wilkinson

Fig. 5.6 Zaatari refugee camp for Syrian refugees in Jordan which only contains a population of 80,000 out of the 1.3 million in the country. It is expected that climate change will contribute to the refugee crisis and the problem of famine. International cooperation is needed to meet this emergency. (Public domain.)

In the world as it is, an estimated 10 million children die each year from starvation or from diseases related to malnutrition.

In the world as it could be, the international community would support programs for agricultural development and famine relief on a much larger scale than at present.

Fig. 5.7 Unsafe drinking water is the largest cause of preventable disease. (Water and the Body.)

In the world as it is, diarrhoea spread by unsafe drinking water kills an estimated 6 million children every year.

In the world as it could be, the installation of safe and adequate water systems and proper sanitation in all parts of the world would have a high priority and would be supported by ample international funds.

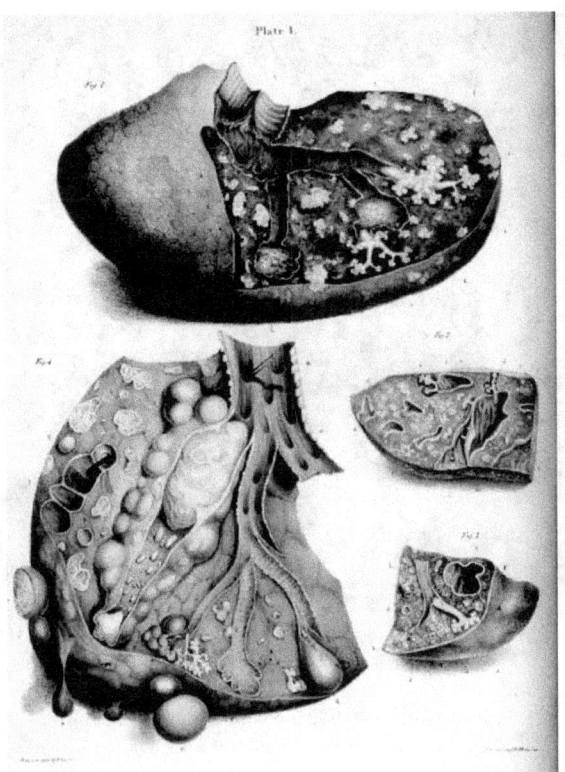

Fig. 5.8 The amount of money needed to effectively combat diseases such as tuberculosis is a tiny fraction of the colossal sums spent on wars and armament. (Public domain.)

In the world as it is, malaria, tuberculosis, AIDS, cholera, schistosomiasis, typhoid fever, typhus, trachoma, sleeping sickness and river blindness cause the illness and death of millions of people each year. For example, it is estimated that 200 million people now suffer from schistosomiasis and that 500 million suffer from trachoma, which often causes blindness. In Africa alone, malaria kills more than a million children every year.

In the world as it could be, these preventable diseases would be controlled by a concerted international effort. The World Health Organization would be given sufficient funds to carry out this project.

Fig. 5.9 A village school. (Rediff.com)

In the world as it is, the rate of illiteracy in the 25 least developed countries is 80%. The total number of illiterates in the world is estimated to be 800 million.

In the world as it could be, the international community would aim at giving all children at least an elementary education. Laws against child labor would prevent parents from regarding very young children as a source of income, thus removing one of the driving forces behind the population explosion. The money invested in education would pay economic dividends after a few years.

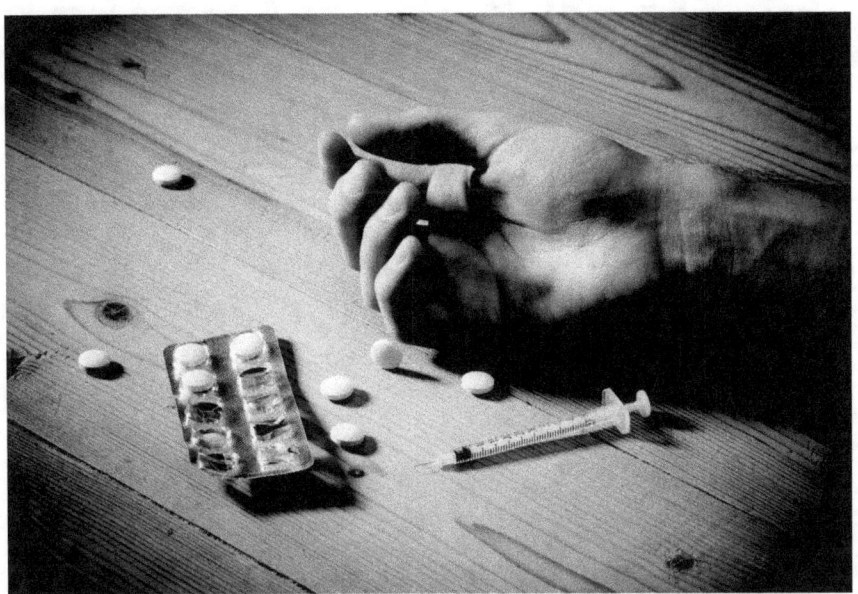

Fig. 5.10 In order to prevent the production of drug-related crops, it is necessary to help poor farmers to find alternatives. (Mynd Works Therapy.)

In the world as it is, opium poppies and other drug-producing plants are grown with little official hindrance in certain parts of Asia, the Middle East, and Latin America. Hard drugs refined from these plants are imported illegally into the developed countries, where they become a major source of high crime rates and human tragedy.

In the world as it could be, all nations would work together in a coordinated world-wide program to prevent the growing, refinement and distribution of harmful drugs.

Some suggestions for further reading

(1) E.J. Hobsbawn, *The Age of Empire, 1875-1914*, Vintage Books, (1989).
(2) L. James, *The Rise and Fall of the British Empire*, St Martin's Press, (1997).
(3) N. Ferguson, *Empire: The Rise and Demise of the British World Order and the Lessons for Global Power*, Basic Books, (2003).
(4) S. Schama, *The Fate of Empire, 1776-2000*, Miramax, (2002).
(5) A.P. Thorton, *The Imperial Idea and Its Enemies: A Study in British Power*, Palgrave Macmillan, (1985).
(6) H. Mejcher, *Imperial Quest for Oil: Iraq, 1910-1928*, Ithaca Books, London, (1976).
(7) P. Sluglett, *Britain in Iraq, 1914-1932*, Ithaca Press, London, (1976).
(8) D.E. Omissi, *British Air Power and Colonial Control in Iraq, 1920-1925*, Manchester University Press, Manchester, (1990).
(9) V.G. Kiernan, *Colonial Empires and Armies, 1815-1960*, Sutton, Stroud, (1998).
(10) R. Solh, *Britain's 2 Wars With Iraq*, Ithaca Press, Reading, (1996).
(11) D. Hiro, *The Longest War: The Iran-Iraq Military Conflict*, Routledge, New York, (1991).
(12) T.E. Lawrence, *A Report on Mesopotamia by T.E. Lawrence*, Sunday Times, August 22, (1920).
(13) D. Fromkin, *A Peace to End All Peace: The Fall of the Ottoman Empire and the Creation of the Modern Middle East*, Owl Books, (2001).
(14) T. Rajamoorthy, *Deceit and Duplicity: Some Reflections on Western Intervention in Iraq*, Third World Resurgence, March-April, (2003).
(15) P. Knightley and C. Simpson, *The Secret Lives of Lawrence of Arabia*, Nelson, London, (1969).
(16) G. Lenczowski, *The Middle East in World Affairs*, Cornell University Press, (1962).
(17) John A. Hobson, *Imperialism; A Study*, (1902).
(18) P. Cain and T. Hopkins, *British Imperialism, 1688-200*, Longman, (2000).
(19) N. Ferguson, *Empire: The Rise and Demise of the British World Order and the Lessons for Global Power*, Basic Books, (2003).
(20) G. Kolko, *Another Century of War*, New Press, (2002).
(21) G. Kolko, *Confronting the Third World: United States Foreign Policy, 1945-1980*, Pantheon Books, (1988).
(22) M.T. Klare, *Resource Wars: The New Landscape of Global Conflict*, Owl Books reprint edition, New York, (2002).

Chapter 6

The Threat of Nuclear War

"The unleashed power of the atom has changed everything except our ways of thinking, and thus we drift towards unparalleled catastrophes."

"I don't know what will be used in the next world war, but the 4th will be fought with stones."

Albert Einstein

Introduction

Today, the greatest threats facing human civilization and the biosphere are catastrophic climate change and nuclear war. Each of these could potentially destroy our civilization, kill most humans, and make most of our planet uninhabitable for most species, including our own.

The peoples of the world must unite and work with dedication to avoid these twin threats.

Fig. 6.1 Saint Paul's Cathedral during the London Blitz. Determined firefighting by citizens saved the cathedral from burning. (Wikipedia.)

Fig. 6.2 A view of Dresden after the firebombing with a statue of "Goodness" in the foreground. (Wikipedia.)

Targeting civilians

The erosion of ethical principles during World War II

When Hitler invaded Poland in September, 1939, US President Franklin Delano Roosevelt appealed to Great Britain, France, and Germany to spare innocent civilians from terror bombing. "The ruthless bombing from the air of civilians in unfortified centers of population during the course of the hostilities", Roosevelt said (referring to the use of air bombardment during World War I) "...has sickened the hearts of every civilized man and woman, and has profoundly shocked the conscience of humanity." He urged "every Government which may be engaged in hostilities publicly to affirm its determination that its armed forces shall in no event, and under no circumstances, undertake the bombardment from the air of civilian populations or of unfortified cities."

Two weeks later, British Prime Minister Neville Chamberlain responded to Roosevelts appeal with the words: "Whatever the lengths to which others may go, His Majesty's Government will never resort to the deliberate attack on women and children and other civilians for purposes of mere terrorism."

Much was destroyed during World War II, and among the casualties of the war were the ethical principles that Roosevelt and Chamberlain announced at its outset. At the time of Roosevelt and Chamberlains declarations, terror bombing of civilians had already begun in the Far East. On 22 and 23 September, 1937, Japanese bombers attacked civilian populations in Nanjing and Canton. The attacks provoked widespread protests. The British Under Secretary of State for Foreign Affairs, Lord Cranborne, wrote: "Words cannot express the feelings of profound horror with which the news of these raids has been received by the whole civilized world. They are often directed against places far from the actual area of hostilities. The military objective, where it exists, seems to take a completely second place. The main object seems to be to inspire terror by the indiscriminate slaughter of civilians..."

On the 25th of September, 1939, Hitler's air force began a series of intense attacks on Warsaw. Civilian areas of the city, hospitals marked with the Red Cross symbol, and fleeing refugees all were targeted in a effort to force the surrender of the city through terror. On the 14th of May, 1940, Rotterdam was also devastated. Between the 7th of September 1940 and the 10th of May 1941, the German Luftwaffe carried out massive air attacks on targets in Britain. By May, 1941, 43,000 British civilians were killed and more than a million houses destroyed.

By the end of the war the United States and Great Britain were bombing of civilians on a far greater scale than Japan and Germany had ever done.

For example, on July 24-28, 1943, British and American bombers attacked Hamburg with an enormous incendiary raid whose official intention was "the total destruction" of the city.

The result was a firestorm that did, if fact, lead to the total destruction of the city. One airman recalled, that "As far as I could see was one mass of fire. A sea of flame has been the description, and thats an understatement. It was so bright that I could read the target maps and adjust the bombsight." Another pilot was "...amazed at the awe-inspiring sight of the target area. It seemed as though the whole of Hamburg was on fire from one end to the other and a huge column of smoke was towering well above us - and we were on 20,000 feet! It all seemed almost incredible and, when I realized that I was looking at a city with a population of two millions, or about that, it became almost frightening to think of what must be going on down there in Hamburg."

Below, in the burning city, temperatures reached 1400°F, a temperature at which lead and aluminum have long since liquefied. Powerful winds sucked new air into the firestorm. There were reports of babies being torn by the high winds from their mothers arms and sucked into the flames. Of the 45,000 people killed, it has been estimated that 50 percent were women and children and many of the men killed were elderly, above military age. For weeks after the raids, survivors were plagued by "...droves of vicious rats, grown strong by feeding on the corpses that were left unburied within the rubble as well as the potatoes and other food supplies lost beneath the broken buildings."

The German cities Kassel, Pforzheim, Mainz, Dresden and Berlin were similarly destroyed, and in Japan, US bombing created firestorms in many cities, for example Tokyo, Kobe and Yokohama. In Tokyo alone, incendiary bombing caused more than 100,000 civilian casualties.

Hiroshima and Nagasaki

On August 6, 1945, at 8:15 in the morning, an atomic bomb was exploded in the air over Hiroshima. The force of the explosion was equivalent to twenty thousand tons of T.N.T.. Out of a city of two hundred and fifty thousand people, almost one hundred thousand were killed by the bomb; and another hundred thousand were hurt.

In some places, near the center of the city, people were completely vaporized, so that only their shadows on the pavement marked the places where they had been. Many people who were not killed by the blast or by burns from the explosion, were trapped under the wreckage of their houses. Unable to move, they were burned to death in the fire which followed.

Some accounts of the destruction of Hiroshima, written by children who

Fig. 6.3 Enrico Fermi (1901–1954). In 1934, he and his team of young Italian physicists split uranium atoms without realizing it. (Public domain.)

survived it, have been collected by Professor Arata Osada. Among them is the following account, written by a boy named Hisato Ito. He was 11 years old when the atomic bomb was exploded over the city:

"On the morning of August 5th (we went) to Hiroshima to see my brother, who was at college there. My brother spent the night with us in a hotel... On the morning of the 6th, my mother was standing near the entrance, talking with the hotel proprietor before paying the bill, while I played with the cat. It was then that a violent flash of blue-white light swept in through the doorway."

"I regained consciousness after a little while, but everything was dark. I had been flung to the far end of the hall, and was lying under a pile of debris caused by the collapse of two floors of the hotel. Although I tried to crawl out of this, I could not move. The fine central pillar, of which the proprietor was so proud, lay flat in front of me. "

"I closed my eyes and was quite overcome, thinking that I was going to die, when I heard my mother calling my name. At the sound of her voice, I opened my eyes; and then I saw the flames creeping close to me. I called frantically to my mother, for I knew that I should be burnt alive if I did not escape at once. My mother pulled away some burning boards and saved me. I shall never forget how happy I felt at that moment - like a bird let out of a cage."

"Everything was so altered that I felt bewildered. As far as my eyes could see, almost all the houses were destroyed and on fire. People passed by, their bodies red, as if they had been peeled. Their cries were pitiful. Others were dead. It was impossible to go farther along the street on account of the bodies, the ruined houses, and the badly wounded who lay about moaning. I did not know what to do; and as I turned to the west, I saw that the flames were drawing nearer.."

"At the waters edge, opposite the old Sentai gardens, I suddenly realized that I had become separated from my mother. The people who had been burned were plunging into the river Kobashi, and then were crying out: 'Its hot! Its hot! They were too weak to swim, and they drowned while crying for help."

In 1951, shortly after writing this account, Hisato Ito died of radiation sickness. His mother died soon afterward from the same cause.

The postwar nuclear arms race

When the news of the atomic bombing of Hiroshima and Nagasaki reached Albert Einstein, his sorrow and remorse were extreme. During the remainder of his life, he did his utmost to promote the cause of peace and to warn humanity against the dangers of nuclear warfare. Together with Bertrand Russell and Joseph Rotblat he helped to found Pugwash Conferences on Science and World Affairs (Nobel Peace Prize 1995), an organization of scientists and other scholars devoted to world peace and to the abolition of nuclear weapons.

When Otto Hahn, the discoverer of fission, heard the news of the destruction of Hiroshima, he and nine other German atomic scientists were being held prisoner at an English country house near Cambridge. Hahn

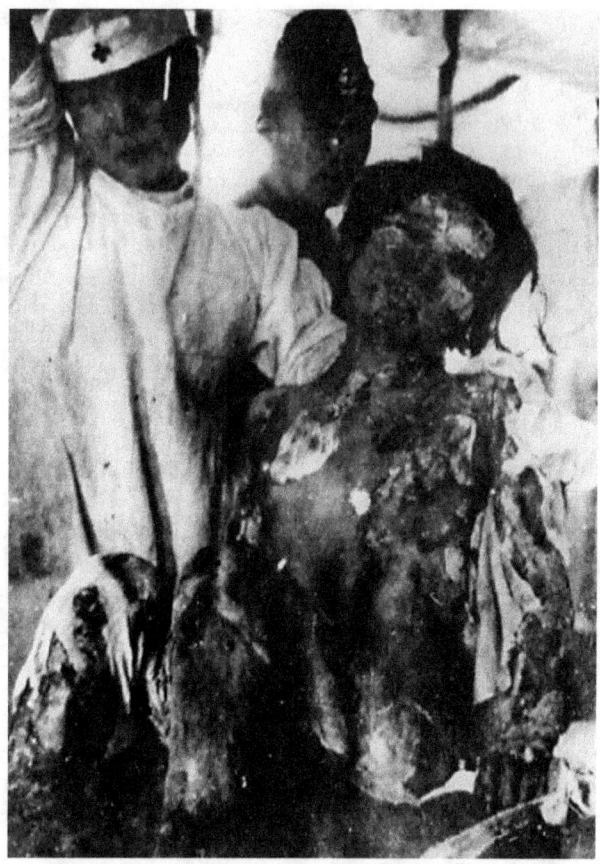

Fig. 6.4 Hiroshima. (duniverso.com.br.)

became so depressed that his colleagues feared that he would take his own life.

World public opinion was also greatly affected by the indiscriminate destruction of human life in Hiroshima and Nagasaki. Shortly after the bombings, the French existentialist author Albert Camus wrote: "Our technical civilization has just reached its greatest level of savagery. We will have to choose, in the more or less near future, between collective suicide and the intelligent use of our scientific conquests. Before the terrifying prospects now available to humanity, we see even more clearly that peace is the only battle worth waging. This is no longer a prayer, but a demand to be made by all peoples to their governments - a demand to choose definitively between hell and reason."

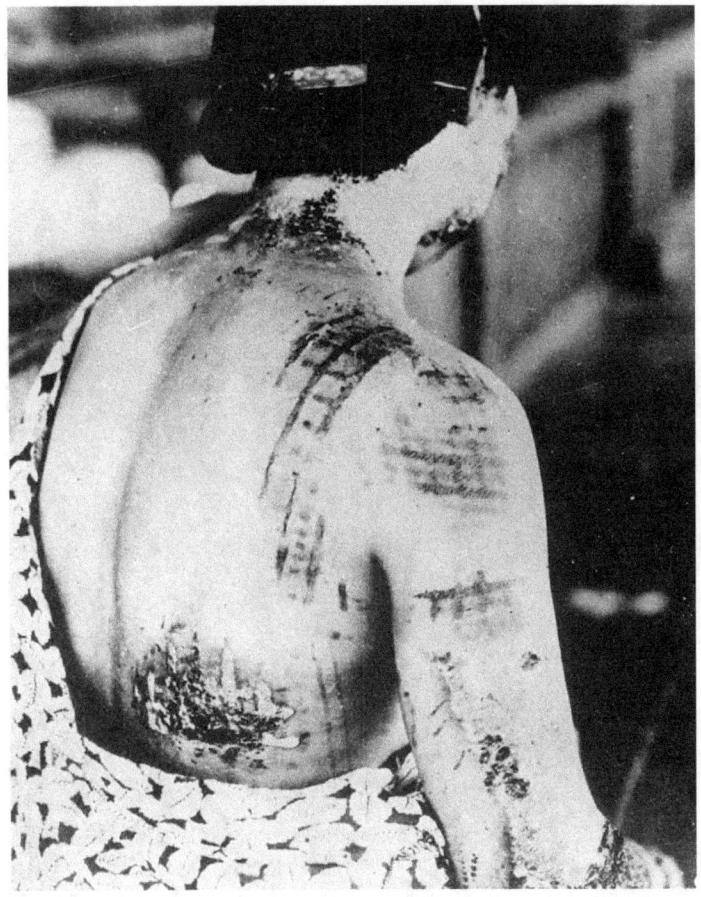

Fig. 6.5 Hiroshima. The greater absorption of thermal energy by dark colors resulted in the clothes pattern, in the tight-fitting areas on this survivor, being burnt into the skin. (Public domain.)

Among the scientists who had worked at Chicago and Los Alamos, there was relief that the war was over; but as descriptions of Hiroshima and Nagasaki became available there were also sharp feelings of guilt. Many scientists who had worked on the bomb project made great efforts to persuade the governments of the United States, England and the Soviet Union to agree to international control of atomic energy; but these efforts met with failure; and the nuclear arms race developed with increasing momentum.

In 1946, the United States proposed the Baruch Plan to internationalize atomic energy, but the plan was rejected by the Soviet Union, which had

Fig. 6.6 Nagasaki before the nuclear explosion and firestorm. (Public domain.)

been conducting its own secret nuclear weapons program since 1943. On August 29, 1949, the USSR exploded its first nuclear bomb. It had a yield equivalent to 21,000 tons of TNT, and had been constructed from Pu-239 produced in a nuclear reactor. Meanwhile the United Kingdom had begun to build its own nuclear weapons.

The explosion of the Soviet nuclear bomb caused feelings of panic in the United States, and President Truman authorized an all-out effort to build superbombs using thermonuclear reactions - the reactions that heat the sun and stars. The idea of using a U-235 fission bomb to trigger a thermonuclear reaction in a mixture of light elements had first been proposed by Enrico

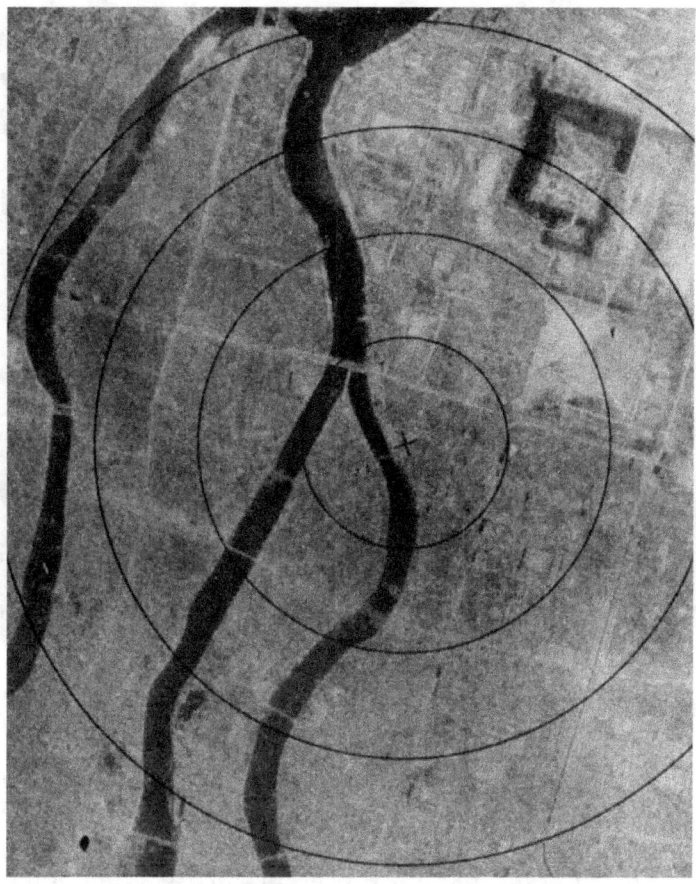

Fig. 6.7 Nagasaki afterwards. (Public domain.)

Fermi in a 1941 conversation with his Chicago colleague Edward Teller. After this conversation, Teller (perhaps the model for Stanley Kubrick's character Dr. Strangelove) became a fanatical advocate of the superbomb.

After Truman's go-ahead, the American program to build thermonuclear weapons made rapid progress, and on October 31, 1952, the first US thermonuclear device was exploded at Eniwetok Atoll in the Pacific Ocean. It had a yield of 10.4 megatons, that is to say it had an explosive power equivalent to 10,400,000 tons of TNT. Thus the first thermonuclear bomb was five hundred times as powerful as the bombs that had devastated Hiroshima and Nagasaki. Lighter versions of the device were soon developed, and these could be dropped from aircraft or delivered by rockets.

Fig. 6.8 The United States exploded a hydrogen bomb near the island of Enewetak in the South Pacific in 1952. The explosive force of the bomb was 500 times greater than the bombs that destroyed Hiroshima and Nagasaki. The Soviet Union tested its first hydrogen bomb in 1953. In March, 1954, the US tested another hydrogen bomb at the Bikini Atoll in the Pacific Ocean. It was 1000 times more powerful than the Hiroshima bomb. The Japanese fishing boat, Lucky Dragon, was 130 kilometers from the Bikini explosion, but radioactive fallout from the test killed one crew member and made all the others seriously ill. (Public domain.)

The Soviet Union and the United Kingdom were not far behind. In 1955 the Soviets exploded their first thermonuclear device, followed in 1957 by the UK. In 1961 the USSR exploded a thermonuclear bomb with a yield of 58 megatons. A bomb of this size, three thousand times the size of the Hiroshima bomb, would be able to totally destroy a city even if it missed it by 50 kilometers. Fall-out casualties would extend to a far greater distance.

In the late 1950s General Gavin, Chief of Army Research and Development in the United States, was asked by the Symington Committee, "If we got into a nuclear war and our strategic air force made an assault in

Fig. 6.9 After discussing the Bikini test and its radioactive fallout with Joseph Rotblat, Lord Russell became concerned for the future of the human gene pool if large numbers of such bombs should ever be used in a war. To warn humanity of the danger, he wrote what came to be known as the Russell-Einstein Manifesto. On July 9, 1955, with Rotblat in the chair, Russell read the Manifesto to a packed press conference. The document contains the words: "Here then is the problem that we present to you, stark and dreadful and inescapable: Shall we put an end to the human race, or shall mankind renounce war?... There lies before us, if we choose, continual progress in happiness, knowledge and wisdom. Shall we, instead, choose death because we cannot forget our quarrels? We appeal as human beings to human beings: Remember your humanity, and forget the rest." Lord Russell devoted much of the remainder of his life to working for the abolition of nuclear weapons. Here he is seen in 1962 in Trafalgar Square, London, addressing a meeting of the Campaign for Nuclear Disarmament. (Public domain.)

force against Russia with nuclear weapons exploded in a way where the prevailing winds would carry them south-east over Russia, what would be the effect in the way of death?"

General Gavin replied: "Current planning estimates run on the order of several hundred million deaths. That would be either way depending on

Fig. 6.10 Albert Einstein wrote: "The unleashed power of the atom has changed everything save our modes of thinking, and we thus drift toward unparalleled catastrophes." He also said, "I don't know what will be used in the next world war, but the 4th will be fought with stones." (Wikimedia.)

which way the wind blew. If the wind blew to the south-east they would be mostly in the USSR, although they would extend into the Japanese area and perhaps down into the Philippine area. If the wind blew the other way, they would extend well back into Western Europe."

Between October 16 and October 28, 1962, the Cuban Missile Crisis occurred, an incident in which the world came extremely close to a full-scale thermonuclear war. During the crisis, President Kennedy and his advisers

Fig. 6.11 Joseph Rotblat devoted the remainder of his life to working for peace and for the abolition of nuclear weapons. He became the president and guiding spirit of the Pugwash Conferences on Science and World Affairs, an organization of scientists and other scholars devoted to these goals. In his 1995 Nobel Peace Prize acceptance speech, Sir Joseph Rotblat (as he soon became) emphasized the same point that had been made in the Russell-Einstein Manifesto - that war itself must be eliminated in order to free civilization from the danger of nuclear destruction. (Pugwash Conferences.)

estimated that the chance of an all-out nuclear war with Russia was 50%. Recently-released documents indicate that the probability of war was even higher than Kennedy's estimate. Robert McNamara, who was Secretary of Defense at the time, wrote later, "We came within a hairbreadth of nuclear war without realizing it... Its no credit to us that we missed nuclear war..."

In 1964 the first Chinese nuclear weapon was tested, and this was followed in 1967 by a Chinese thermonuclear bomb with a yield of 3.3 megatons. France quickly followed suit testing a fission bomb in 1966 and a

Fig. 6.12 To the insidious argument that "the end justifies the means", Mahatma Gandhi answered firmly: "They say 'means are after all means. I would say 'means are after all everything. As the means, so the end. Indeed the Creator has given us control (and that very limited) over means, none over end... The means may be likened to a seed, and the end to a tree; and there is the same inviolable connection between the means and the end as there is between the seed and the tree. Means and end are convertible terms in my philosophy of life." In other words, if evil means are used, the end achieved will be contaminated by the means used to achieve it. Gandhi's insight can be applied to the argument that the nuclear bombings that destroyed Hiroshima and Nagasaki helped to end World War II and were therefore justified. In fact, these terrible events lead to a nuclear arms race that still casts an extremely dark shadow over the future of human civilization. (Public domain.)

thermonuclear bomb in 1968. In all about thirty nations contemplated building nuclear weapons, and many made active efforts to do so.

Because the concept of deterrence required an attacked nation to be able to retaliate massively even though many of its weapons might be destroyed by a preemptive strike, the production of nuclear warheads reached insane heights, driven by the collective paranoia of the Cold War. More than 50,000 nuclear warheads were produced worldwide, a large number of them thermonuclear. The collective explosive power of these warheads was equivalent to 20,000,000,000 tons of TNT, i.e. 4 tons for every man, woman and child on the planet, or, expressed differently, a million times the explosive power of the bomb that destroyed Hiroshima.

The end of the Cold War

In 1985, Michael Gorbachev (1931-) became the General Secretary of the Communist Party of the Soviet Union. Gorbachev had become convinced by his conversations with scientists that the policy of nuclear confrontation between the United States and the USSR was far too dangerous to be continued over a long period of time. If continued, sooner or later, through accident of miscalculation, it would result in a disaster of unprecedented proportions. Gorbachev also believed that the USSR was in need of reform, and he introduced two words to characterize what he felt was needed: *glasnost* (openness) and *perestroika* (reconstruction).

In 1986, US President Ronald Reagan met Mikhail Gorbachev in Reykjavik, Iceland. The two leaders hoped that they might find ways of reducing the danger that a thermonuclear Third World War would be fought between their two countries. Donald Reagan, the White House Chief of Staff, was present at the meeting, and he records the following conversation: "At one point in time Gorbachev said 'I would like to do away with all nuclear weapons. And Reagan hit the table and said 'Well why didn't you say so in the first place! Thats exactly what I want to do! And if you want to do away with all the weapons, Ill agree to do away with all the weapons. Of course well do away with all the weapons. 'Good, [said Gorbachev] 'Thats great, but you must confine SDI to the laboratory. 'No I wont, said Reagan. 'No way. SDI continues. I told you that I am never going to give up SDI." The SDI program, which seemingly prevented Presidents Reagan and Gorbachev from reaching an agreement to completely eliminate their nuclear weapons was Reagan's "Star Wars" program which (in violation of the ABM Treaty) proposed to set up a system of of radar, satellites and missiles to shoot down attacking missiles.

Gorbachev s reforms effectively granted self-government to the various parts of the Soviet Union, and he himself soon resigned from his post as its leader, since the office was no longer meaningful. Most of the newly-independent parts of the old USSR began to introduce market economies, and an astonished world witnessed a series of unexpected and rapid changes: On September 10, 1989 Hungarian government opened its border for East

German refugees; on November 9, 1989 Berlin Wall was reopened; on December 22, 1989 Brandenburg Gate was opened; and on October 3, 1990 Germany was reunited. The Cold War was over!

The Non-Proliferation Treaty

During the Cold War, a number of international treaties attempting to reduce the global nuclear peril had been achieved after much struggle. Among these, the 1968 Nuclear Non-Proliferation Treaty (NPT) has special importance. The NPT was designed to prevent the spread of nuclear weapons beyond the five nations that already had them; to provide assurance that "peaceful" nuclear activities of non-nuclear-weapon states would not be used to produce such weapons; to promote peaceful use of nuclear energy to the greatest extent consistent with non-proliferation of nuclear weapons; and finally, to ensure that definite steps towards complete nuclear disarmament would be taken by all states, as well steps towards comprehensive control of conventional armaments (Article VI).

The non-nuclear-weapon states insisted that Article VI be included in the treaty as a price for giving up their own ambitions. The full text of Article VI is as follows: "Each of the Parties to the Treaty undertakes to pursue negotiations in good faith on effective measures relating cessation of the nuclear arms race at an early date and to nuclear disarmament, and on a Treaty on general and complete disarmament under strict international control."

The NPT has now been signed by 187 countries and has been in force as international law since 1970. However, Israel, India, Pakistan, and Cuba have refused to sign, and North Korea, after signing the treaty, withdrew from it in 1993. Israel began producing nuclear weapons in the late 1960s (with the help of a reactor provided by France) and the country is now believed to possess 100-150 of them, including neutron bombs. Israels policy is one of "nuclear opacity" - i.e., visibly possessing nuclear weapons while denying their existence.

South Africa, with the help of Israel and France, also produced nuclear weapons, which it tested in the Indian Ocean in 1979. In 1991 however, South Africa signed the NPT and destroyed its nuclear weapons.

India produced what it described as a "peaceful nuclear explosion" in 1974. By 1989 Indian scientists were making efforts to purify the lithium-6 isotope, a key component of the much more powerful thermonuclear bombs. In 1998, India conducted underground tests of nuclear weapons, and is now believed to have roughly 60 warheads, constructed from Pu-239 produced in "peaceful" reactors.

Pakistan's efforts to obtain nuclear weapons were spurred by India's 1974 "peaceful nuclear explosion". Zulfiquar Ali Bhutto, who initiated Pakistan's program, first as Minister of Fuel, Power and Natural Resources, and later as President and Prime Minister, declared: "There is a Christian Bomb, a Jewish Bomb and a Hindu Bomb. There must be an Islamic Bomb! We will get it even if we have to starve - even if we have to eat grass!" As early as 1970, the laboratory of Dr. Abdul Qadeer Khan, (a metallurgist who was to become Pakistan's leading nuclear bomb maker) had been able to obtain from a Dutch firm the high-speed ultracentrafuges needed for uranium enrichment. With unlimited financial support and freedom from auditing requirements, Dr. Khan purchased restricted items needed for nuclear weapon construction from companies in Europe and the United States. In the process, Dr. Khan became an extremely wealthy man. With additional help from China, Pakistan was ready to test five nuclear weapons in 1998. The Indian and Pakistani nuclear bomb tests, conducted in rapid succession, presented the world with the danger that these devastating bombs would be used in the conflict over Kashmir. Indeed, Pakistan announced that if a war broke out using conventional weapons, Pakistan's nuclear weapons would be used "at an early stage".

In Pakistan, Dr. A.Q. Khan became a great national hero. He was presented as the person who had saved Pakistan from attack by India by creating Pakistan's own nuclear weapons. In a Washington Post article[a] Pervez Hoodbhoy wrote: "Nuclear nationalism was the order of the day as governments vigorously promoted the bomb as the symbol of Pakistan's high scientific achievement and self-respect, and as the harbinger of a new Muslim era." Similar manifestations of nuclear nationalism could also be seen in India after India's 1998 bomb tests.

Early in 2004, it was revealed that Dr. Khan had for years been selling nuclear secrets and equipment to Lybia, Iran and North Korea. However, observers considered that it was unlikely that Khan would be tried for these offenses, since a trial might implicate Pakistan's army as well as two of its former prime ministers. Furthermore, Dr. Khan has the strong support of Pakistan's Islamic fundamentalists. Recent assassinations emphasize the precariousness of Pakistan's government. There is a danger that it may be overthrown by Islamic fundamentalists, who would give Pakistan's nuclear weapons to terrorist organizations. This type of danger is a general one associated with nuclear proliferation. As more and more countries obtain nuclear weapons, it becomes increasingly likely that one of them will undergo a revolution, during the course of which nuclear weapons will fall into the hands of subnational organizations.

---

[a] 1 February, 2004.

Article VIII of the Non-Proliferation Treaty provides for a conference to be held every five years to make sure that the NPT is operating as intended. In the 1995 NPT Review Conference, the lifetime of the treaty was extended indefinitely, despite the general dissatisfaction with the bad faith of the nuclear weapon states: They had dismantled some of their warheads but had taken no significant steps towards complete nuclear disarmament. The 2000 NPT Review Conference made it clear that the nuclear weapons states could not postpone indefinitely their commitment to nuclear disarmament by linking it to general and complete disarmament, since these are separate and independent goals of Article VI. The Final Document of the conference also contained 13 Practical Steps for Nuclear Disarmament, including ratification of a Comprehensive Test Ban Treaty (CTBT), negotiations on a Fissile Materials Cutoff Treaty, the preservation and strengthening of the Anti-Ballistic Missile (ABM) Treaty, greater transparency with regard to nuclear arsenals, and making irreversability a principle of nuclear reductions. Another review conference is scheduled for 2010, a year that marks the 55th anniversary of the destruction of Hiroshima and Nagasaki.

Something must be said about the concept of irreversability mentioned in the Final Document of the 2000 NPT Review Conference. Nuclear weapons can be destroyed in a completely irreversible way by getting rid of the special isotopes which they use. In the case of highly enriched uranium (HEU), this can be done by mixing it thoroughly with ordinary unenriched uranium. In natural uranium, the rare fissile isotope U-235 is only 0.7%. The remaining 99.3% consists of the common isotope, U-238, which under ordinary circumstances cannot undergo fission. If HEU is mixed with a sufficient quantity of natural uranium, so that the concentration of U-235 falls below 20%, it can no longer be used in nuclear weapons.

Getting rid of plutonium irreversibly is more difficult, but it could be cast into large concrete blocks and dumped into extremely deep parts of the ocean (e.g. the Japan Trench) where recovery would be almost impossible. Alternatively, it could be placed in the bottom of very deep mine shafts, which could afterwards be destroyed by means of conventional explosives. None of the strategic arms reduction treaties, neither the SALT treaties nor the 2002 Moscow Treaty, incorporate irreversability.

The recent recommendation by four distinguished German statesmen that all short-range nuclear weapons be destroyed is particularly interesting [13]. The strongest argument for the removal of US tactical nuclear weapons from Europe is the danger of collapse of the NPT. The 2005 NPT Review Conference was a disaster, and there is a danger that at the 2010 Review Conference, the NPT will collapse entirely because of the discriminatory position of the nuclear weapon states (NWS) and their failure to honor their commitments under Article VI. NATOs present nuclear weapon policy also violates the NPT, and correcting this violation would help to save the 2010 Review Conference from failure.

At present, the air forces of the European countries in which the US nuclear weapons are stationed perform regular training exercises in which they learn how to deliver the weapons. This violates the spirit, and probably also the letter, of Article IV, which prohibits the transfer of nuclear weapons from an NWS to a non-NWS. The "nuclear sharing" proponents maintain that such transfers would only happen in an emergency; but there is nothing in the NPT saying that the treaty would not hold under all circumstances. Furthermore, NATO would be improved, rather than damaged, by giving up "nuclear sharing". If President Obama wishes to fulfill his campaign promises [14] - if he wishes to save the NPT - a logical first step would be to remove US tactical nuclear weapons from Europe.

Flaws in the concept of nuclear deterrence

Before discussing other defects in the concept of deterrence, it must be said very clearly that the idea of "massive nuclear retaliation" is completely unacceptable from an ethical point of view. The doctrine of retaliation, performed on a massive scale, violates not only the principles of common human decency and common sense, but also the ethical principles of every major religion. Retaliation is especially contrary to the central commandment of Christianity which tells us to love our neighbor, even if he or she is far away from us, belonging to a different ethnic or political group, and even if our distant neighbor has seriously injured us. This principle has a fundamental place not only in in Christianity but also in Buddhism. "Massive retaliation" completely violates these very central ethical principles, which are not only clearly stated and fundamental but also very practical, since they prevent escalatory cycles of revenge and counter-revenge.

Contrast Christian ethics with estimates of the number of deaths that would follow a US nuclear strike against Russia: Several hundred million deaths. These horrifying estimates shock us not only because of the enormous magnitude of the expected mortality, but also because the victims would include people of every kind: women, men, old people, children and infants, completely irrespective of any degree of guilt that they might have. As a result of such an attack, many millions of people in neutral countries would also die. This type of killing has to be classified as genocide.

When a suspected criminal is tried for a wrongdoing, great efforts are devoted to clarifying the question of guilt or innocence. Punishment only follows if guilt can be proved beyond any reasonable doubt. Contrast this with the totally indiscriminate mass slaughter that results from a nuclear attack!

It might be objected that disregard for the guilt or innocence of victims is a universal characteristic of modern war, since statistics show that, with time, a larger and larger percentage of the victims have been civilians, and especially children. For example, the air attacks on Coventry during World War II, or the fire bombings of Dresden and Tokyo, produced massive casualties which involved all segments of the population with complete disregard for the question of guilt or innocence. The answer, I think, is that modern war has become generally unacceptable from an ethical point of view, and this unacceptability is epitomized in nuclear weapons.

The enormous and indiscriminate destruction produced by nuclear weapons formed the background for an historic 1996 decision by the International Court of Justice in the Hague. In response to questions put to it by WHO and the UN General Assembly, the Court ruled that "the threat and use of nuclear weapons would generally be contrary to the rules of international law applicable in armed conflict, and particularly the principles and rules of humanitarian law." The only *possible* exception to this general rule might be "an extreme circumstance of self-defense, in which the very survival of a state would be at stake". But the Court refused to say that even in this extreme circumstance the threat or use of nuclear weapons would be legal. It left the exceptional case undecided. In addition, the World Court added unanimously that "there exists an obligation to pursue in good faith *and bring to a conclusion* negotiations leading to nuclear disarmament in all its aspects under strict international control."

This landmark decision has been criticized by the nuclear weapon states as being decided "by a narrow margin", but the structuring of the vote made the margin seem more narrow than it actually was. Seven judges voted against Paragraph 2E of the decision (the paragraph which states that the threat or use of nuclear weapons would be generally illegal, but which mentions as a possible exception the case where a nation might be defending itself from an attack that threatened its very existence.) Seven judges voted for the paragraph, with the President of the Court, Muhammad Bedjaoui of Algeria casting the deciding vote. Thus the Court adopted it, seemingly by a narrow margin. But three of the judges who voted against 2E did so because they believed that no possible exception should be mentioned! Thus, if the vote had been slightly differently structured, the result would have be ten to four.

Of the remaining four judges who cast dissenting votes, three represented nuclear weapons states, while the fourth thought that the Court ought not to have accepted the questions from WHO and the UN. However Judge Schwebel from the United States, who voted against Paragraph 2E, nevertheless added, in a separate opinion, "It cannot be accepted that the use of nuclear weapons on a scale which would - or could - result in the deaths of many millions in indiscriminate inferno and by far-reaching fallout, have pernicious effects in space and time, and render uninhabitable much of the earth, could be lawful." Judge Higgins from the UK, the first woman judge in the history of the Court, had problems with the word "generally" in Paragraph 2E and therefore voted against it, but she thought that a more profound analysis might have led the Court to conclude in favor of illegality in all circumstances. Judge Fleischhauer of Germany said in his separate opinion, "The nuclear weapon is, in many ways, the negation of the humanitarian considerations underlying the law applicable in armed conflict and the principle of neutrality. The nuclear weapon cannot distinguish between civilian and military targets. It causes immeasurable suffering. The radiation released by it is unable to respect the territorial integrity of neutral States."

President Bedjaoui, summarizing the majority opinion, called nuclear weapons "the ultimate evil", and said "By its nature, the nuclear weapon, this blind weapon, destabilizes humanitarian law, the law of discrimination in the use of weapons... The ultimate aim of every action in the field of nuclear arms will always be nuclear disarmament, an aim which is no longer utopian and which all have a duty to pursue more actively than ever."

Thus the concept of nuclear deterrence is not only unacceptable from the standpoint of ethics; it is also contrary to international law. The World Courts 1996 advisory Opinion unquestionably also represents the opinion of the majority of the worlds peoples. Although no formal plebiscite has been taken, the votes in numerous resolutions of the UN General Assembly speak very clearly on this question. For example the New Agenda Resolution (53/77Y) was adopted by the General Assembly on 4 December 1998 by a massively affirmative vote, in which only 18 out of the 170 member states voted against the resolution.[b] The New Agenda Resolution proposes numerous practical steps towards complete nuclear disarmament, and it calls on the Nuclear-Weapon States "to demonstrate an unequivocal commitment to the speedy and total elimination of their nuclear weapons and without delay to pursue in good faith and bring to a conclusion negotiations leading to the elimination of these weapons, thereby fulfilling their obligations under Article VI of the Treaty on the Non-Proliferation of Nuclear Weapons (NPT)". Thus, in addition to being ethically unacceptable and contrary to international law, nuclear weapons also contrary to the principles of democracy.

Having said these important things, we can now turn to some of the other defects in the concept of nuclear deterrence. One important defect is

---

[b]Of the 18 countries that voted against the New Agenda resolution, 10 were Eastern European countries hoping for acceptance into NATO, whose votes seem to have been traded for increased probability of acceptance.

that nuclear war may occur through accident or miscalculation - through technical defects or human failings. This possibility is made greater by the fact that despite the end of the Cold War, thousands of missiles carrying nuclear warheads are still kept on a "hair-trigger" state of alert with a quasi-automatic reaction time measured in minutes. There is a constant danger that a nuclear war will be triggered by error in evaluating the signal on a radar screen. For example, the BBC reported recently that a group of scientists and military leaders are worried that a small asteroid entering the earths atmosphere and exploding could trigger a nuclear war if mistaken for a missile strike.

A number of prominent political and military figures (many of whom have ample knowledge of the system of deterrence, having been part of it) have expressed concern about the danger of accidental nuclear war. Colin S. Grey[c] expressed this concern as follows: "The problem, indeed the enduring problem, is that we are resting our future upon a nuclear deterrence system concerning which we cannot tolerate even a single malfunction." General Curtis E. LeMay[d] has written, "In my opinion a general war will grow through a series of political miscalculations and accidents rather than through any deliberate attack by either side." Bruce G. Blair[e] has remarked that "It is obvious that the rushed nature of the process, from warning to decision to action, risks causing a catastrophic mistake ... This system is an accident waiting to happen."

"But nobody can predict that the fatal accident or unauthorized act will never happen", Fred Ikle of the Rand Corporation has written, "Given the huge and far-flung missile forces, ready to be launched from land and sea on on both sides, the scope for disaster by accident is immense... In a matter of seconds - through technical accident or human failure - mutual deterrence might thus collapse."

---

[c]Chairman, National Institute for Public Policy.

[d]Founder and former Commander in Chief of the United States Strategic Air Command.

[e]Brookings Institute.

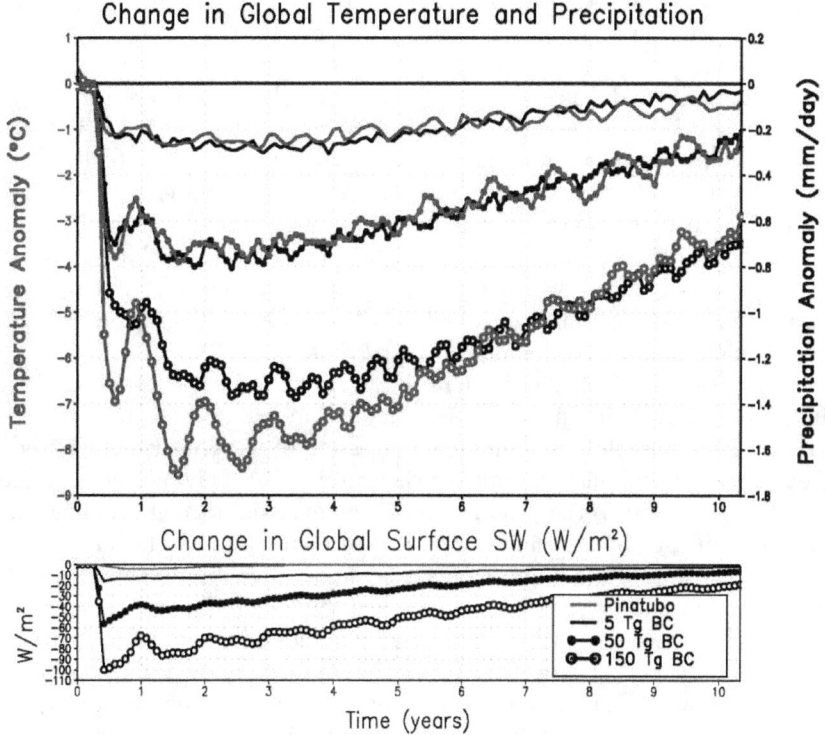

Fig. 6.13 Recent studies by atmospheric scientists have shown that the smoke from burning cities produced by even a limited nuclear war would have a devastating effect on global agriculture. The studies show that the smoke would rise to the stratosphere, where it would spread globally and remain for a decade, blocking sunlight and destroying the ozone layer. Because of the devastating effect on global agriculture, darkness from even a small nuclear war (e.g. between India and Pakistan) would result in an estimated billion deaths from famine. (O. Toon, A. Robock and R. Turco, "The Environmental Consequences of Nuclear War", Physics Today, vol. 61, No. 12, 2008, p. 37-42.)

Another serious failure of the concept of nuclear deterrence is that it does not take into account the possibility that atomic bombs may be used by terrorists. Indeed, the threat of nuclear terrorism has today become one of the most pressing dangers that the world faces, a danger that is particularly acute in the United States.

Since 1945, more than 3,000 metric tons (3,000,000 kilograms) of highly enriched uranium and plutonium have been produced - enough for several hundred thousand nuclear weapons. Of this, roughly a million kilograms are in Russia, inadequately guarded, in establishments where the technicians are poorly paid and vulnerable to the temptations of bribery. There is a continuing danger that these fissile materials will fall into the hands of terrorists, or organized criminals, or irresponsible governments. Also, an extensive black market for fissile materials, nuclear weapons components etc. has recently been revealed in connection with the confessions of Pakistan's bomb-maker, Dr. A.Q. Khan. Furthermore, if Pakistan's less-than-stable government should be overthrown, complete nuclear weapons could fall into the hands of terrorists.

On November 3, 2003, Mohamed ElBaradei, Director General of the International Atomic Energy Agency, made a speech to the United Nations in which he called for "limiting the processing of weapons-usable material (separated plutonium and high enriched uranium) in civilian nuclear programmes - as well as the production of new material through reprocessing and enrichment - by agreeing to restrict these operations to facilities exclusively under international control." It is almost incredible, considering the dangers of nuclear proliferation and nuclear terrorism, that such restrictions were not imposed long ago. Nuclear reactors used for "peaceful" purposes unfortunately also generate fissionable isotopes of plutonium, neptunium and americium. Thus all nuclear reactors must be regarded as ambiguous in function, and all must be put under strict international control. One might ask, in fact, whether globally widespread use of nuclear energy is worth the danger that it entails.

The Italian nuclear physicist Francesco Calogero, who has studied the matter closely, believes that terrorists could easily construct a simple gun-type nuclear bomb if they were in possession of a critical mass of highly enriched uranium. In such a simple atomic bomb, two grapefruit-sized subcritical portions of HEU are placed at opposite ends of the barrel of an artillery piece and are driven together by means of a conventional explosive. Prof. Calogero estimates that the fatalities produced by the explosion of such a device in the center of a large city could exceed 100,000.

We must remember the remark of U.N. Secretary General Kofi Annan after the 9/11/2001 attacks on the World Trade Center. He said, "*This time* it was not a nuclear explosion". The meaning of his remark is clear: If the world does not take strong steps to eliminate fissionable materials and nuclear weapons, it will only be a matter of time before they will be used in terrorist attacks on major cities. Neither terrorists nor organized criminals can be deterred by the threat of nuclear retaliation, since they have no territory against which such retaliation could be directed. They blend invisibly into the general population. Nor can a "missile defense system" prevent terrorists from using nuclear weapons, since the weapons can be brought into a port in any one of the hundreds of thousands of containers that enter on ships each year, a number far too large to be checked exhaustively.

In this dangerous situation, the only logical thing for the world to do is to get rid of both fissile materials and nuclear weapons as rapidly as possible. We must acknowledge that the idea of nuclear deterrence is a dangerous fallacy, and acknowledge that the development of military systems based on nuclear weapons has been a terrible mistake, a false step that needs to be reversed. If the most prestigious of the nuclear weapons states can sincerely acknowledge their mistakes and begin to reverse them, nuclear weapons will seem less glamorous to countries like India, Pakistan, North Korea and Iran, where they now are symbols of national pride and modernism.

Civilians have for too long played the role of passive targets, hostages in the power struggles of politicians. It is time for civil society to make its will felt. If our leaders continue to enthusiastically support the institution of war, if they will not abolish nuclear weapons, then let us have new leaders.

Establishment opinion shifts towards nuclear abolition

Today there are indications that the establishment is moving towards the point of view that the peace movement has always held: - that nuclear weapons are essentially genocidal, illegal and unworthy of civilization; and that they must be completely abolished as quickly as possible. There is a rapidly-growing global consensus that a nuclear-weapon-free world can and must be achieved in the very near future.

One of the first indications of the change was the famous Wall Street Journal article by Schultz, Perry, Kissinger and Nunn advocating complete abolition of nuclear arms [1]. This was followed quickly by Mikhail Gorbachev's supporting article, published in the same journal [2], and a statement by distinguished Italian statesmen [3]. Meanwhile, in October 2007, the Hoover Institution had arranged a symposium entitled "Reykjavik Revisited; Steps Towards a World Free of Nuclear Weapons" [4].

In Britain, Sir Malcolm Rifkind, Lord Hurd and Lord Owen (all former Foreign Secretaries) joined the former NATO Secretary General Lord Robertson as authors of an article in The Times advocating complete abolition of nuclear weapons [5]. The UK's Secretary of State for Defense, Des Brown, speaking at a disarmament conference in Geneva, proposed that the UK "host a technical conference of P5 nuclear laboratories on the verification of nuclear disarmament before the next NPT Review Conference in 2010" to enable the nuclear weapon states to work together on technical issues.

In February, 2008, the Government of Norway hosted an international conference on "Achieving the Vision of a World Free of Nuclear Weapons" [7]. A week later, Norway's Foreign Minister, Jonas Gahr Støre, reported the results of the conference to a disarmament meeting in Geneva [8]. On July 11, 2008 , speaking at a Pugwash Conference in Canada, Norway's Defense Minister, Anne-Grete Strøm-Erichsen, reiterated her country's strong support for the complete abolition of nuclear weapons [9].

In July 2008, Barack Obama said in his Berlin speech, "It is time to secure all loose nuclear materials; to stop the spread of nuclear weapons; and to reduce the arsenals from another era. This is the moment to begin the work of seeking the peace of a world without nuclear weapons."

Later that year, in September, Vladimir Putin said, "Had I been told just two or three years ago I wouldn't believe that it would be possible, but I believe that it is now quite possible to liberate humanity from nuclear weapons..."

Other highly-placed statesmen added their voices to the growing consensus: Australia's Prime Minister, Kevin Rudd, visited the Peace Museum at Hiroshima, where he made a strong speech advocating nuclear abolition. He later set up an International Commission on Nuclear Non-Proliferation and Disarmament co-chaired by Australia and Japan [10].

On January 9, 2009, four distinguished German statesmen (Richard von Weizäcker, Helmut Schmidt, Egon Bahr and Hans-Dietrich Genscher) published an article entitled "Towards a Nuclear-Free World: a German View" in the International Herald Tribune [12]. Among the immediate steps recommended in the article are the following:

- The vision of a nuclear-weapon-free world... must be rekindled.
- Negotiations aimed at drastically reducing the number of nuclear weapons must begin...
- The Nuclear Non-Proliferation Treaty (NPT) must be greatly reinforced.
- America should ratify the Comprehensive Nuclear Test-Ban Treaty.
- All short-range nuclear weapons must be destroyed.
- The Anti-Ballistic Missile (ABM) Treaty must be restored. Outer space may only be used for peaceful purposes.

Going to zero

On December 8-9, 2008, approximately 100 international leaders met in Paris to launch the Global Zero Campaign [11]. They included Her Majesty Queen Noor of Jordan, Norway's former Prime Minister Gro Harlem Brundtland, former UK Foreign Secretaries Sir Malcolm Rifkind, Margaret Beckett and David Owen, Ireland's former Prime Minister Mary Robinson, UK philanthropist Sir Richard Branson, former UN Under-Secretary-General Jayantha Dhanapala, and Nobel Peace Prize winners President Jimmy Carter, President Mikhail Gorbachev, Archbishop Desmond Tutu and Prof. Muhammad Yunus. The concrete steps advocated by Global Zero include:

- Deep reductions to Russian-US arsenals, which comprise 96% of the world's 27,000 nuclear weapons.
- Russia and the United States, joined by other nuclear weapons states, cutting arsenals to zero in phased and verified reductions.
- Establishing verification systems and international management of the fuel cycle to prevent future development of nuclear weapons.

The Global Zero website [11] contains a report on a new public opinion poll covering 21 nations, including all of the nuclear weapons states. The poll showed that public opinion overwhelmingly favors an international agreement for eliminating all nuclear weapons according to a timetable. It was specified that the agreement would include monitoring. The average in all countries of the percent favoring such an agreement was 76%. A few results of special interest mentioned in the report are Russia 69%; the United States, 77%; China, 83%; France, 86%, and Great Britain, 81%.

In his April 5, 2009 speech in Prague the newly-elected U.S. President Barack Obama said: "To reduce our warheads and stockpiles, we will negotiate a new strategic arms reduction treaty with Russia this year. President Medvedev and I will begin this process in London, and we will seek an agreement by the end of the year that is sufficiently bold. This will set the stage for further cuts, and we will seek to involve all nuclear weapon states in this endeavor... To achieve a global ban on nuclear testing, my administration will immediately and aggressively pursue U.S. ratification of the Comprehensive Test Ban Treaty."

A few days later, on April 24, 2009, the European Parliament recommended complete nuclear disarmament by 2020. An amendment introducing the "Model Nuclear Weapons Convention" and the "Hiroshima-Nagasaki Protocol" as concrete tools to achieve a nuclear weapons free world by 2020 was approved with a majority of 177 votes against 130. The Nuclear Weapons Convention is analogous to the conventions that have successfully banned chemical and biological weapons.

The role of public opinion

Public opinion is extremely important for the actual achievement of complete nuclear abolition. In the first place, the fact that the public is overwhelmingly against the retention of nuclear weapons means that the continuation of nuclear arsenals violates democratic principles. Secondly, the weapons are small enough to be easily hidden. Therefore the help of "whistle-blowers" will be needed to help inspection teams to make sure that no country violates its agreement to irreversibly destroy every atomic bomb. What is needed is a universal recognition that nuclear weapons are an absolute evil, and that their continued existence is a threat to human civilization and to the life of every person on the planet.

Our aim must be to delegitimize nuclear weapons, in much the same way that unnecessary greenhouse gas emissions have recently been delegitimized, or cigarette smoking delegitemized, or racism delegitimized. This should be an easy task because of the essentially genocidal nature of nuclear weapons. For half a century, ordinary people have been held as hostages, never knowing from day to day whether their own lives and the lives of those they love would suddenly be sacrificed on the alter of thermonuclear nationalism and power politics. We must let the politicians know that we are no longer willing to be hostages; and we must also accept individual responsibility for reporting violations of international treaties, although our own nation might be the violator.

Most of us grew up in schools where we were taught that duty to our nation was the highest duty; but the times we live in today demand a change of heart, a higher loyalty to humanity as a whole. If the mass media cooperate in delegitimizing nuclear weapons, if educational systems cooperate and if religions [f] cooperate, the change of heart that we need - the global ethic that we need - can quickly be achieved.

Complete abolition of nuclear weapons

Although the Cold War has ended, the danger of a nuclear catastrophe is greater today than ever before. There are almost 16,000 nuclear weapons in the world today, of which more than 90 percent are in the hands of Russia and the United States. About 2,000 of these weapons are on hair-trigger alert, meaning that whoever is in charge of them has only a few minutes to decide whether the signal indicating an attack is real, or an error. The most important single step in reducing the danger of a disaster would be to take all weapons off hair-trigger alert.

Incidents in which global disaster is avoided by a hair's breadth are constantly occurring. For example, on the night of 26 September, 1983, Lt. Col. Stanislav Petrov, a young software engineer, was on duty at a

---

[f] As an example of the role that religions can play, we can consider the Buddhist organization Soka Gakkai International (SGI), which has 12 million members throughout the world. SGI's President Daisaku Ikeda has declared nuclear weapons to be an absolute evil and for more than 50 years the organization has worked for their abolition.

Fig. 6.14 Women Strike for Peace during the Cuban Missile Crisis in 1962. (Public domain.)

surveillance center near Moscow. Suddenly the screen in front of him turned bright red. An alarm went off. Its enormous piercing sound filled the room. A second alarm followed, and then a third, fourth and fifth, until the noise was deafening. The computer showed that the Americans had launched a strike against Russia. Petrov's orders were to pass the information up the chain of command to Secretary General Yuri Andropov. Within minutes, a nuclear counterattack would be launched. However, because of certain inconsistent features of the alarm, Petrov disobeyed orders and reported it as a computer error, which indeed it was. Most of us probably owe our lives to his brave and coolheaded decision and his knowledge of software systems. Narrow escapes such as this show us clearly that in the long run, the survival of our civilization will require the complete abolition of nuclear weapons.

Although their number has been substantially reduced from its Cold War maximum, the total explosive power of todays weapons is equivalent to roughly half a million Hiroshima bombs. To multiply the tragedy of Hiroshima and Nagasaki by a factor of half a million changes the danger qualitatively. What is threatened today is the complete breakdown of human society.

A nuclear war between Russia and the United States would be a catastrophe of unimaginable proportions, from which human civilization would hardly recover. Recently, the danger of such a catastrophic war has increased because of large-scale NATO exercises near to the Russian border. In a situation of tension, a nuclear war between the two superpowers could occur through human error, technical failure, or escalation, although neither government planned for it or wished for it.

As the number of nuclear weapon states grows larger, there is an increasing chance that a revolution will occur in one of them, putting nuclear weapons into the hands of terrorist groups or organized criminals. Today, for example, Pakistans less-than-stable government might be overthrown, and Pakistans nuclear weapons might end in the hands of terrorists. The weapons might then be used to destroy one of the worlds large coastal cities, having been brought into the port by one of numerous container ships that dock every day. Such an event might trigger a large-scale nuclear conflagration.

Today, the world is facing a grave danger from the reckless behavior of the government of the United States, which recently arranged a coup that overthrew the elected government of Ukraine. Although Victoria Nulands December 13, 2013 speech talks much about democracy, the people who carried out the coup in Kiev can hardly be said to be democracy's best representatives. Many belong to the Svoboda Party, which had its roots in the Social-National Party of Ukraine (SNPU). The name was an intentional reference to the Nazi Party in Germany.

It seems to be the intention of the US to establish NATO bases in Ukraine, no doubt armed with nuclear weapons. In trying to imagine how the Russians feel about this, we might think of the US reaction when a fleet of ships sailed to Cuba in 1962, bringing Soviet nuclear weapons. In the confrontation that followed, the world was bought very close indeed to an all-destroying nuclear war. Does not Russia feel similarly threatened by the thought of hostile nuclear weapons on its very doorstep? Can we not learn from the past, and avoid the extremely high risks associated with the similar confrontation in Ukraine today?

In general, aggressive interventions, in Iran, Syria, Ukraine, the Korean Peninsula and elsewhere, all present dangers for uncontrollable escalation into large and disastrous conflicts, which might potentially threaten the survival of human civilization.

Few politicians or military figures today have any imaginative understanding of what a war with thermonuclear weapons would be like. Recent studies have shown that in a nuclear war, the smoke from firestorms in burning cities would rise to the stratosphere where it would remain for a decade, spreading throughout the world, blocking sunlight, blocking the hydrological cycle and destroying the ozone layer. The effect on global agriculture would be devastating, and the billion people who are chronically undernourished today would be at risk. Furthermore, the tragedies of Chernobyl and Fukushima remind us that a nuclear war would make large areas of the world permanently uninhabitable because of radioactive contamination. A full-scale thermonuclear war would be the ultimate ecological catastrophe. It would destroy human civilization and much of the biosphere.

One can gain a small idea of the terrible ecological consequences of a nuclear war by thinking of the radioactive contamination that has made large areas near to Chernobyl and Fukushima uninhabitable, or the testing of hydrogen bombs in the Pacific, which continues to cause cancer, leukemia and birth defects in the Marshall Islands more than half a century later.

The United States tested a hydrogen bomb at Bikini in 1954. Fallout from the bomb contaminated the island of Rongelap, one of the Marshall Islands 120 kilometers from Bikini. The islanders experienced radiation illness, and many died from cancer. Even today, half a century later, both people and animals on Rongelap and other nearby islands suffer from birth defects. The most common defects have been "jelly fish babies", born with no bones and with transparent skin. Their brains and beating hearts can be seen. The babies usually live a day or two before they stop breathing.

A girl from Rongelap describes the situation in the following words: "I cannot have children. I have had miscarriages on seven occasions... Our culture and religion teach us that reproductive abnormalities are a sign that women have been unfaithful. For this reason, many of my friends keep quiet about the strange births that they have had. In privacy they give birth, not to children as we like to think of them, but to things we could only describe as octopuses, apples, turtles and other things in our experience. We do not have Marshallese words for these kinds of babies, because they were never born before the radiation came."

The Republic of the Marshall Islands is suing the nine countries with nuclear weapons at the International Court of Justice at The Hague, arguing they have violated their legal obligation to disarm. The Guardian reports that "In the unprecedented legal action, comprising nine separate cases brought before the ICJ on Thursday, the Republic of the Marshall Islands accuses the nuclear weapons states of a 'flagrant denial of human justice. It argues it is justified in taking the action because of the harm it suffered as a result of the nuclear arms race.

The Pacific chain of islands, including Bikini Atoll and Enewetak, was the site of 67 nuclear tests from 1946 to 1958, including the Bravo shot, a 15-megaton device equivalent to a thousand Hiroshima blasts, detonated in 1954. The Marshallese islanders say they have been suffering serious health and environmental effects ever since.

The island republic is suing the five 'established nuclear weapons states recognized in the 1968 nuclear non-proliferation treaty (NPT), the US, Russia (which inherited the Soviet arsenal), China, France and the UK, as well as the three countries outside the NPT who have declared nuclear arsenals: India, Pakistan and North Korea, and the one undeclared nuclear weapons state, Israel. The Republic of the Marshall Islands is not seeking monetary compensation, but instead it seeks to make the nuclear weapon states comply with their legal obligations under Article VI of the Nuclear Nonproliferation Treaty and the 1996 ruling of the International Court of Justice.

The Nuclear Age Peace Foundation (NAPF) is a consultant to the Marshall Islands on the legal and moral issues involved in bringing this case. David Krieger, President of NAPF, upon hearing of the motion to dismiss the case by the U.S. responded, "The U.S. government is sending a terrible message to the world, that is, that U.S. courts are an improper venue for resolving disputes with other countries on U.S. treaty obligations. The U.S. is, in effect, saying that whatever breaches it commits are all right if it says so. That is bad for the law, bad for relations among nations, bad for nuclear non-proliferation and disarmament, and not only bad, but extremely dangerous for U.S. citizens and all humanity."

The RMI has appealed the U.S. attempt to reject its suit in the U.S, Federal Court, and it will continue to sue the nine nuclear nations in the International Court of Justice. Whether or not the suits succeed in making the nuclear nations comply with international law, attention will be called to the fact the nine countries are outlaws. In vote after vote in the United Nations General Assembly, the peoples of the world have shown how deeply they long to be free from the menace of nuclear weapons. Ultimately, the tiny group of power-hungry politicians must yield to the will of the citizens whom they are at present holding as hostages.

It is a life-or-death question. We can see this most clearly when we look far ahead. Suppose that each year there is a certain finite chance of a nuclear catastrophe, let us say 2 percent. Then in a century the chance of survival will be 13.5 percent, and in two centuries, 1.8 percent, in three centuries, 0.25 percent, in 4 centuries, there would only be a 0.034 percent chance of survival and so on. Over many centuries, the chance of survival would shrink almost to zero. Thus by looking at the long-term future, we can clearly see that if nuclear weapons are not entirely eliminated, civilization will not survive.

Civil society must make its will felt. A thermonuclear war today would be not only genocidal but also omnicidal. It would kill people of all ages, babies, children, young people, mothers, fathers and grandparents, without any regard whatever for guilt or innocence. Such a war would be the ultimate ecological catastrophe, destroying not only human civilization but also much of the biosphere. Each of us has a duty to work with dedication to prevent it.

One important possibility for progress on the seemingly intractable issue of nuclear disarmament would be for a nation or group of nations to put forward a proposal for a Nuclear Weapons Convention for direct vote on the floor of the UN General Assembly. It would almost certainly be adopted by a massive majority. I believe that such a step would be a great achievement, even if bitterly opposed by some of the nuclear weapons states. When the will of the majority of the worlds peoples is clearly expressed in an international treaty, even if the treaty functions imperfectly, the question of legality is clear. Everyone can see which states are violating international law. In time, world public opinion will force the criminal states to conform with international law.

In the case of a Nuclear Weapons Convention, world public opinion would have especially great force. It is generally agreed that a full-scale nuclear war would have disastrous effects, not only on belligerent nations but also on neutral countries. Mr. Javier Pérez de Cuéllar , former Secretary-General of the United Nations, emphasized this point in one of his speeches: "I feel", he said, "that the question may justifiably be put to the leading nuclear powers: by what right do they decide the fate of humanity? From Scandinavia to Latin America, from Europe and Africa to the Far East, the destiny of every man and woman is affected by their actions. No one can expect to escape from the catastrophic consequences of a nuclear war on the fragile structure of this planet. ..."

"No ideological confrontation can be allowed to jeopardize the future of humanity. Nothing less is at stake: todays decisions affect not only the present; they also put at risk succeeding generations. Like supreme arbiters, with our disputes of the moment, we threaten to cut off the future and to extinguish the lives of innocent millions yet unborn. There can be no greater arrogance. At the same time, the lives of all those who lived before us may be rendered meaningless; for we have the power to dissolve in a conflict of hours or minutes the entire work of civilization, with all the brilliant cultural heritage of humankind.

"...In a nuclear age, decisions affecting war and peace cannot be left to military strategists or even to governments. They are indeed the responsibility of every man and woman. And it is therefore the responsibility of all of us... to break the cycle of mistrust and insecurity and to respond to humanity's yearning for peace."

The eloquent words of Javier Pérez de Cuéllar express the situation in which we now find ourselves: Accidental nuclear war, nuclear terrorism, insanity of a person in a position of power, or unintended escalation of a conflict, could at any moment plunge our beautiful world into a catastrophic thermonuclear war which might destroy not only human civilization but also much of the biosphere.

A model Nuclear Weapons Convention already exists. It was drafted in 1996 and updated in 2007 by three NGOs: International Association of Lawyers Against Nuclear Arms, International Network of Engineers and Scientists Against Nuclear Proliferation and International Physicians for the Prevention of Nuclear War. The Nuclear Weapons Convention (NWC) can be downloaded in many languages from the website of Unfold Zero. It could be put to a direct vote at the present session of the UN General Assembly. The mechanism for doing this could exactly parallel the method by which the Arms Trade Treaty was adopted in 2013. The UN Ambassador of Costa Rica could send a copy of the NWC to Secretary General Ban Ki-moon, asking him, on behalf of Costa Rica, Mexico and Austria to put it to a swift vote in the General Assembly.

There is strong evidence that the NWC would be passed by a large majority. For example, Humanitarian Initiative Joint Statement of 2015 was endorsed by 159 governments. Furthermore, the consensus document of the NPT Review Conference of 2010, endorsed by 188 state parties, contains the following sentence: "The Conference expresses its deep concern at the humanitarian consequences of any use of nuclear weapons and reaffirms the need for all States at all times to comply with applicable international law, including international humanitarian law".

We can expect that the adoption of a Nuclear Weapons Convention will be opposed by the states that currently possess these weapons. One reason for this is the immense profits that suppliers make by "modernizing" nuclear arsenals. For example, the Arms Control Association states "The U.S. military is in the process of modernizing all of its existing strategic delivery systems and refurbishing the warheads they carry to last for the next 30-50 years." It adds "Three independent estimates put the expected total cost over the next 30 years at as much as $1 trillion." We should notice that these plans for long-term retention of nuclear weapons are blatant violations of Article VI of the NPT.

Money is often the motive for crimes, and in this case, a vast river of money is driving us in the direction of a catastrophic nuclear war. If we wait for the approval of the nuclear weapon states, we will have to wait forever, and the general public, whose active help we need in abolishing nuclear weapons, will feel more and more helpless and powerless. To prevent this, we need concrete progress rather than endless delay.

There are strong precedents for the adoption of the NWC against the opposition of powerful states. The Arms Trade Treaty is one precedent, the International Criminal Court is another and the Ottawa Treaty is a third.

The adoption of an Arms Trade Treaty is a great step forward; the adoption of the ICC, although its operation is imperfect, is also a great step forward, and likewise, the Antipersonnel Land-Mine Convention is a great step forward. In my opinion, the adoption of a Nuclear Weapons Convention, even in the face of powerful opposition, would also be a great step forward. When the will of the majority of the worlds peoples is clearly expressed in an international treaty, even if the treaty functions imperfectly, the question of legality is clear. Everyone can see which states are violating international law. In time, world public opinion will force the criminal states to conform to the law.

Fig. 6.15 Fireball of the Tsar Bomba (RDS-220), the largest weapon ever detonated (1961). Fission-fusion-fission bombs of almost unlimited power can be constructed by adding a layer of inecpensive ordinary uranium outside a core containing a fission-fusion bomb. Such a bomb would completely destroy a city even if it missed the target by 50 kilometers. (Fair use: https://en.wikipedia.org/wiki/Tsar_Bomba)

In the world as it is, the nuclear weapons now stockpiled are sufficient to kill everyone on earth several times over. Nuclear technology is spreading, and many politically unstable countries have recently acquired nuclear weapons or may acquire them soon. Even terrorist groups or organized criminals may acquire such weapons, and there is an increasing danger that they will be used.

In the world as it could be, both the manufacture and the possession of nuclear weapons would be prohibited. The same would hold for other weapons of mass destruction.

Suggestions for further reading

(1) A. Robock, L. Oman, G. L. Stenchikov, O. B. Toon, C. Bardeen, and R. Turco, "Climatic consequences of regional nuclear conflicts", Atmospheric Chemistry and Physics, Vol. 7, p. 2003-2012, 2007.

(2) M. Mills, O. Toon, R. Turco, D. Kinnison, R. Garcia, "Massive global ozone loss predicted following regional nuclear conflict", Proceedings of the National Academy of Sciences (USA), vol. 105(14), pp. 5307-12, Apr 8, 2008.

(3) O. Toon, A. Robock, and R. Turco, "The Environmental Consequences of Nuclear War", Physics Today, vol. 61, No. 12, p. 37-42, 2008.

(4) R. Turco, O. Toon, T. Ackermann, J. Pollack, and C. Sagan, "Nuclear Winter: Global consequences of multiple nuclear explosions", Science, Vol. 222, No. 4630, pp. 1283-1292, December 1983.

(5) A. Robock, L. Oman, G. Stenchikov, "Nuclear winter revisited with a modern climate model and current nuclear arsenals: Still catastrophic consequences", Journal of Geophysical Research - Atmospheres, Vol. 112, No. D13, p. 4 of 14, 2007.

(6) I. Helfand, "An Assessment of the Extent of Projected Global Famine Resulting From Limited, Regional Nuclear War", International Physicians for the Prevention of Nuclear War, Physicians for Social Responsibility, Leeds, MA, 2007.

(7) George P. Schultz, William J. Perry, Henry A. Kissinger and Sam Nunn, "A World Free of Nuclear Weapons", The Wall Street Journal, January 4, 2007, page A15 and January 15, 2008, page A15.

(8) Mikhail Gorbachev, "The Nuclear Threat", The Wall Street Journal, January 30, 2007, page A15.

(9) Massimo DAlema, Gianfranco Fini, Giorgio La Malfa, Arturo Parisi and Francesco Calogero, "For a World Free of Nuclear Weapons", Corriere Della Sera, July 24, 2008.

(10) Hoover Institution, "Reykjavik Revisited; Steps Towards a World Free of Nuclear Weapons", October, 2007.

(11) Douglas Hurd, Malcolm Rifkind, David Owen and George Robertson, "Start Worrying and Learn to Ditch the Bomb", The Times, June 30, 2008.

(12) Des Brown, Secretary of State for Defense, UK, "Laying the Foundations for Multilateral Disarmament", Geneva Conference on Disarmament, February 5, 2008.

(13) Government of Norway, International Conference on "Achieving the Vision of a World Free of Nuclear Weapons", Oslo, Norway, February 26-27, 2008.

(14) Jonas Gahr Støre, Foreign Minister, Norway, "Statement at the Conference on Disarmament", Geneva, March 4, 2008.
(15) Anne-Grete Strøm-Erichsen, Defense Minister, Norway, "Emerging Opportunities for Nuclear Disarmament", Pugwash Conference, Canada, July 11, 2008.
(16) Kevin Rudd, Prime Minister, Australia, "International Commission on Nuclear Non-Proliferation and Disarmament", Media Release, July 9, 2008.
(17) Global Zero, www.globalzero.org/paris-conference
(18) Helmut Schmidt, Richard von Weizäcker, Egon Bahr and Hans-Dietrich Genscher, "Towards a Nuclear-Free World: a German View", International Herald Tribune, January 9, 2009.
(19) Hans M. Kristensen and Elliot Negin, "Support Growing for Removal of U.S. Nuclear Weapons from Europe", Common Dreams Newscenter, first posted May 6, 2005.
(20) David Krieger, "President-elect Obama and a World Free of Nuclear Weapons", Nuclear Age Peace Foundation Website, 2008.
(21) J.L. Henderson, *Hiroshima*, Longmans (1974).
(22) A. Osada, *Children of the A-Bomb, The Testament of Boys and Girls of Hiroshima*, Putnam, New York (1963).
(23) M. Hachiya, M.D., *Hiroshima Diary*, The University of North Carolina Press, Chapel Hill, N.C. (1955).
(24) M. Yass, *Hiroshima*, G.P. Putnams Sons, New York (1972).
(25) R. Jungk, *Children of the Ashes*, Harcourt, Brace and World (1961).
(26) B. Hirschfield, *A Cloud Over Hiroshima*, Baily Brothers and Swinfin Ltd. (1974).
(27) J. Hersey, *Hiroshima*, Penguin Books Ltd. (1975).
(28) R. Rhodes, *Dark Sun: The Making of the Hydrogen Bomb*, Simon and Schuster, New York, (1995)
(29) R. Rhodes, *The Making of the Atomic Bomb*, Simon and Schuster, New York, (1988).
(30) D.V. Babst et al., *Accidental Nuclear War: The Growing Peril*, Peace Research Institute, Dundas, Ontario, (1984).
(31) S. Britten, *The Invisible Event: An Assessment of the Risk of Accidental or Unauthorized Detonation of Nuclear Weapons and of War by Miscalculation*, Menard Press, London, (1983).
(32) M. Dando and P. Rogers, *The Death of Deterrence*, CND Publications, London, (1984).
(33) N.F. Dixon, *On the Psychology of Military Incompetence*, Futura, London, (1976).
(34) D. Frei and C. Catrina, *Risks of Unintentional Nuclear War*, United Nations, Geneva, (1982).

(35) H. LEtang, *Fit to Lead?*, Heinemann Medical, London, (1980).
(36) SPANW, *Nuclear War by Mistake - Inevitable or Preventable?*, Swedish Physicians Against Nuclear War, Lulea, (1985).
(37) J. Goldblat, *Nuclear Non-proliferation: The Why and the Wherefore*, (SIPRI Publications), Taylor and Francis, (1985).
(38) IAEA, *International Safeguards and the Non-proliferation of Nuclear Weapons*, International Atomic Energy Agency, Vienna, (1985).
(39) J. Schear, ed., *Nuclear Weapons Proliferation and Nuclear Risk*, Gower, London, (1984).
(40) D.P. Barash and J.E. Lipton, *Stop Nuclear War! A Handbook*, Grove Press, New York, (1982).
(41) C.F. Barnaby and G.P. Thomas, eds., *The Nuclear Arms Race: Control or Catastrophe*, Francis Pinter, London, (1982).
(42) L.R. Beres, *Apocalypse: Nuclear Catastrophe in World Politics*, Chicago University press, Chicago, IL, (1980).
(43) F. Blackaby et al., eds., *No-first-use*, Taylor and Francis, London, (1984).
(44) NS, ed., *New Statesman Papers on Destruction and Disarmament* (NS Report No. 3), New Statesman, London, (1981).
(45) H. Caldicot, *Missile Envy: The Arms Race and Nuclear War*, William Morrow, New York, (1984).
(46) R. Ehrlich, *Waging the Peace: The Technology and Politics of Nuclear Weapons*, State University of New York Press, Albany, NY, (1985).
(47) W. Epstein, *The Prevention of Nuclear War: A United Nations Perspective*, Gunn and Hain, Cambridge, MA, (1984).
(48) W. Epstein and T. Toyoda, eds., *A New Design for Nuclear Disarmament*, Spokesman, Nottingham, (1975).
(49) G.F. Kennan, *The Nuclear Delusion*, Pantheon, New York, (1983).
(50) R.J. Lifton and R. Falk, *Indefensible Weapons: The Political and Psychological Case Against Nuclearism*, Basic Books, New York, (1982).
(51) J.R. Macy, *Despair and Personal Power in the Nuclear Age*, New Society Publishers, Philadelphia, PA, (1983).
(52) A.S. Miller et al., eds., *Nuclear Weapons and Law*, Greenwood Press, Westport, CT, (1984).
(53) MIT Coalition on Disarmament, eds., *The Nuclear Almanac: Confronting the Atom in War and Peace*, Addison-Wesley, Reading, MA, (1984).
(54) UN, *Nuclear Weapons: Report of the Secretary-General of the United Nations*, United Nations, New York, (1980).
(55) IC, *Proceedings of the Conference on Understanding Nuclear War*, Imperial College, London, (1980).
(56) B. Russell, *Common Sense and Nuclear Warfare*, Allen and Unwin, London, (1959).

(57) F. Barnaby, *The Nuclear Age*, Almqvist and Wiksell, Stockholm, (1974).
(58) D. Albright, F. Berkhout and W. Walker, *Plutonium and Highly Enriched Uranium 1996: World Inventories, Capabilities and Policies*, Oxford University Press, Oxford, (1997).
(59) G.T. Allison et al., *Avoiding Nuclear Anarchy: Containing the Threat of Loose Russian Nuclear Weapons and Fissile Material*, MIT Press, Cambridge MA, (1996).
(60) B. Bailin, *The Making of the Indian Atomic Bomb: Science, Secrecy, and the Post-colonial State*, Zed Books, London, (1998).
(61) G.K. Bertsch and S.R. Grillot, (Eds.), *Arms on the Market: Reducing the Risks of Proliferation in the Former Soviet Union*, Routledge, New York, (1998).
(62) P. Bidawi and A. Vanaik, *South Asia on a Short Fuse: Nuclear Politics and the Future of Global Disarmament*, Oxford University Press, Oxford, (2001).
(63) F.A. Boyle, *The Criminality of Nuclear Deterrence: Could the U.S. War on Terrorism Go Nuclear?*, Clarity Press, Atlanta GA, (2002).
(64) G. Burns, *The Atomic Papers: A Citizens Guide to Selected Books and Articles on the Bomb, the Arms Race, Nuclear Power, the Peace Movement, and Related Issues*, Scarecrow Press, Metuchen NJ, (1984).
(65) L. Butler, *A Voice of Reason*, The Bulletin of Atomic Scientists, **54**, 58-61, (1998).
(66) R. Butler, *Fatal Choice: Nuclear Weapons and the Illusion of Missile Defense*, Westview Press, Boulder CO, (2001).
(67) R.P. Carlisle (Ed.), *Encyclopedia of the Atomic Age*, Facts on File, New York, (2001).
(68) G.A. Cheney, *Nuclear Proliferation: The Problems and Possibilities*, Franklin Watts, New York, (1999).
(69) A. Cohen, *Israel and the Bomb*, Colombia University Press, New York, (1998).
(70) S.J. Diehl and J.C. Moltz, *Nuclear Weapons and Nonproliferation: A Reference Handbook*, ABC-Clio Information Services, Santa Barbara CA, (2002).
(71) H.A. Feiveson (Ed.), *The Nuclear Turning Point: A Blueprint for Deep Cuts and De-Alerting of Nuclear Weapons*, Brookings Institution Press, Washington D.C., (1999).
(72) R. Forsberg et al., *Nonproliferation Primer: Preventing the Spread of Nuclear, Chemical and Biological Weapons*, MIT Press, Cambridge, (1995).
(73) R. Hilsman, *From Nuclear Military Strategy to a World Without War: A History and a Proposal*, Praeger Publishers, Westport, (1999).

(74) International Physicians for the Prevention of Nuclear War and The Institute for Energy and Environmental Research *Plutonium: Deadly Gold of the Nuclear Age*, International Physicians Press, Cambridge MA, (1992).

(75) R.W. Jones and M.G. McDonough, *Tracking Nuclear Proliferation: A Guide in Maps and Charts, 1998*, The Carnegie Endowment for International Peace, Washington D.C., (1998).

(76) R.J. Lifton and R. Falk, *Indefensible Weapons: The Political and Psychological Case Against Nuclearism*, Basic Books, New York, (1982).

(77) R.E. Powaski, *March to Armageddon: The United States and the Nuclear Arms Race, 1939 to the Present*, Oxford University Press, (1987).

(78) J. Rotblat, J. Steinberger and B. Udgaonkar (Eds.), *A Nuclear-Weapon-Free World: Desirable? Feasible?*, Westview Press, (1993).

(79) The United Methodist Council of Bishops, *In Defense of Creation: The Nuclear Crisis and a Just Peace*, Graded Press, Nashville, (1986).

(80) U.S. Congress Office of Technology Assessment (Ed.), *Dismantling the Bomb and Managing the Nuclear Materials*, U.S. Government Printing Office, Washington D.C., (1993).

(81) P. Boyer, *By the Bombs Early Light: American Thought and Culture at the Dawn of the Atomic Age*, University of North Carolina Press, (1985).

(82) A. Makhijani and S. Saleska, *The Nuclear Power Deception: Nuclear Mythology From Electricity 'Too Cheap to Meter to 'Inherently Safe Reactors*, Apex Press, (1999).

(83) C. Perrow, *Normal Accidents: Living With High-Risk Technologies*, Basic Books, (1984).

(84) P. Rogers, *The Risk of Nuclear Terrorism in Britain*, Oxford Research Group, Oxford, (2006).

(85) MIT, *The Future of Nuclear Power: An Interdisciplinary MIT Study*, http://web.mit.edu/nuclearpower, (2003).

(86) Z. Mian and A. Glaser, *Life in a Nuclear Powered Crowd*, INES Newsletter No. 52, 9-13, April, (2006).

(87) E. Chivian, and others (eds.), *Last Aid: The Medical Dimensions of Nuclear War*, W.H. Freeman, San Fransisco, (1982).

(88) Medical Associations Board of Science and Education, *The Medical Effects of Nuclear War*, Wiley, (1983).

Chapter 7

Facing a Set of Linked Problems

"Questioning growth is deemed the act of lunatics, idealists and revolutionaries. But question it we must. The myth of growth has failed us. It has failed the billion people who still live on less than $2 a day. It has failed the fragile ecological systems on which we depend for survival. It has failed spectacularly, in its own terms, to provide economic stability and to secure people's jobs."

Professor Tim Jackson

Introduction

During the 21st century, the world will be facing a number of extremely pressing problems. Instead of debating which of them is the most serious, we should recognize that the problems are interconnected and that we must solve them simultaneously. This chapter will explore some of the links:

Links between human rights and climate change

The earth's atmosphere and climate are shared by all nations. At the United Nations summit meeting on climate in Copenhagen in December 2009 (COP15), leaders of the developing countries rightly stated that the huge per-capita greenhouse gas emissions of industrialized nations constitute a violation of human rights. In many developing countries, millions of lives are already threatened by climate changes produced by carbon emissions from the industrialized countries. Examples of this are famine-producing droughts in Africa, and rising ocean levels that threaten the existence of island nations. The future will hold much more severe threats unless the citizens of industrialized nations are willing to change their lifestyles. However, China, still classified as a developing country, has now become the world's largest emitter of CO_2. In general, the developing countries too have a responsibility to respect the human rights of others by restricting greenhouse gas emissions.

Links between poverty and population growth

This link was explored in Chapter 4. In countries where population is growing rapidly, it is impossible for governments to create the necessary infrastructure rapidly enough to keep up with growth. Schools, health care, transportation, housing, sanitation, and job creation all lag behind the relentless growth in numbers. Thus population growth produces poverty and misery.

There is also a reciprocal link that leads to a vicious circle: Poverty is observed to produce rapid population growth. When the level of education is low, many women are unaware of the techniques that can be used to control family size, or they cannot afford them, or the women have no jobs outside their homes and their status is so low that it is their husbands who decide the number of children. Furthermore, in many poor countries, child labor is not effectively prohibited, and there is no social security system or pension system to take care of old people. The children are expected to do this. Thus many couples produce large numbers of children because they believe that they will gain an economic advantage from them. The vicious circle driven by the reciprocal links between population growth and poverty is called the "demographic trap", a name given to it by the demographer J.L. Kayfetz.

Links between energy scarcity, population growth, climate change and famine

At present a child dies from starvation every five seconds - six million children die from hunger every year. If adults are included, the number is much larger. It is estimated that hunger is involved, directly or indirectly, in the deaths of 35 million people each year. Over a billion people in today's world are chronically undernourished. There is a threat that unless prompt and well-informed action is taken by the international community, the tragic loss of life that is already being experienced will increase to unimaginable proportions.

As glaciers melt in the Himalayas, threatening the summer water supplies of India and China; as ocean levels rise, drowning the fertile rice-growing river deltas of Asia; as aridity begins to decrease the harvests of Africa, North America and Europe; as populations grow; as aquifers are overdrawn; as cropland is lost to desertification and urban growth; and as energy prices become prohibitively high, making modern energy-intensive agricultural methods impossible, the billion people who now are undernourished but still survive, might not survive. They might become the victims of a famine whose proportions could exceed anything that the world has previously experienced.

The threat of severe future food shortages can only be avoided by prompt and well-informed policy changes. These include stabilization of populations, stabilization of climate; stabilization of aquifers; conservation of soils; protection of cropland; restriction of the use of grain for motor fuels; investment in agriculture and agricultural research; and establishment of a global food bank. It will also be helpful to avoid transportation of food over long distances, since this requires large inputs of energy. Finally, the consumption of meat will need to be reduced, since meat production is a very inefficient way of utilizing food calories from grain. (Livestock production is also a major source of greenhouse gases, accounting for more emissions than the entire transportation sector. See "Livestock's Long Shadow", Food and Agricultural Organization of the United Nations, Rome, 2006).

Links between population growth and war

Population growth is placing increasing pressure on global resources. Many of today's conflicts in the Middle East, Central Asia, Latin America and

Africa can be seen in this light. There is a danger that the era of resource scarcity that the world is now entering will be characterized by bitter wars for the possession of increasingly scarce oil, water and metals.

In his book, *Resource Wars: The New Landscape of Global Conflict* (2002), Michael T. Klare[a] shows that many recent wars can be interpreted as struggles for the control of natural resources. In order that such conflicts should not become more frequent in the future, a cooperative attitude towards resources is needed. The global community must face resource scarcity with solidarity. Furthermore, it is vital that the global population be rapidly stabilized.

Links between poverty, disease and war

Another problem facing the world today is the resurgence of infectious disease. Examples of this are pandemics of HIV/AIDS, malaria, and drug-resistant tuberculosis.

Clearly these pandemics are linked to poverty, both as causes of poverty and as its effects. They are also linked to the problem of war.

Today, the world spends roughly 1.5 trillion (million million) US dollars each year on armaments. This amount of money is almost too large to imagine, and if we instead used it constructively, almost all of the problems facing the world today could be solved. In particular, a tiny fraction of the money wasted (or worse than wasted) on armaments could drastically reduce the number of deaths from malnutrition and preventable disease.

Links between war and environmental degradation

Warfare during the 20th century has not only caused the loss of 175 million lives (primarily civilians) - it has also caused the greatest ecological catastrophes in human history. The damage takes place even in times of peace. Studies by Joni Seager, a geographer at the University of Vermont, conclude that "a military presence anywhere in the world is the single most reliable predictor of ecological damage".

Modern warfare destroys environments to such a degree that it has been described as an "environmental holocaust." For example, herbicides use in the Vietnam War killed an estimated 6.2 billion board-feet of hardwood trees in the forests north and west of Saigon, according to the American Association for the Advancement of Science. Herbicides such as Agent Orange also made enormous areas of previously fertile land unsuitable for

---

[a]Michael Klare is the Five College Professor of Peace and World Security Studies, based at Hampshire College in Amherst Massachusetts, but also lecturing at Amherst, Mount Holyoke, Smith and the University of Massachusetts.

agriculture for many years to come.[b] In Vietnam and elsewhere in the world, valuable agricultural land has also been lost because land mines or the remains of cluster bombs make it too dangerous for farming.

During the Gulf War of 1990, the oil spills amounted to 150 million barrels, 650 times the amount released into the environment by the notorious Exxon Valdez disaster. During the Gulf War an enormous number of shells made of depleted uranium were fired. When the dust produced by exploded shells is inhaled it often produces cancer, and it will remain in the environment of Iraq for decades.

Radioactive fallout from nuclear tests pollutes the global environment and causes many thousands of cases of cancer, as well as birth abnormalities. Most nuclear tests have been carried out on lands belonging to indigenous peoples.

The danger of a catastrophic nuclear war casts a dark shadow over the future of our species. It also casts a very black shadow over the future of the global environment. Here the dark shadow can be interpreted literally: It is the shadow cast by smoke from burning cities, which would rise to the stratosphere, spread globally, and remain for a decade, destroying not only agriculture, but also normal vegetation and the ozone layer. It the case of a large-scale nuclear war, the resulting famine would mean that most humans and animals would die from starvation. Details of this threat to terrestrial life are given in the references at the end of Chapter 6.

Links between climate change and nuclear weapons

Climate change is linked to the danger of nuclear war because of the widespread (but false) belief that nuclear power generation is an answer to global warming. If many nations throughout the world decide to build power-generating reactors, the number of countries possessing nuclear weapons will increase dramatically because it is almost impossible to distinguish between civilian and military nuclear programs.

By reprocessing spent nuclear fuel rods, using ordinary chemical means, a nation with a power reactor can obtain a weapons-usable isotope of plutonium, Pu-239. Even when such reprocessing is performed under international control, the uncertainty as to the amount of Pu-239 obtained is large enough so that the operation might superficially seem to conform to regulations while still supplying enough Pu-239 to make many bombs.

---

[b] Agent Orange also produced cancer, birth abnormalities and other serious forms of illness both in the Vietnamese population and among the foreign soldiers fighting in Vietnam.

The enrichment of uranium[c] is also linked to reactor use. Many reactors of modern design make use of low enriched uranium (LEU) as a fuel. Nations operating such a reactor may claim that they need a program for uranium enrichment in order to produce LEU for fuel rods. However, by operating their ultracentrifuges a little longer, they can easily produce highly enriched uranium (HEU), i.e., uranium containing a high percentage of the rare isotope U-235, and therefore usable in weapons.

The widely held belief that global warming can be avoided by switching to nuclear power is false. In a carefully documented book "Nuclear Power is Not the Answer to Global Warming or Anything Else", the Australian physician Helen Caldicott points out that if a detailed accounting of CO_2 emissions is made during all the phases of nuclear power generation, including both construction and decommissioning of the plant, together with mining, transportation and refinement of the uranium ore, the CO_2 emissions are seen to be comparable with those produced by a coal-fired power plant.

Known reserves of uranium are only sufficient for the generation of 8×10^{20} joules of electrical energy,[d] i.e., about 25 TWy, i.e. less than enough to supply the world with power for two years. It is sometimes argued that a larger amount of electricity could be obtained from the same amount of uranium through the use of fast breeder reactors, but this would involve totally unacceptable proliferation risks. In fast breeder reactors, the fuel rods consist of highly enriched uranium. Around the core, is an envelope of natural uranium. The flux of fast neutrons from the core is sufficient to convert a part of the U-238 in the envelope into Pu-239, a fissionable isotope of plutonium.

Fast breeder reactors are prohibitively dangerous from the standpoint of nuclear proliferation because both the highly enriched uranium from the fuel rods and the Pu-239 from the envelope are directly weapons-usable. It would be impossible, from the standpoint of equity, to maintain that some nations have the right to use fast breeder reactors, while others do not. If all nations used fast breeder reactors, the number of nuclear weapons states would increase drastically.

In the countries where it is presently used, nuclear power generation is heavily subsidized, and were it not for these subsidies, it would not be able to compete with wind energy or solar energy. It is vital that the subsidies be shifted from nuclear power to the development of various forms of renewable energy.

---

[c]i.e. Production of uranium with a higher percentage of U-235 than is found in natural uranium.

[d]Craig, J.R., Vaugn, D.J. and Skinner, B.J., *Resources of the Earth: Origin, Use and Environmental Impact, Third Edition*, page 210.

The mass media have been much better in making the general public aware of threats from climate change than they have been in calling attention to the great threat posed by the continued existence of nuclear weapons. However, a discussion of nuclear dangers ought to be linked to the climate debate because a nuclear war, even a small one, would seriously affect the global climate for a period of approximately ten years. (See www.nucleardarkness.org).

Links between nuclear weapons and the rights of indigenous peoples

In a recent article, Alyn Ware[e] has written:

"Indigenous peoples around the world have had their sovereignty infringed, their territories destroyed and their health impacted by the nuclear arms race. Uranium mining, nuclear weapons testing and nuclear waste dumping have all been done on indigenous territories, including the Shoshone lands in the United States, Maoha islands in French-occupied Polynesia, Ugihur lands in Lop Nor (China), Pitjantjatjara lands in Maralinga (Australia), Marshallese Islands in the Pacific, and the Arctic territories of the Kazakhs, Sami, Vepsians, Karalians, Aluet, Nentses and Komi..."

"The testing of nuclear weapons on indigenous territories has mostly stopped, but has left a legacy of environmental contamination and transgenerational health effects from the radiation released in nuclear activities. Meanwhile the deployment of nuclear weapons on submarines and the testing of nuclear weapons-carrying missiles continues."

For example, between 1946 and 1958, the US carried out 67 nuclear tests in the Marshall Islands. The total power of the bombs exploded in these tests was equivalent to 7000 Hiroshima-size bombs. Even today the Marshall Islanders still suffer from exceptionally high rates of cancer and birth defects as a result of the tests. In Australia, UK nuclear tests were carried out in desert regions known to be inhabited by indigenous peoples, while Soviet nuclear tests were also carried out on tribal lands.

Links between sustainability, education and global citizenship

As the economist Richard Florida has pointed out (see Chapter 2), cultural activities do not make a heavy impact on the global environment. He sees the transition to a sustainable society as being linked with a shift of human efforts away from the manufacture of goods and towards creative

---

[e]Alyn Ware is the Global Coordinator for Parliamentarians for Nuclear Nonproliferation and Disarmament.

non-industrial activities. This transition, which is already starting to take place, is linked with the development of global citizenship. The growth of human culture involves cooperation across national boundaries. Thus as we move towards sustainability we are simultaneously moving towards global citizenship and a global ethic. Among living organisms on earth, the evolution of a complex culture is unique to humans, and it is responsible for our success as a species. Human cultural evolution will increasingly become synonymous with global cooperation and a global ethic.

Links between poverty and war

The problem of eliminating global inequality and poverty is linked in several ways with the problem of eliminating war.

In the first place, war is one of the greatest sources of poverty. For example, much of the enormous third world debt is due to arms purchases, which make no constructive contribution to development. Indeed, the pervasiveness of small arms in parts of Africa makes armed conflict there so endemic that development is all but impossible. Wars also destroy infrastructure and damage ecology on a large scale. One can think, for example of the destruction of power plants and water purification plants in Iraq during the first Gulf war or the use of defoliants during the Viet Nam War. The defoliants destroyed forests and made large areas of land unsuitable for agriculture.

Victims of land mines, poison gas, bombings or small arms can be crippled for life and can become economic burdens for their societies. Soldiers are taken away from useful occupations during wars, and if killed or severely wounded, they can no longer help their families. The treatment of war casualties imposes a great burden on medical facilities. Industries making munitions are diverted from useful activities. Thus, in a variety of ways, war is one of the most important sources of poverty.

A second link between poverty and the problem of war has to do with the contrasts between rich and poor nations as a hindrance to the development of international law and governance. If we survey the nations of the world, we can find many within which good government has been achieved, together with some measure of internal peace and happiness. Some very large countries function as coherent (although not ideal) social units, with freedom from internal wars. Each of these countries is so large and has such an inhomogeneous population that the problem of achieving internal peace within them is not qualitatively different from the problem of achieving peace throughout the entire world. The same methods that each of these enormous countries uses within itself could be used globally - for example education for social cohesion, and systems of laws that act on individuals.

However, plans for strong government at a global level are blocked by enormous contrasts between rich and poor nations. Rich nations fear that with a strong world government, they would lose the advantages that they now have.

The European Union has already encountered the problem of economic inequality as a barrier to efforts to strengthen and enlarge the federation. The richer nations of the EU fear that they will have to pay high taxes to support economic progress in the poorer parts of the Union. Nevertheless, the EU is attempting to solve these problems, motivated by the conviction that whatever its defects, some degree of political union is needed to rule out the possibility that the horrors of World Wars I and II will ever be repeated. The EU is extremely interesting since it gives us a model of what needs to be done globally. Of course the global contrasts between rich and poor nations are far greater than those found within the EU, but these global North-South contrasts also can be and must be eliminated.

Market forces in ecology

Most analysts agree that our economic system needs to be modified in such a way that it will stop leading us towards ecological disaster, and instead encourage the development of a stable and sustainable global society. This theme is strongly presented in Paul Hawken's excellent book, *The Ecology of Commerce*, (Collins Business, 2005). It is also a central element in Lester Brown's far-sighted and eloquent book, *Plan B 3.0* (W.W. Norton, 2008). Both Brown and Hawken argue that income taxes should be reduced, and that taxes should instead be placed on ecologically damaging activities, such as burning fossil fuels.

The widely read book *Natural Capitalism* (1999), coauthored by Paul Hawken, Amory Lovens and Hunter Lovens, views the global economy as being a part of a larger economy that also includes natural ecosystems. The three authors believe that we should attribute value to human intelligence and culture, to hydrocarbons and minerals, as well as to such things as trees and microscopic fungi. Amory Lovens has advocated an approach that he calls the "feebate", in which taxes on unsustainable products such as Sports Utility Vehicles (SUV's) are pooled to support the development and purchase of sustainable alternatives, such as hybrid electric vehicles.

Pricing that includes externalities

The concept of externalities was developed by the English economist Arthur Cecil Pigou, (1877-1959), a Professor of Political Economics at Cambridge

University. He invented the idea that the free market can fail to give the true price to a product or service because costs to society or to the environment are not included. For example, the price of a package of cigarettes might not include the cost to society of the illnesses and deaths that result from smoking. Similarly, the social and environmental costs of industrial activities that cause pollution are usually neglected. In his major book, *Wealth and Welfare* (1912, 1920), Pigou introduced the idea that such "market failures" could be corrected by governments through a combination of taxation and subsidies. Such taxes are called "Pigovian taxes" in his honor. Today they are also often called "green taxes".

Green taxes have already been introduced in many countries, for example in Denmark; but there is a need for the introduction of a greatly expanded and universalized system of green taxes throughout the world. This must be done so thoroughly that the price of every item in the global economy will include its true environmental and social cost. Only then will we be able to rely on the free market to lead us in an intelligent and non-destructive direction. Unless all nations throughout the world incorporate social and ecological features in their economic systems, globalization and free trade put the enlightened economies into unfair competition with the less enlightened ones; fairly treated labor may be brought into competition with child labor or slave labor; and countries with green taxes may be forced by free trade to compete with countries with no environmental legislation. Thus globalization and free trade will become fair and beneficial only when protection of labor becomes universal throughout the world, and only when pricing that includes externals becomes universal.

Payment for tropical forests

We mentioned that rainforests are under assault. An area of tropical forest the size of Greece is lost each year. It is difficult for the governments of developing countries to resist economic pressures from beef or soybean farmers, from palm oil plantations and from logging companies. For this reason it has been proposed that the intact forests should be assigned a value commensurate with the services that they perform for the world by sequestering carbon and preserving biodiversity. According to this proposal, the wealthy industrial nations should pay a yearly fee to tropical nations that preserve their rainforests intact as a reward for the global services that the forests perform.

Reformed economics

Reform is urgently needed in the teaching of economics and business: Classical economics developed during the 18th and 19th centuries, when the global supply of land and raw materials seemed unlimited (at least within the foreseeable future), and when the only limitation to economic development was the shortage of capital. This might be called an "open world" situation, a situation in which growth became the Holy Grail of all economists. Today we are in a "closed world" situation. The possibility of development is limited, not by the availabilty of capital, but by shrinking supplies of arable land, water and non-renewable resources, and by the finite carrying capacity of the global environment. Nevertheless economics continues to be taught along classical lines and for this reason, economists continue to worship growth. We urgently need to introduce biology and ecology into the education of economists. The economics of growth must be replaced by equilibrium economics, where considerations of ecology, carrying capacity, and sustainability are given their proper weight, and where the quality of life of future generations has as much importance as present profits.

Throughout the history of industrialism, there have been individual leaders of industry who have shown a strong sense of social responsibility. For example, Robert Owen (1771-1858) was not only a successful factory owner but also a pioneer of the Cooperative Movement and of trade unions. Andrew Carnegie (1835-1919) kept very little of his vast fortune for himself and used by far the greater part to establish universities and libraries. More recently, Ted Turner gave exactly one billion dollars (about a third of his fortune) to the United Nations. When asked to comment, he said "It's only money." He then added "I hope that Bill Gates is watching". Bill Gates was indeed watching, and Gates has since become a public benefactor on an extremely large scale, establishing a program aimed at vaccinating every child in Africa. He presently spends only part of his time on Microsoft, and devotes the remainder to his many philanthropies.

In 2006, the Nobel Peace Prize was awarded to Grameen Bank and to its founder Prof. Muhammad Yunus for making small loans available to rural populations, first in Bangladesh and later in many other countries. The borrowers are too poor to qualify for credit from conventional banks.

Yunus' experiments with microcredit began in 1974, during a famine, when he lent 27 dollars to a group of 42 families to allow them to make small items for sale. Today Grameen Bank (literally "Bank of the Villages") has more than 2000 branches. It makes loans without collateral to solidarity and self-help groups, relying on a collective sense of honor to enforce repayment. Yunus and Grameen Bank have especially concentrated on making microcredit available to women, believing that loans to groups of women will give greater benefits to families than loans made to men.

Grameen Bank is an example of a business whose purpose is not to make a profit but rather to provide a social service. Recently Grameen combined with the French firm Groupe Dannone to form a corporation called "Grameen Dannone Foods Social Business Enterprise in Bangladesh". The new corporation is run along business lines, but rather than making a profit, it aims at a social goal - providing affordable healthy daily nutrition to low-income, nutritionally deprived populations in Bangladesh. Here, as in Grameen Bank, the driving force is not individual self-interest (as recommended by Adam Smith) but rather a social goal.

Fig. 7.1 Prof. Muhammad Yunus, who shared the 2006 Nobel Peace Prize with Grameen Bank. He pioneered the concept of businesses that aim at social goals rather than profit. (Wikipedia.)

These examples are not completely isolated. There is today a general movement towards corporate social responsibility (CSR). Several major oil companies now have have divisions devoted to renewable energy technologies. It is to be hoped that this is not just "window dressing", and that the oil companies will devote a significant fraction of their profits to developing renewable energy. It would be very desirable for large pharmaceutical companies to exhibit social responsibility by giving tropical diseases more attention, and by making medicines (for example against AIDS) available in the developing world at affordable prices.

Some of the ideals of CSR were formalized in the ten principles of the UN Global Compact (a forum of corporations and NGOs launched by UN Secretary General Kofi Annan in 2000). Today the Global Compact has more than 3000 corporate members and includes more than 1000 NGOs. It's ten principles are as follows:

- **Human Rights:** Businesses should
  1) support and respect the protection of internationally proclaimed human rights and
  2) make sure that they are not complicit in human rights abuses.

- **Labor Standards:** Businesses should uphold
  3) the freedom of association and the effective recognition of the right to collective bargaining,
  4) the elimination of all forms of forced or compulsory labor,
  5) the effective abolition of child labor; and
  6) the elimination of discrimination in employment and occupation.

- **Environment:** Businesses should
  7) support a precautionary approach to environmental challenges,
  8) undertake initiatives to promote environmental responsibility; and
  9) encourage the development and diffusion of environmentally friendly technologies.

- **Anti-Corruption:** Businesses should
  10) work against corruption in all its forms, including extortion and bribery.

Of course, the fact that a corporation has added its name to the UN Global Compact does not mean that it always follows the ten principles, any more than membership in the Boy Scouts guarantees that a young man is trustworthy, loyal, helpful, friendly, courteous, kind,..etc. Nevertheless, the ten principles have a normative value, and one can hope that they will someday be incorporated into the education of economists and businessmen.

Economists, industrialists and business leaders have a duty to the peoples of the world and to the global environment in much the same way that physicians have a sacred duty to the welfare of their patients. Therefore the education of economists and industrialists ought to emphasize ethical and ecological principles. Like doctors, economists and industrialists carry matters of life and death in their hands: Think of the 10 million children who die each year from poverty-related causes; think of the wholesale extinction of species; think of global warming; think of the risk of a catastrophic future famine caused by population growth, by energy shortages, by climate change and by ecological degradation. We urgently need to introduce biology, ecology and ethics into the education of economists. The economics of growth must be replaced by equilibrium economics, where considerations of ecology, carrying capacity, and sustainability are given proper weight, and where the quality of life of future generations has as much importance as present profits.

Not only economists, but also students of business administration should be made conscious of the negative effects of globalization as well as the positive ones, and they should consider the measures that will be needed to correct the negative effects. Students of business administration should be helped to develop an attitude of responsibility towards the less developed countries of the world, so that if they later become administrators in multinational corporations, they will choose generous and enlightened policies rather than exploitative ones.

The economic impact of war and preparation for war should be included in the training of economists. Both the direct and indirect costs of war should be studied, for example the effect of unimaginably enormous military budgets in reducing the money available to solve pressing problems posed by the resurgence of infectious disease (e.g. AIDS, and drug-resistant forms of malaria and tuberculosis); the problem of population stabilization; food problems; loss of arable land; future energy problems; the problem of finding substitutes for vanishing nonrenewable resources, and so on. Many of these problems were discussed at a recent conference of economists in Copenhagen, but the fact that all such global emergencies could be adequately addressed with a fraction of the money wasted on military budgets was not discussed.

Finally, economics curricula should include the problems of converting war-related industries to peaceful ones - the problem of beating swords into plowshares. It is often said that our economies are dependent on arms industries. If this is so, it is an unhealthy dependence, analogous to drug addiction, since arms industries do not contribute to future-oriented infrastructure. The problem of conversion is an important one. It is the economic analog of the problem of ending a narcotics addiction, and it ought to be given proper weight in the education of economists.

Present economic problems seen as an opportunity

The economic recession that began with the US subprime mortgage crisis of 2007 and 2008 can be seen as an opportunity. It is thought to be temporary, but it is a valuable warning of irreversible long-term changes that will come later in the 21st century when the absolute limits of economic growth are reached. Already today we are faced with the problems of preventing unemployment and simultaneously building the infrastructure of an ecologically sustainable society.

Today's economists believe that growth is required for economic health; but at some point during this century, growth will no longer be possible. If no changes have been made in our economic system when this happens, we will be faced with massive unemployment.Two changes are needed to prevent this:

(1) Labor must be moved to tasks related to ecological sustainability. These include development of renewable energy, reforestation, soil and water conservation, replacement of private transportation by public transport. Health and family planning services must also be made available to all.
(2) Opportunities for employment must be shared among those in need of work, even if this means reducing the number of hours that each person works each week and simultaneously reducing the use of luxury goods, unnecessary travel, conspicuous consumption and so on. It will be necessary for governments to introduce laws reducing the length of the working week, thus ensuring that opportunities for employment are shared equally.

We have the chance, already today, to make these changes in our economic system. The completely unregulated free market alone has proved to be inadequate in a situation where economic growth has slowed or halted, as is very apparent in the context of the present financial crisis. But halfway through the 21st century, economic growth will be halted permanently by ecological constraints and vanishing resources. We must construct a steady-state economic system - one that can function without growth. Our new economic system needs to have a social and ecological conscience, it needs to be responsible, and it needs to have a farsighted global ethic. We have the opportunity to anticipate and prevent future shocks by working today to build a new economic system.

Fig. 7.2 A design by William Morris (1834-1896). Together with John Ruskin (1819-1900) and others in the Arts and Crafts Movement, Morris criticized the Industrial Revolution and division of labor for destroying craftsmanship, traditions of design, traditional skills, and pride in work. His Utopian book, "News from Nowhere", is a plea for a return to cooperative workshops where good design and craftsmanship would flourish. Ruskin's book, "Unto This Last" (which later greatly influenced Gandhi), points out that the pleasure of warm friendships with coworkers in small cooperative communities is not given sufficient weight by the economic systems of industrial societies. (Public domain.)

The introduction of Pigovian taxes by one country may make it less able to compete with other countries that do not include externalities in their pricing. Until such reforms become universal, free trade may give unfair advantages to countries which give the least attention to social and

Fig. 7.3 Professor Tim Jackson is the Economics Commissioner for Britain's Sustainable Development Commission. In March, 2009, he published a book entitled "Prosperity without growth? The transition to a sustainable economy". 'Questioning growth' is deemed the act of lunatics, idealists and revolutionaries", Jackson states, "But question it we must. The myth of growth has failed us. It has failed the billion people who still live on less than $2 a day. It has failed the fragile ecological systems on which we depend for survival. It has failed spectacularly, in its own terms, to provide economic stability and to secure people's jobs." (Twitter.)

environmental ethics. Thus free trade and globalization will become fair and beneficial only when ethical economic practices become universal.

Suggestions for further reading

(1) R.L. Heilbroner, *The Worldly Philosophers*, 5th edition, Simon and Schuster, (1980).
(2) R. Harrod, *Life of John Maynard Keynes*, Harcourt, Brace, New York, (1951).
(3) J.M. Keynes, *Economic Consequences of the Peace*, Harcourt, Brace, New York, (1920).
(4) J.M. Keynes, *Essays in Persuasion*, Harcourt, Brace, New York, (1951).
(5) J.M. Keynes, *The General Theory of Employment, Interest and Money*, Harcourt, Brace, New York, (1964).
(6) R. Lekachman, *The Age of Keynes*, Random House, New York, (1966).
(7) R. Florida, *The Rise of the Creative Class*, Basic Books, (2002).
(8) Herman Daly, *Steady-State Economics: Second Edition with New Essays*, Island Press, (1991).
(9) Herman Daly, *Economics in a Full World*, Scientific American, Vol. 293, Issue 3, September, (2005).
(10) Herman Daly and John Cobb, *For the Common Good*, Beacon Press, Boston, (1989).
(11) H.E. Daly, *Sustainable Growth - An Impossibility Theorem*, Development, **3**, 45-47, (1990).
(12) H.E. Daly and K.N. Townsend, (editors), *Valuing the Earth. Economics, Ecology, Ethics*, MIT Press, Cambridge, Massachusetts, (1993)
(13) E.O. Wilson, *The Diversity of Life*, Allen Lane, The Penguin Press, (1992).
(14) Lester R. Brown et. al.,*Saving the Planet. How to Shape an Environmentally Sustainable Global Economy*, W.W. Norton, New York, (1991).
(15) L.R. Brown, *Building a Sustainable Society*, W.W. Norton, (1981).
(16) L.R. Brown, and P. Shaw, *Six Steps to a Sustainable Society*, Worldwatch Paper 48, Worldwatch Institute, Washington D.C., (1982).
(17) Muhammad Yunus, *Banker to the Poor; Microcredit and the Battle Against World Poverty*, (2003).
(18) UN Global Compact, http://www.unglobalcompact.org (2007).
(19) UN Millennium Development Goals http://www.un.org/millenniumgoals/ (2007).
(20) Amartya Sen, *Poverty and Famine; An Essay on Entitlement and Deprivation*, Oxford Univeersity Press, (1981).
(21) Amartya Sen, *Development as Freedom*, Oxford University Press, (1999).
(22) Amartya Sen, *Inequality Reexamined*, Harvard University Press, (1992).
(23) Paul F. Knitter and Chandra Muzaffar, editors, *Subverting Greed; Re-*

ligious Perspectives on the Global Economy, Orbis Books, Maryknoll, New York, (2002).
(24) Edy Korthals Altes, *The Contribution of Religions to a Just and Sustainable Economic Development*, in F. David Peat, editor, *The Pari Dialogues, Volume 1*, Pari Publishing, (2007).
(25) Hendrik Opdebeeck, *Globalization Between Market and Democracy*, in F. David Peat, editor, *The Pari Dialogues, Volume 1*, Pari Publishing, (2007).
(26) Paul Hawken *The Ecology of Commerce; A Declaration of Sustainability*, Collins Business, (2005).
(27) Luther Standing Bear, *Land of the Spotted Eagle*, Houghton Mifflin, (1933).
(28) T. Gyatso, HH the Dalai Lama, *Ancient Wisdom, Modern World: Ethics for the New Millennium*, Abacus, London, (1999).
(29) T. Gyatso, HH the Dalai Lama, *How to Expand Love: Widening the Circle of Loving Relationships*, Atria Books, (2005).
(30) J. Rotblat and D. Ikeda, *A Quest for Global Peace*, I.B. Tauris, London, (2007).
(31) M. Gorbachev and D. Ikeda, *Moral Lessons of the Twentieth Century*, I.B. Tauris, London, (2005).
(32) D. Krieger and D. Ikeda, *Choose Hope*, Middleway Press, Santa Monica CA 90401, (2002).
(33) P.F. Knitter and C. Muzaffar, eds., *Subverting Greed: Religious Perspectives on the Global Economy*, Orbis Books, Maryknoll, New York, (2002).
(34) S. du Boulay, *Tutu: Voice of the Voiceless*, Eerdmans, (1988).
(35) Earth Charter Initiative *The Earth Charter*, www.earthcharter.org
(36) P.B. Corcoran, ed., *The Earth Charter in Action*, KIT Publishers, Amsterdam, (2005).
(37) R. Costannza, ed., *Ecological Economics: The Science and Management of Sustainability*, Colombia University Press, New York, (1991).
(38) A. Peccei, *The Human Quality*, Pergamon Press, Oxford, (1977).
(39) A. Peccei, *One Hundred Pages for the Future*, Pergamon Press, New York, (1977).
(40) E. Pestel, *Beyond the Limits to Growth*, Universe Books, New York, (1989).
(41) Pope Francis I, Laudato si', https://laudatosi.com/watch
(42) John S. Avery, *The Need for a New Economic System*, Irene Publishing, Sparsnäs Sweden, (2016).

(43) John S. Avery, *Collected Essays, Volumes 1-3*, Irene Publishing, Sparsnäs Sweden, (2016).
(44) John S. Avery, *Space-Age Science and Stone-Age Politics*, Irene Publishing, Sparsnäs Sweden, (2016).
(45) John S. Avery, *Science and Society*, World Scientific, (2016).
(46) Aspen Institute for Humanistic Studies, Program in International Affairs, *The Planetary Bargain*, Aspen, Colorado, (1975).
(47) W. Berry, *Home Economics*, North Point Press, San Francisco, (1987).
(48) E. Eckholm, *Planting for the Future: Forestry for Human Needs*, Worldwatch Paper 26, Worldwatch Institute, Washington D.C., (1979).
(49) R. Goodland, H. Daly, S. El Serafy and B. von Droste (editors), *Environmentally Sustainable Economic Development: Building on Brundtland*, UNESCO, Paris, (1991).
(50) F. Hirsch, *Social Limits to Growth*, Harvard University Press, Cambridge, (1976).
(51) W. Leontief, et al., *The Future of the World Economy*, Oxford University Press, (1977).
(52) M. Lipton, *Why Poor People Stay Poor*, Harvard University Press, (1977).
(53) J. McHale, and M.C. McHale, *Basic Human Needs: A Framework for Action*, Center for Integrative Studies, Huston, (1977).
(54) D.L. Meadows, *Alternatives to Growth*, Ballinger, Cambridge, (1977).
(55) D.H. Meadows, *The Global Citizen*, Island Press, Washington D.C., (1991).
(56) D.L. Meadows, and D.H. Meadows (editors), *Toward Global Equilibrium*, Wright-Allen Press, Cambridge, Mass., (1973).
(57) L.W. Milbrath, *Envisioning a Sustainable Society*, State University of New York Press, Albany, (1989).
(58) R.E. Miles, *Awakening from the American Dream: The Social and Political Limits to Growth*, Universe Books, New York, (1976).
(59) S. Postel, and L. Heise, *Reforesting the Earth* , Worldwatch Paper 83, Worldwatch Institute, Washington D.C., (1988).
(60) M. Sagoff, *The Economy of the Earth*, Cambridge University Press, (1988).
(61) E.F. Schumacher, *Small is Beautiful: Economics As If People Mattered*, Harper and Row, New York, (1973).
(62) World Bank, *World Development Report*, Oxford University Press, New York, (published annually).
(63) G.P. Zachary, *A 'Green Economist' Warns Growth May Be Overrated*, The Wall Street Journal, June 25, (1996).

Chapter 8

Outlawing War

"We must stand guard against the acquisition of unwarranted influence, whether sought or unsought, by the military-industrial complex. The potential for the disastrous rise of misplaced power exists and will persist."

Dwight David Eisenhower

"Naturally, the common people don't want war, neither in Russia nor in England, nor for that matter in Germany. That is understood. But, after all, it is the leaders of the country that determine the policy, and it is always a simple matter to drag the people along, whether it is a democracy or a fascist dictatorship... All you have to do is to tell them that they are being attacked, and denounce the peacemakers for lack of patriotism and exposing the country to danger. It works the same in any country."

Herman Goering, interviewed in Spandau Prison

Introduction

The enormous destructive power of modern weapons has made the war-system obsolete. Nevertheless, because of institutional and cultural inertia, and because of a tendency towards tribalism in inherited human nature, nationalism persists. But today, the absolutely sovereign nation-state has become a dangerous anachronism.

Probably our best hope for the future lies in developing the United Nations into a federation, with the power to make laws that are binding on individuals.

Outlawing war

The United Nations Charter

The Second World War was terrible enough to make world leaders resolve to end the institution of war once and for all, and the United Nations was set up for this purpose.

Article 2 of the UN Charter requires that "All members shall refrain in their international relations from the threat or use of force against the territorial integrity or political independence of any state."

This requirement is somewhat qualified by Article 51, which says that "Nothing in the present Charter shall impair the inherent right of individual or collective self-defense if an armed attack occurs against a Member of the United Nations, until the Security Council has taken measures necessary to maintain international peace and security."

Thus, in general, war is illegal under the UN Charter. Self-defense against an armed attack is permitted, but only for a limited time, until the Security Council has had time to act.

The Nuremberg Principles

At the end of the Second World War, when the full extent of the atrocities that had been committed by the Nazi's became known, it was decided to prosecute Nazi leaders for crimes against peace, war crimes, and crimes against humanity (such as extermination camps). There was disagreement about how such trials should be held, but after some debate between the Allied countries, it was agreed that 24 Nazi officials and military leaders would be tried by an International Military Tribunal in Nuremberg, Germany, a former center of Nazi politics. There were originally 24 defendants, but two of them committed suicide. One was presumed dead but was nevertheless tried in absentia. Of the twenty-one remaining defendants, eleven were given the death penalty, eight were sentenced to long prison terms, and three were acquitted. Similar trials also took place in Japan.

In 1946 the United Nations General Assembly unanimously affirmed "the principles of international law recognized by the Charter of the Nuremberg Tribunal and the judgment of the Tribunal". The General Assembly also established an International Law Commission to formalize the Nuremberg Principles, and the result was the following list:

- **Principle I**: Any person who commits an act which constitutes a crime under international law is responsible, and therefore liable to punishment.
- **Principle II**: The fact that internal law does not impose a penalty for an act which constitutes a crime under international law does not relieve the person who committed the act from responsibility under international law.
- **Principle III**: The fact that the person who committed an act which constitutes a crime under international law acted as Head of State or responsible government official does not relieve him from responsibility under international law.
- **Principle IV**: The fact that a person acted pursuant to order of his Government or of a superior does not relieve him of responsibility under international law, provided that a moral choice was in fact possible for him.
- **Principle V**: Any person charged with a crime under international law has the right to a fair trial on the facts and law.
- **Principle VI**: The crimes hereinafter set out are punishable as crimes under international law:

    a. Crimes against peace: (i) Planning, preparation, initiation or waging of war of aggression or a war in violation of international treaties, agreements or assurances; (ii) Participation in a common plan or conspiracy for the accomplishment of any of the acts mentioned under (i).

    b. War crimes: Violations of the laws or customs of war which include, but are not limited to, murder, ill-treatment of prisoners of war or persons on the seas, killing of hostages, plunder of public or private property, wanton destruction of cities, towns or villages, or devastation not justified by military necessity.

    c. Crimes against humanity: Atrocities and offenses, including but not limited to, murder, extermination, deportation, imprisonment, torture, rape, or other inhumane acts committed against any civilian population, or persecutions on political, racial or religious grounds, whether or not in violation of the laws of the country where perpetrated.

- **Principle VII**: Complicity in the commission of a crime against peace, a war crime, or a crime against humanity as set forth in Principle VI is a crime under international law.

The Nuremberg Principles are being used today as the basis for the International Criminal Court's trials of individuals accused of genocide and war crimes in Rwanda and in the former Yugoslavia.

The Principles throw an interesting light onto the status of soldiers. According to the Nuremberg Principles, it is not only the right, but also the duty of individuals to make moral and legal judgments concerning wars in which they are asked to fight. If a soldier participates in an illegal war (and all wars, apart from actions of the UN Security Council, are now illegal) then the soldier is liable to prosecution for violating international law. The fact that he or she was acting under orders is not an excuse. The training of soldiers is designed to remove the burdens of moral and legal responsibility from a soldier's individual shoulders; but the Nuremberg Principles are designed to put these burdens squarely back where they belong - on the shoulders of the individual.

Perpetual (illegal) war

Eisenhower warns against the military-industrial complex

The two world wars of the 20th Century involved a complete reordering of the economies of the belligerent countries, and a dangerous modern phenomenon was created - the military-industrial complex.

In his farewell address (January 17, 1961) US President Dwight David Eisenhower warned of the dangers of the war-based economy that World War II had forced his nation to build: "...We have been compelled to create an armaments industry of vast proportions", Eisenhower said, "...Now this conjunction of an immense military establishment and a large arms industry is new in American experience. The total influence - economic, political, even spiritual - is felt in every city, every state house, every office in the federal government. ...We must not fail to comprehend its grave implications. Our toil, resources and livelihood are all involved; so is the very structure of our society. ... We must stand guard against the acquisition of unwarranted influence, whether sought or unsought, by the military-industrial complex. The potential for the disastrous rise of misplaced power exists and will persist. We must never let the weight of this combination endanger our democratic processes. We should take nothing for granted."

Fig. 8.1 Eisenhower's Farewell Address deserves to be read by everyone who feels concern for the future of the world. The power of the military-industrial complex, against which the retiring US President warned, is one of the main reasons why the institution of war persists, although everyone realizes that war is responsible for much of the suffering of humanity. (Public domain.)

This farsighted speech by Eisenhower deserves to be studied by every student of economics. As the retiring president pointed out, the military-industrial complex is a threat both to peace and to democracy. It is not unique to the United States but exists in many countries. The world today spends roughly a trillion (i.e. a million million) US dollars each year on armaments. It is obvious that very many people make their living from war, and therefore it is correct to speak of war as a social, political and economic institution. The military-industrial complex is one of the main reasons why war persists, although everyone realizes that war is the cause of much of the suffering of humanity. We know that war is madness, but it persists. We know that it threatens the survival of our species, but it persists, entrenched in the attitudes of historians, newspaper editors and television producers, entrenched in the methods by which politicians finance their campaigns, and entrenched in the financial power of arms manufacturers - entrenched also in the ponderous and costly hardware of war, the fleets of warships, bombers, tanks, nuclear missiles and so on.

As we try to introduce ethical elements into economics, we should remember the existence of this "devil's dynamo". It perpetuates the institution of war although the peoples of the world long for peace, and despite the fact that human civilization itself is threatened by all-destroying modern weapons. What is to be done? More attention to this problem is needed, and more research, more public awareness.

Einstein and Freud: Why war?

"The unleashed power of the atom has changed everything except our ways of thinking, and thus we drift towards unparalleled catastrophes."

"I don't know what will be used in the next world war, but the 4th will be fought with stones."

Albert Einstein (1879-1955)

Besides being one of the greatest physicists of all time, Albert Einstein was a lifelong pacifist, and his thoughts on peace can speak eloquently to us today. We need his wisdom today, when the search for peace has become vital to our survival as a species.

Family background

Albert Einstein was born in Ulm, Germany, in 1879. He was the son of middle-class, irreligious Jewish parents, who sent him to a Catholic school. Einstein was slow in learning to speak, and at first his parents feared that he might be retarded; but by the time he was eight, his grandfather could say in a letter: "Dear Albert has been back in school for a week. I just love that boy, because you cannot imagine how good and intelligent he has become."

Remembering his boyhood, Einstein himself later wrote: "When I was 12, a little book dealing with Euclidian plane geometry came into my hands at the beginning of the school year. Here were assertions, as for example the intersection of the altitudes of a triangle in one point, which, though by no means self-evident, could nevertheless be proved with such certainty that any doubt appeared to be out of the question. The lucidity and certainty made an indescribable impression on me."

When Albert Einstein was in his teens, the factory owned by his father and uncle began to encounter hard times. The two Einstein families moved to Italy, leaving Albert alone and miserable in Munich, where he was supposed to finish his course at the gymnasium. Einstein's classmates had given him the nickname "Beidermeier", which means something like "Honest John"; and his tactlessness in criticizing authority soon got him into trouble. In Einstein's words, what happened next was the following: "When I was in the seventh grade at the Lutpold Gymnasium, I was summoned by my home-room teacher, who expressed the wish that I leave the school. To my remark that I had done nothing wrong, he replied only, 'Your mere presence spoils the respect of the class for me'."

Einstein left gymnasium without graduating, and followed his parents to Italy, where he spent a joyous and carefree year. He also decided to change his citizenship. "The over-emphasized military mentality of the German State was alien to me, even as a boy", Einstein wrote later. "When my father moved to Italy, he took steps, at my request, to have me released from German citizenship, because I wanted to be a Swiss citizen."

Einstein's letter to Freud: Why war?

Because of his later fame as a physicist, Einstein was asked to make many speeches. and in all these he condemned violence and nationalism, urging that these be replaced by and international cooperation and law under an effective international authority. He also wrote many letters and articles pleading for peace and for the renunciation of militarism and violence.

Einstein believed that the production of armaments is damaging, not only economically, but also spiritually. In 1930 he signed a manifesto for world disarmament sponsored by the Woman's International League for Peace and Freedom. In December of the same year, he made his famous statement in New York that if two percent of those called for military service were to refuse to fight, governments would become powerless, since they could not imprison that many people. He also argued strongly against compulsory military service and urged that conscientious objectors should be protected by the international community. He argued that peace, freedom of individuals, and security of societies could only be achieved through disarmament, the alternative being "slavery of the individual and annihilation of civilization".

In letters, and articles, Einstein wrote that the welfare of humanity as a whole must take precedence over the goals of individual nations, and that we cannot wait until leaders give up their preparations for war. Civil society, and especially public figures, must take the lead. He asked how decent and self-respecting people can wage war, knowing how many innocent people will be killed.

In 1931, the International Institute for Intellectual Cooperation invited Albert Einstein to enter correspondence with a prominent person of his own choosing on a subject of importance to society. The Institute planned to publish a collection of such dialogues. Einstein accepted at once, and decided to write to Sigmund Freud to ask his opinion about how humanity could free itself from the curse of war. A translation from German of part of the long letter that he wrote to Freud is as follows:

"Dear Professor Freud, The proposal of the League of Nations and its International Institute of Intellectual Cooperation at Paris that I should invite a person to be chosen by myself to a frank exchange of views on any problem that I might select affords me a very welcome opportunity of conferring with you upon a question which, as things are now, seems the most important and insistent of all problems civilization has to face. This is the problem: Is there any way of delivering mankind from the menace of war? It is common knowledge that, with the advance of modern science, this issue has come to mean a matter of life or death to civilization as we know it; nevertheless, for all the zeal displayed, every attempt at its solution has ended in a lamentable breakdown."

"I believe, moreover, that those whose duty it is to tackle the problem professionally and practically are growing only too aware of their impotence to deal with it, and have now a very lively desire to learn the views of men who, absorbed in the pursuit of science, can see world-problems in the perspective distance lends. As for me, the normal objective of my thoughts affords no insight into the dark places of human will and feeling. Thus in the enquiry now proposed, I can do little more than seek to clarify the question at issue and, clearing the ground of the more obvious solutions, enable you to bring the light of your far-reaching knowledge of man's instinctive life upon the problem.."

"As one immune from nationalist bias, I personally see a simple way of dealing with the superficial (i.e. administrative) aspect of the problem: the setting up, by international consent, of a legislative and judicial body to settle every conflict arising between nations... But here, at the outset,

Fig. 8.2 "Why War?" (Public domain.)

I come up against a difficulty; a tribunal is a human institution which, in proportion as the power at its disposal is... prone to suffer these to be deflected by extrajudicial pressure..."

Freud replied with a long and thoughtful letter in which he said that a tendency towards conflict is an intrinsic part of human emotional nature, but that emotions can be overridden by rationality, and that rational behavior is the only hope for humankind. The full exchange between Einstein and Freud can be found on the following link:
http://www.freud.org.uk/file-uploads/files/WHY

A few more things that Einstein said about peace:

"We cannot solve our problems with the same thinking that we used when we created them."

"It has become appallingly obvious that our technology has exceeded our humanity."

"Peace cannot be kept by force; it can only be achieved by understanding."

"The world is a dangerous place to live; not because of the people who are evil, but because of the people who don't do anything about it."

"Insanity: doing the same thing over and over again and expecting to get different results."

"Nothing will end war unless the people themselves refuse to go to war."

"Past thinking and methods did not prevent world wars. Future thinking must prevent war."

"You cannot simultaneously prevent and prepare for war."

"Never do anything against conscience, even if the state demands it."

"Taken as a whole, I would believe that Gandhi's views were the most enlightened of all political men of our time."

"Without ethical culture, there is no salvation for humanity."

Federalism and global governance

It is becoming increasingly clear that the concept of the absolutely sovereign nation-state is a dangerous anachronism in a world of thermonuclear weapons, instantaneous communication, and economic interdependence. Probably our best hope for the future lies in developing the United Nations into a World Federation. The strengthened United Nations should have a legislature with the power to make laws that are binding on individuals, and the ability to arrest and try individual political leaders for violations of these laws. The World Federation should also have the power of taxation, and the military and legal powers necessary to guarantee the human rights of ethnic minorities within nations.

Making the United Nations into a World Federation

A federation of states is, by definition, a limited union where the federal government has the power to make laws that are binding on individuals, but where the laws are confined to interstate matters, and where all powers not expressly delegated to the federal government are retained by the individual states. In other words, in a federation each of the member states runs its own internal affairs according to its own laws and customs; but in certain agreed-on matters, where the interests of the states overlap, authority is specifically delegated to the federal government.

Since the federal structure seems well suited to a world government with limited and carefully-defined powers that would preserve as much local autonomy as possible, it is worthwhile to look at the histories of a few of the federations. There is much that we can learn from their experiences.

The success of federations

Historically, the federal form of government has proved to be extremely robust and successful. Many of today's nations are federations of smaller, partially autonomous, member states. Among these nations are Argentina, Australia, Austria, Belgium, Brazil, Canada, Germany, India, Mexico, Russia, Spain, South Africa and the United States.

The Swiss Federation is an interesting example, because its regions speak three different languages: German, French and Italian. In 1291, citizens of Uri, Schwyz and Unterwalden, standing on the top of a small mountain called Rütli, swore allegiance to the first Swiss federation with the words "we will be a one and only nation of brothers". During the 14th century, Luzern, Zürich, Glarus, Zug and Bern also joined. Later additions during the 15th and 16th centuries included Fribourg, Solothurn, Basel, Schaffhausen and Appenzell. In 1648 Switzerland declared itself to be an independent nation, and in 1812, the Swiss Federation declared its neutrality. In 1815, the French-speaking regions Valais, Neuchatel and Genéve were added, giving Switzerland its final boundaries.

In some ways, Switzerland is a very advanced democracy, and many issues are decided by the people of the cantons in direct referendums. On the other hand, Switzerland was very late in granting votes to women (1971), and it was only in 1990 that a Swiss federal court forced Appenzell Innerrhoden to comply with this ruling. Switzerland was also very late in joining the United Nations (10 September, 2002).

The Federal Constitution of United States of America is one of the most important and influential constitutions in history. It later formed a model for many other governments, especially in South America. The example of the United States is especially interesting because the original union of states formed by the Articles of Confederation in 1777 proved to be too weak, and it had to be replaced eleven years later by a federal constitution.

During the revolutionary war against England the 13 former colonies sent representatives to a Continental Congress, and on May 10, 1776, the Congress authorized each of the colonies to form its own local provincial government. On July 4, 1776 it published a formal Declaration of

Independence. The following year, the Congress adopted the Articles of Confederation defining a government of the new United States of America. The revolutionary war continued until 1783, when the Treaty of Paris was signed by the combatants, ending the war and giving independence to the United States. However, the Articles of Confederation soon proved to be too weak. The main problem with the Articles was that laws of the Union acted on its member states rather than on individual citizens.

In 1887, a Constitutional Convention was held in Philadelphia with the aim of drafting a new and stronger constitution. In the same year, Alexander Hamilton began to publish the Federalist Papers, a penetrating analysis of the problems of creating a workable government uniting a number of semi-independent states. The key idea of the Federalist Papers is that the coercion of states is neither just nor feasible, and that a government uniting several states must function by acting on individuals. This central idea was incorporated into the Federal Constitution of the United States, which was adopted in 1788. Another important feature of the new Constitution was that legislative power was divided between the Senate, where the states had equal representation regardless of their size, and the House of Representatives, where representation was proportional to the populations of the states. The functions of the executive, the legislature and the judiciary were separated in the Constitution, and in 1789 a Bill of Rights was added.

George Mason, one of the architects of the federal constitution of the United States, believed that "such a government was necessary as could directly operate on individuals, and would punish those only whose guilt required it", while James Madison (another drafter of the U.S. federal constitution) remarked that the more he reflected on the use of force, the more he doubted "the practicability, the justice and the efficacy of it when applied to people collectively, and not individually". Finally, Alexander Hamilton, in his Federalist Papers, discussed the Articles of Confederation with the following words: "To coerce the states is one of the maddest projects that was ever devised... Can any reasonable man be well disposed towards a government which makes war and carnage the only means of supporting itself - a government that can exist only by the sword? Every such war must involve the innocent with the guilty. The single consideration should be enough to dispose every peaceable citizen against such a government... What is the cure for this great evil? Nothing, but to enable the... laws to operate on individuals, in the same manner as those of states do."

Because the states were initially distrustful of each other and jealous of their independence, the powers originally granted to the US federal government were minimal. However, as it evolved, the Federal Government of the United States gradually became stronger, and bit by bit it became involved in an increasingly wide range of activities.

The formation of the federal government of Australia is interesting because it illustrates the power of ordinary citizens to influence the large-scale course of events. In the 19th century, the six colonies British that were later to be welded into the Commonwealth of Australia imposed tariffs on each other, so that citizens living near the Murray River (for example) would have to stop and pay tolls each time they crossed the river. The tolls, together with disagreements over railways linking the colonies, control of river water and other common concerns, finally became so irritating that citizens' leagues sprang up everywhere to demand federation. By the 1890's such federation leagues could be found in cities and towns throughout the continent. In 1893, the citizens' leagues held a conference in Corowa, New South Wales, and proposed the "Corowa Plan", according to which a Constitutional Convention should be held. After this, the newly drafted constitution was to be put to a referendum in all of the colonies. This would be the first time in history that ordinary citizens would take part in the nation-building process. In January, 1895, the Carawa Plan was adopted by a meeting of Premiers in Hobart, and finally, despite the apathy and inaction of many politicians, the citizens had their way: The first Australian federal election was held March, 1901, and on May 9, 1901, the Federal Parliament of Australia opened. Australia was early in granting votes for women (1903). Its voting system has evolved gradually. Today there is a system of compulsory voting by citizens for both the Australian House of Representatives and the Australian Senate.

The successes and problems of the European Union provide invaluable experience as we consider the measures that will be needed to make the United Nations into a federation. On the whole, the EU has been an enormous success, demonstrating beyond question that it is possible to begin with a very limited special-purpose federation and to gradually expand it, judging at each stage whether the cautiously taken steps have been successful. The European Union has today made war between its member states virtually impossible. This goal, now achieved, was in fact the vision that inspired the leaders who initiated the European Coal and Steel Community in 1950.

The European Union is by no means without its critics or without problems, but, as we try to think of what is needed for United Nations reform, these criticisms and problems are just as valuable to us as are the successes of the EU.

Countries that have advanced legislation protecting the rights of workers or protecting the environment complain that their enlightened laws will be nullified if everything is reduced to the lowest common denominator in the EU. This complaint is a valid one, and two things can be said about it: Firstly, diversity is valuable, and therefore it may be undesirable to homogenize legislation, even if uniform rules make trade easier. Secondly, if certain rules are to be made uniform, it is the most enlightened environmental laws or labor laws that ought to be made the standard, rather than the least enlightened ones. Similar considerations would hold for a reformed and strengthened United Nations.

Another frequently heard complaint about the EU is that it takes decision-making far away from the voters, to a remote site where direct political will of the people can hardly be felt. This criticism is also very valid. Often, in practice, the EU has ignored or misunderstood one of the basic ideas of federalism: A federation is a compromise between the desirability of local self-government, balanced against the necessity of making central decisions on a few carefully selected issues. As few issues as possible should taken to Bruxelles, but there are certain issues that are so intrinsically transnational in their implications that they must be decided centrally. This is the principle of subsidiarity, so essential for the proper operation of federations - local government whenever possible, and only a few central decisions when absolutely necessary. In applying the principle of subsidiarity to a world government of the future, one should also remember that UN reform will take us into new and uncharted territory. Therefore it is prudent to grant only a few carefully chosen powers, one at a time, to a reformed and strengthened UN, to see how these work, and then to cautiously grant other powers, always bearing in mind that wherever possible, local decisions are the best.

Weakness of the U.N. Charter and steps towards a World Federation

Laws must be made binding on individuals

Among the weaknesses of the present U.N. Charter is the fact that it does not give the United Nations the power to make laws which are binding on individuals. At present, in international law, we treat nations as though they were persons: We punish entire nations by sanctions when the law is broken, even when only the leaders are guilty, even though the burdens of the sanctions fall most heavily on the poorest and least guilty of the citizens, and even though sanctions often have the effect of uniting the citizens of a country behind the guilty leaders. To be effective, the United Nations needs a legislature with the power to make laws which are binding on individuals, and the power to to arrest individual political leaders for flagrant violations of international law.

The present United Nations Charter is similar to the United States Articles of Confederation, a fatally weak union that lasted only eleven years, from 1777 to 1788. Like it, the UN attempts to act by coercing states. Although the United Nations Charter has lasted almost sixty years and has been enormously valuable, its weaknesses are also apparent, like those of the Articles. One can conclude that the proper way to reform the United Nations is to make it into a full federation, with the power to make and enforce laws that are binding on individuals.

The International Criminal Court, which was established when the Rome Treaty came into force in 2002, is a step in the right direction. The ICC's jurisdiction extends only to the crime of genocide, crimes against humanity, war crimes, and (at some time in the future) the crime of aggression. In practice, the ICC is open to the criticisms that it is often unable to enforce its rulings and that it lacks impartiality. Nevertheless, the establishment of the ICC is a milestone in humanity's efforts to replace the brutal military force of powerful governments by the rule of law. For the first time in history, individuals are being held responsible for violating international laws.

The voting system of the U.N. General Assembly must be reformed

Another weakness of the present United Nations Charter is the principle of "one nation one vote" in the General Assembly. This principle seems to establish equality between nations, but in fact it is very unfair: For example it gives a citizen of China or India less than a thousandth the voting power of a citizen of Malta or Iceland. A reform of the voting system is clearly needed. (A recent and detailed discussion of these issues has been given by Dr. Francesco Stipo, Reference 1.)

One possible plan (proposed by Bertrand Russell) would be for final votes to be cast by regional blocks, each block having one vote. The blocks might be: 1) Latin America 2) Africa 3) Europe 4) North America 5) Russia and Central Asia 6) China 7) India and Southeast Asia 8) The Middle East and 9) Japan, Korea and Oceania.

Today, Ambassadors and Permanent Representatives at the United Nations are appointed by national governments. However, in the long-term future, this system may evolve into a more democratic one, where citizens will vote directly for their representatives, as they do in many federations, such as Australia, Germany, the United States and the European Union.

The United Nations must be given the power to impose taxes

If the UN is to become an effective World Federation, it will need a reliable source of income to make the organization less dependent on wealthy countries, which tend to give support only to those interventions of which they approve. A promising solution to this problem is the so-called "Tobin tax", named after the Nobel-laureate economist James Tobin of Yale University. Tobin proposed that international currency exchanges should be taxed at a rate between 0.1 and 0.25 percent. He believed that even this extremely low rate of taxation would have the beneficial effect of damping speculative transactions, thus stabilizing the rates of exchange between currencies. When asked what should be done with the proceeds of the tax, Tobin said, almost as an afterthought, "Let the United Nations have it."

The volume of money involved in international currency transactions is so enormous that even the tiny tax proposed by Tobin would provide the United Nations with between 100 billion and 300 billion dollars annually. By strengthening the activities of various UN agencies, the additional income would add to the prestige of the United Nations and thus make the organization more effective when it is called upon to resolve international political conflicts.

The budgets of UN agencies, such as the World Health Organization, the Food and Agricultural Organization, UNESCO and the UN Development Programme, should not just be doubled but should be multiplied by a factor of at least twenty. With increased budgets the UN agencies could sponsor research and other actions aimed at solving the world's most pressing problems - AIDS, drug-resistant infections diseases, tropical diseases, food insufficiencies, pollution, climate change, alternative energy strategies, population stabilization, peace education, as well as combating poverty, malnutrition, illiteracy, lack of safe water and so on. Scientists would would be less tempted to find jobs with arms-related industries if offered the chance to work on idealistic projects. The United Nations could be given its own television channel, with unbiased news programs, cultural programs, and "State of the World" addresses by the UN Secretary General.

Besides the Tobin tax, other measure have been proposed to increase the income of the United Nations. For example, it has been proposed that income from resources of the sea bed be given to the UN, and that the UN be given the power to tax carbon dioxide emissions. All of the proposals for giving the United Nations an adequate income have been strongly opposed by a few nations that wish to control the UN through its purse strings, especially by the United States, which has threatened to withdraw from the UN if a Tobin tax is introduced. However, it is absolutely essential for the future development of the United Nations that the organization be given the power to impose taxes. No true government can exist without this power. It is just as essential as is the power to make and enforce laws that are binding on individuals.

The United Nations must be given a standing military force

At present, when the United Nations is called upon to meet an emergency, such as preventing genocide, an *ad hoc* force must be raised, and the time required to do this often means that the emergency action is fatally delayed. The UN should immediately be given a standing force of volunteers from all nations, ready to meet emergencies. The members of this force would owe their primary loyalty to the UN, and one of its important duties would be to prevent gross violations of human rights.

In the perspective of a longer time-frame, we need to work for a world where national armies will be very much reduced in size, where the United Nations will have a monopoly on heavy armaments, and where the manufacture or possession of nuclear weapons, as well as the export of arms and ammunition from industrialized countries to the developing countries, will be prohibited. (See reference 3).

Looking towards the future, we can foresee a time when the United Nations will have the power to make and enforce international laws which are binding on individuals. Under such circumstances, true police action will be possible, incorporating all of the needed safeguards for lives and property of the innocent.

One can hope for a future world where public opinion will support international law to such an extent that a new Hitler or Saddam Hussein or a future Milosevic will not be able to organize large-scale resistance to arrest - a world where international law will be seen by all to be just, impartial and necessary - a well-governed global community within which each person will owe his or her ultimate loyalty to humanity as a whole.

The veto power in the Security Council must be eliminated

We should remember that the UN Charter was drafted and signed before the first nuclear bomb was dropped on Hiroshima; and it also could not anticipate the extraordinary development of international trade and communication which characterizes the world today. The five permanent members of the Security Council, China, France, Russia, the United Kingdom and the United States, were the victors of World War II, and were given special privileges by the Charter as it was established in 1945, among these the power to veto UN actions on security issues. In practice, the veto power of the P5 nations has made the UN ineffective, and it has become clear that changes are needed. If the Security Council is retained in a World Federation, the veto power must be eliminated.

Subsidiarity

The need for international law must be balanced against the desirability of local self-government. Like biological diversity, the cultural diversity of humankind is a treasure to be carefully guarded. A balance or compromise between these two desirable goals can be achieved by granting only a few carefully chosen powers to a World Federation with sovereignty over all other issues retained by the member states. This leaves us with a question: Which issues should be decided centrally, and which locally?

The present United Nations Charter contains guarantees of human rights, but there is no effective mechanism for enforcing these guarantees. In fact there is a conflict between the parts of the Charter protecting human rights and the concept of absolute national sovereignty. Recent history has given us many examples of atrocities committed against ethnic minorities by leaders of nation-states, who claim that sovereignty gives them the right to run their internal affairs as they wish, free from outside interference. One feels that it ought to be the responsibility of the international community to prevent gross violations of human rights, such as genocide; and if this is in conflict with the concept of national sovereignty, then sovereignty must yield.

In the future, overpopulation and famine are likely to become increasingly difficult and painful problems in several parts of the world. Since various cultures take widely different attitudes towards birth control and family size, the problem of population stabilization seems to be one which should be decided locally. At the same time, aid for local family planning programs, as well as famine relief, might appropriately come from global agencies, such as WHO and FAO. With respect to large-scale migration, it would be unfair for a country which has successfully stabilized its own population, and which has eliminated poverty within its own borders, to be forced to accept a flood of migrants from regions of high fertility. Therefore the extent of immigration should be among those issues to be decided locally.

Security, and controls on the manufacture and export of armaments will require an effective authority at the global level.

The steps needed to convert the United Nations into a World Federation can be taken cautiously, one at a time. Having see the results of of a particular step, one can move on to the next. The establishment of the International Criminal Court is an important first step towards a system of international laws that acts on individuals. Another important step would be to give the UN a much larger and more reliable source of income. The establishment of a standing UN emergency military force is another step that ought to be taken in the near future.

Obstacles to a World Federation

It is easy to write down what is needed to convert the United Nations into a World Federation. But will not the necessary steps towards a future world of peace and law be blocked by the powerholders of today? Not everyone wants peace. Not everyone wants international law.[a]

The United Nations was established at the end of the most destructive war the world had ever seen, and its horrors were fresh in the minds of the delegates to the 1945 San Francisco Conference. The main purpose of the Charter that they drafted was to put an end to the institution of war. It was hoped that as a consequence, the UN would also end the colonial era, since war is needed to maintain the unequal relationships of colonialism. Neither of these things happened. War is still with us, and war is still used to maintain the intolerable economic inequalities of neocolonialism. The fact that military might is still used by powerful industrialized nations to maintain economic hegemony over less developed countries has been amply documented by Professor Michael Klare in his books on Resource Wars.

Today 2.7 billion people live on less than $2 a day - 1.1 billion on less than $1 per day. 18 million of our fellow humans die each year from poverty-related causes. In 2006, 1.1 billion people lacked safe drinking water, and waterborne diseases killed an estimated 1.8 million people. The developing countries are also the scene of a resurgence of other infectious diseases, such as malaria, drug-resistant tuberculosis and HIV/AIDS.[b]

Meanwhile, in 2011, world military budgets reached a total of 1.7 trillion dollars (i.e. 1.7 million million dollars). This amount of money is almost too large to be imagined. The fact that it is being spent means that many people are making a living from the institution of war. Wealthy and powerful lobbies from the military-industrial complex are able to influence mass media and governments. Thus the institution of war persists, although we know very well that it is a threat to civilization and that it responsible for much of the suffering that humans experience.

Today's military spending of almost two trillion US dollars per year would be more than enough to finance safe drinking water for the entire world, and to bring primary health care and family planning advice to all. If used constructively, the money now wasted (or worse than wasted) on the institution of war could also help the world to make the transition from fossil fuel use to renewable energy systems.

---

[a]The interested reader can find the "Hague Invasion Act" described on the Internet.

[b]It would be wrong to attribute poverty in the developing world entirely to war, and to exploitation by the industrialized countries. Rapid population growth is also a cause of poverty. Nevertheless, the enormous contrast between the rich and poor parts of the world is partly the result of unfair trade agreements imposed by means of "regime change" and "nation building", i.e. interference backed by military force.

The way in which some industrialized countries maintain their control over less developed nations can be illustrated by the "resource curse", i.e. the fact that resource-rich developing countries are no better off economically than those that lack resources, but are cursed with corrupt and undemocratic governments. This is because foreign corporations extracting local resources under unfair agreements exist in a symbiotic relationship with corrupt local officials.

As long as enormous gaps exist between the rich and poor nations of the world, the task turning the United Nations into an equitable and just federation will be blocked. Thus we are faced with the challenge of breaking the links between poverty and war. Civil society throughout the world must question the need for colossal military budgets, since, according to the present UN Charter, as well as the Nuremberg Principles, war is a violation of international law, except when sanctioned by the Security Council. By following this path we can free the world from the intolerable suffering caused by poverty and from the equally intolerable suffering caused by war.

Governments of large nations compared with global government

The problem of achieving internal peace over a large geographical area is not insoluble. It has already been solved. There exist today many nations or regions within each of which there is internal peace, and some of these are so large that they are almost worlds in themselves. One thinks of China, India, Brazil, Australia, the Russian Federation, the United States, and the European Union. Many of these enormous societies contain a variety of ethnic groups, a variety of religions and a variety of languages, as well as striking contrasts between wealth and poverty. If these great land areas have been forged into peaceful and cooperative societies, cannot the same methods of government be applied globally?

Today there is a pressing need to enlarge the size of the political unit from the nation-state to the entire world. The progress of science has created this need, but science has also given us the means to enlarge the political unit: Our almost miraculous modern communications media, if properly used, have the power to weld all of humankind into a single supportive and cooperative society.

Fig. 8.3 Painting shells in a shell filling factory during World War I. Unimaginable sums of money, which could be used to solve the world's urgent problems, are instead thrown to the all-devouring monster of war. (Public domain.)

In the world as it is, 1.7 trillion US dollars are spent each year on armaments.

In the world as it could be, the enormous sums now wasted on war would be used to combat famine, poverty, illiteracy, preventable disease and climate change.

.

Fig. 8.4 Scientists and engineers would much prefer to use their education and abilities in a constructive way. UNESCO could give them the opportunity to do this. (Public domain.)

In the world as it is, 40% of all research funds are used for projects related to armaments.

In the world as it could be, research in science and engineering would be redirected towards solving the urgent problems now facing humanity, such as the development of better methods for treating tropical diseases, new energy sources, and new agricultural methods. An expanded UNESCO would replace national military establishments as the patron of science and engineering.

Fig. 8.5 It is impossible to overstate the importance of the International Criminal Court. At last, international law is acting on individuals rather than states. Today the ICC functions very imperfectly, but we can try to improve its impartiality and expand its jurisdiction. (Public domain.)

In the world as it is, there is no generally enforcible system of international law, although the International Criminal Court is a step in the right direction.

In the world as it could be, the General Assembly of the United Nations would have the power to make international laws. These laws would be binding for all citizens of the world community, and the United Nations would enforce its laws by arresting or fining individual violators, even if they were heads of states. However, the laws of the United Nations would be restricted to international matters, and each nation would run its own internal affairs according to its own laws.

Fig. 8.6 The legacy of arms from old conflicts contributes to new conflicts. (Public domain.)

In the world as it is, armaments exported from the industrial countries to the Third World amount to a value of roughly 17 billion dollars per year. This trade in arms increases the seriousness and danger of conflicts in the less developed countries, and diverts scarce funds from their urgent needs.

In the world as it could be, international trade in arms would be strictly limited by enforcible laws.

Fig. 8.7 The flag of the United Nations. Today, loyalty to one's own nation must be supplemented by a higher loyalty to humanity as a whole. (Public Domain.)

In the world as it is, each nation considers itself to be "sovereign". In other words, every country considers that it can do whatever it likes, without regard for the welfare of the world community. This means that at the international level we have anarchy.

In the world as it could be, the concept of national sovereignty would be limited by the needs of the world community. Each nation would decide most issues within its own boundaries, but would yield some of its sovereignty in international matters.

Fig. 8.8 Emblem of the United Nations. One can hope that, with a reformed voting system, the United Nations will become more effective. (Public domain.)

In the world as it is, the system of giving "one nation one vote" in the United Nations General Assembly means that Monaco, Liechtenstein, Malta and Andorra have as much voting power as China, India, the United States and Russia combined. For this reason, UN resolutions are often ignored.

In the world as it could be, the voting system of the General Assembly would be reformed. One possible plan would be for final votes to be cast by regional blocks, each block having one vote. The blocks might be. 1) Latin America 2) Africa 3) Europe 4) North America 5) Russia and Central Asia 6) China 7) India and Southeast Asia 8) The Middle East and 9) Japan, Korea and Oceania.

Fig. 8.9 James Tobin at the Council of Economic Advisers. Approximately age 44. Tobin was Sterling Professor of Economics at Yale University, member of the Council of Economic Advisers, and Nobel Laureate in Economics. He proposed that international currency transactions be taxed at a small fraction of a procent, and that the proceeds be given to support the United Nations. (Public domain.)

In the world as it is, the United Nations has no reliable means of raising revenues.

In the world as it could be, the United Nations would have the power to tax international business transactions, such as exchange of currencies. Each member state would also pay a yearly contribution, and failure to pay would mean loss of voting rights.

Fig. 8.10 The Nuremberg Principles place responsibility on the shoulders of the individual. No one can be excused from responsibility just by saying "I was following the orders of my government". (Sniggle.net.)

In the world as it is, young men are forced to join national armies, where they are trained to kill their fellow humans. Often, if they refuse for reasons of conscience, they are thrown into prison.

In the world as it could be, national armies would be very much reduced in size. A larger force of volunteers would be maintained by the United Nations to enforce international laws. The United Nations would have a monopoly on heavy armaments, and the manufacture or possession of nuclear weapons would be prohibited.

Suggestions for further reading

(1) Y. Nakash, *The Shi'is of Iraq*, Princeton University Press, (1994).
(2) D. Fromkin, *A Peace to End All Peace: The Fall of the Ottoman Empire and the Creation of the Modern Middle East*, Owl Books, (2001).
(3) S.K. Aburish, *Saddam Hussein: The Politics of Revenge*, Bloomsbury, London, (2001).
(4) M. Muffti, *Sovereign Creations: Pan-Arabism and Political Order in Syria and Iraq*, Cornell University Press, (1996).
(5) C. Clover, *Lessons of the 1920 Revolt Lost on Bremer*, Financial Times, November 17, (2003).
(6) J. Kifner, *Britain Tried First. Iraq Was No Picnic Then*, New York Times, July 20, (2003).
(7) J. Feffer, B. Egrenreich and M.T. Klare, *Power Trip: US Unilateralism and Global Strategy After September 11*, Seven Stories Press, (2003).
(8) J.D. Rockefeller, *Random Reminiscences of Men and Events*, Doubleday, New York, (1909).
(9) M.B. Stoff, *Oil, War and American Security: The Search for a National Policy on Oil, 1941-1947*, Yale University Press, New Haven, (1980).
(10) W.D. Muscable, *George F. Kennan and the Making of American Foreign Policy*, Princeton University Press, Princeton, (1992).
(11) J. Stork, *Middle East Oil and the Energy Crisis*, Monthly Review, New York, (1976).
(12) F. Benn, *Oil Diplomacy in the Twentieth Century*, St. Martin's Press, New York, (1986).
(13) R. Sale, *Saddam Key in Early CIA Plot*, United Press International, April 10, (2003).
(14) K. Roosevelt, *Countercoup: The Struggle for the Control of Iran*, McGraw-Hill, New York, (1979).
(15) J. Fitchett and D. Ignatius, *Lengthy Elf Inquiry Nears Explosive Finish*, International Herald Tribune, February 1, (2002).
(16) M.T. Klare, *Resource Wars: The New Landscape of Global Conflict*, Owl Books reprint edition, New York, (2002).
(17) M. Klare, *Bush-Cheney Energy Strategy: Procuring the Rest of the World's Oil*, Foreign Policy in Focus, (Interhemispheric Resource Center/Institute for Policy Studies/SEEN), Washington DC and Silver City NM, January, (2004).
(18) M. Klare, *Endless Military Superiority*, The Nation magazine, July 15, (2002).
(19) M.T. Klare, *Geopolitics Reborn: The Global Struggle Over Oil and Gas Pipelines*, Current History, December issue, 428-33, (2004).

(20) P. Grose, *Allen Dulles: The Life of a Gentleman Spy*, Houghton Mifflin, Boston, (1994).
(21) S. Warren, *Exxon's Profit Surged in 4th Quarter*, Wall Street Journal, February 12, (2004).
(22) R. Suskind, *The Price of Loyalty: George W. Bush, the White House and the Education of Paul O'Neill*, Simon and Schuster, New York, (2004).
(23) D. Morgan and D.B. Ottaway, *In Iraqi War Scenario, Oil is Key Issue as U.S. Drillers Eye Huge petroleum Pool*, Washington Post, September 15, (2002).
(24) D. Rose, *Bush and Blair Made Secret Pact for Iraqi War*, The Observer, April 4, (2004).
(25) E. Vulliamy, P. Webster and N.P. Walsh, *Scramble to Carve Up Iraqi Oil Reserves Lies Behind US Diplomacy*, The Observer, October 6, (2002).
(26) Y. Ibrahim, *Bush's Iraq Adventure is Bound to Backfire*, International Herald Tribune, November 1, (2002).
(27) P. Beaumont and F. Islam, *Carve-Up of Oil Riches Begins*, The Observer, November 3, (2002).
(28) M. Dobbs, *US Had Key Role in Iraq Buildup*, Washington Post, December 30, (2002).
(29) R. Sale, *Saddam Key in Early CIA Plot*, United Press International, April 10, (2003).
(30) R. Morris, *A Tyrant Forty Years in the Making*, New York Times, March 14, (2003).
(31) H. Batatu, *The Old Social Classes and the Revolutionary Movements of Iraq*, Princeton University Press, (1978).
(32) D.W. Riegel, Jr., and A.M. D'Amato, *US Chemical and Biological Warfare-Related Dual Use Exports to Iraq and their Possible Impact on the Health Consequences of the Persian Gulf War*, Report to US Senate ("The Riegel Report"), May 25, (1994).
(33) P.E. Tyler, *Officers Say US Aided Iraq in War Despite Use of Gas*, New York Times, August 18, (2002).
(34) D. Priest, *Rumsfeld Visited Baghdad in 1984 to Reassure Iraqis, Documents Show*, Washington Post, December 19, (2003).
(35) S. Zunes, *Saddam's Arrest Raises Troubling Questions*, Foreign Policy in Focus, (http://www.globalpolicy.org/), December (2003).
(36) D. Leigh and J. Hooper, *Britain's Dirty Secret*, Guardian, March 6, (2003).
(37) J. Battle, (Ed.), *Shaking Hands With Saddam Hussein: The US Tilts Towards Iraq, 1980-1984*, National Security Archive Electronic Briefing Book No. 82, February 25, (2003).

(38) J.R. Hiltermann, *America Didn't Seem to Mind Poison Gas*, International Herald Tribune, January 17, (2003).
(39) D. Hiro, *Iraq and Poison Gas*, Nation, August 28, (2002).
(40) T. Weiner, *Iraq Uses Techniques in Spying Against its Former Tutor, the US*, Philadelphia Inquirer, February 5, (1991).
(41) S. Hussein and A. Glaspie, *Excerpts From Iraqi Document on Meeting with US Envoy*, The New York Times, International, September 23, (1990).
(42) D. Omissi, *Baghdad and British Bombers*, Guardian, January 19, (1991).
(43) D. Vernet, *Postmodern Imperialism*, Le Monde, April 24, (2003).
(44) J. Buchan, *Miss Bell's Lines in the Sand*, Guardian, March 12, (2003).
(45) C. Tripp, *Iraq: The Imperial Precedent*, Le Monde Diplomatique, January, (2003).
(46) G.H.W. Bush and B. Scowcroft, *Why We Didn't Remove Saddam*, Time, 2 March, (1998).
(47) J.A. Baker III, *The Politics of Diplomacy: Revolution, War and Peace, 1989-1992*, G.P. Putnam's Sons, New York, (1995).
(48) H. Thomas, *Preventive War Sets Serious Precedent*, Seattle Post Intelligencer, March 20, (2003).
(49) R.J. Barnet, *Intervention and Revolution: The United States in the Third World*, World Publishing, (1968).
(50) T. Bodenheimer and R. Gould, *Rollback: Right-wing Power in U.S. Foreign Policy*, South End Press, (1989).
(51) G. Guma, *Uneasy Empire: Repression, Globalization, and What We Can Do*, Toward Freedom, (2003).
(52) W. Blum, *A Brief History of U.S. Interventions: 1945 to the Present*, Z magazine, June, (1999).
(53) W. Blum, *Killing Hope: U.S. Military and CIA Intervention Since World War II*
(54) J.M. Cypher, *The Iron Triangle: The New Military Buildup*, Dollars and Sense magazine, January/February, (2002).
(55) L. Meyer, *The Power of One*, (World Press Review), Reforma, Mexico City, August 5, (1999).
(56) W. Hartung, F. Berrigan and M. Ciarrocca, *Operation Endless Deployment: The War With Iraq Is Part of a Larger Plan for Global Military Dominance*, The Nation magazine, October 21, (2002).
(57) I. Ramonet, *Servile States*, Le Monde diplomatique, Fromkin Paris, October (2002), World Press Review, December, (2002).
(58) J.K. Galbraith, *The Unbearable Costs of Empire*, American Prospect magazine, November, (2002).
(59) G. Monbiot, *The Logic of Empire*, The Guardian, August 6, (2002),

World Press Review, October, (2002).
(60) W.R. Pitt, *The Greatest Sedition is Silence*, Pluto Press, (2003).
(61) J. Wilson, *Republic or Empire?*, The Nation magazine, March 3, (2003).
(62) W.B. Gallie, *Understanding War: Points of Conflict*, Routledge, London, (1991).
(63) R. Falk and S.S. Kim, eds., *The War System: An Interdisciplinary Approach*, Westview, Boulder, CO, (1980).
(64) J.D. Clarkson and T.C. Cochran, eds., *War as a Social Institution*, Colombia University Press, New York, (1941).
(65) S. Melman, *The Permanent War Economy*, Simon and Schuster, (1974). Morgan
(66) H. Mejcher, *Imperial Quest for Oil: Iraq, 1910-1928*, Ithaca Books, London, (1976).
(67) D. Hiro, *The Longest War: The Iran-Iraq Military Conflict*, Routledge, New York, (1991).
(68) M. Klare, *Bush-Cheney Energy Strategy: Procuring the Rest of the World's Oil*, Foreign Policy in Focus, (Interhemispheric Resource Center/Institute for Policy Studies/SEEN), Washington DC and Silver City NM, January, (2004).
(69) J. Fitchett and D. Ignatius, *Lengthy Elf Inquiry Nears Explosive Finish*, International Herald Tribune, February 1, (2002).
(70) T. Rajamoorthy, *Deceit and Duplicity: Some Reflections on Western Intervention in Iraq*, Third World Resurgence, March-April, (2003).
(71) P. Knightley and C. Simpson, *The Secret Lives of Lawrence of Arabia*, Nelson, London, (1969).
(72) G. Lenczowski, *The Middle East in World Affairs*, Cornell University Press, (1962).
(73) D. Rose, *Bush and Blair Made Secret Pact for Iraq War*, Observer, April 4, (2004).
(74) B. Gellman, *Allied Air War Struck Broadly in Iraq; Officials Acknowledge Strategy Went Beyond Purely Military Targets*, Washington Post, June 23, (1991).
(75) M. Fletcher and M. Theodoulou, *Baker Says Sanctions Must Stay as Long as Saddam Holds Power*, Times, May 23, (1991).
(76) J. Pienaar and L. Doyle, *UK Maintains Tough Line on Sanctions Against Iraq*, Independent, May 11, (1991).
(77) B. Blum (translator), *Ex-National Security Chief Brzezinski Admits: Afghan Islamism Was Made in Washington*, Nouvel Observateur, January 15, (1998).
(78) G. Vidal, *Dreaming War: Blood for Oil and the Bush-Cheney Junta*, Thunder's Mouth Press, (2002).

(79) H. Thomas, *Preventive War Sets Serious Precedent*, Seattle Post-Intelligencer, March 20, (2003).
(80) C. Johnson, *The Sorrows of Empire: Militarism, Secrecy, and the End of the Republic*, Henry Hold and Company, New York, (2004).
(81) C. Johnson, *Blowback: The Costs and Consequences of American Empire*, Henry Hold and Company, New York, (2000).
(82) M. Parenti, *Against Empire: The Brutal Realities of U.S. Global Domination*, City Lights Books, 261 Columbus Avenue, San Francisco, CA94133, (1995).
(83) E. Ahmad, *Confronting Empire*, South End Press, (2000).
(84) W. Greider, *Fortress America*, Public Affairs Press, (1998).
(85) J. Pilger, *Hidden Agendas*, The New Press, (1998).
(86) S.R. Shalom, *Imperial Alibis*, South End Press, (1993).
(87) C. Boggs (editor), *Masters of War: Militarism and Blowback in the Era of American Empire*, Routledge, (2003).
(88) J. Pilger, *The New Rulers of the World*, Verso, (2992).
(89) G. Vidal, *Perpetual War for Perpetual Peace: How We Got To Be So Hated*, Thunder's Mouth Press, (2002).
(90) W. Blum, *Rogue State: A Guide to the World's Only Superpower*, Common Courage Press, (2000).
(91) M. Parenti, *The Sword and the Dollar*, St. Martin's Press, 175 Fifth Avenue, New York, NY 10010, (1989).
(92) T. Bodenheimer and R. Gould, *Rollback: Right-wing Power in U.S. Foreign Policy*, South End Press, (1989).
(93) G. Guma, *Uneasy Empire: Repression, Globalization, and What We Can Do*, Toward Freedom, (2003).
(94) W. Blum, *A Brief History of U.S. Interventions: 1945 to the Present*, Z magazine, June, (1999).
(95) W. Blum, *Killing Hope: U.S. Military and CIA Intervention Since World War II*
(96) J.M. Cypher, *The Iron Triangle: The New Military Buildup*, Dollars and Sense magazine, January/February, (2002).
(97) L. Meyer, *The Power of One*, (World Press Review), Reforma, Mexico City, August 5, (1999).
(98) C. Johnson, *Time to Bring the Troops Home*, The Nation magazine, May 14, (2001).
(99) W. Hartung, F. Berrigan and M. Ciarrocca, *Operation Endless Deployment: The War With Iraq Is Part of a Larger Plan for Global Military Dominance*, The Nation magazine, October 21, (2002).
(100) C. Johnson, *The Sorrows of Empire: Militarism, Secrecy, and the End of the Republic*, Henry Hold and Company, New York, (2004).
(101) C. Johnson, *Blowback: The Costs and Consequences of American*

Empire, Henry Hold and Company, New York, (2000).
(102) I. Ramonet, *Servile States*, Le Monde diplomatique, Paris, October (2002), World Press Review, December, (2002).
(103) J.K. Galbraith, *The Unbearable Costs of Empire*, American Prospect magazine, November, (2002).
(104) G. Monbiot, *The Logic of Empire*, The Guardian, August 6, (2002), World Press Review, October, (2002).
(105) W.R. Pitt and S. Ritter, *War on Iraq*, Context Books
(106) W.R. Pitt, *The Greatest Sedition is Silence*, Pluto Press, (2003).
(107) J. Wilson, *Republic or Empire?*, The Nation magazine, March 3, (2003).
(108) R. Dreyfuss, *Just the Beginning: Is Iraq the Opening Salvo in a War to Remake the World?*, The American Prospect magazine, April, (2003).
(109) D. Moberg, *The Road From Baghdad: The Bush Team Has Big Plans For the 21st Century. Can the Rest of the World Stop Them?*, These Times magazine, May, (2003).
(110) J.M. Blair, *The Control of Oil*, Random House, New York, (1976).
(111) R.S. Foot, S.N. MacFarlane and M. Mastanduno, *US Hegemony and International Organizations: The United States and Multilateral Institutions*, Oxford University Press, (2003).
(112) P. Bennis and N. Chomsky, *Before and After: US Foreign Policy and the September 11th Crisis*, Olive Branch Press, (2002).
(113) J. Garrison, *America as Empire: Global Leader or Rouge Power?*, Berrett-Koehler Publishers, (2004).
(114) A.J. Bacevich, *American Empire: The Realities and Consequences of US Diplomacy*, Harvard University Press, (2002).
(115) D.R. Francis, *Hidden Defense Costs Add Up to Double Trouble*, Christian Science Monator, February 23, (2004).
(116) A. Sampson, *The Seven Sisters: The Great Oil Companies of the World and How They Were Made*, Hodder and Staughton, London, (1988).
(117) D. Yergin, *The Prize*, Simon and Schuster, New York, (1991).
(118) E. Abrahamian, *Iran Between Two Revolutions*, Princeton University Press, Princeton, (1982).

Chapter 9

The Evolution of Cooperation

"Alas, two souls are living in my breast!"

Goethe's Faust

"No man is an island, entire of itself; every man is a piece of the continent, a part of the main. If a clod be washed away by the sea, Europe is the less, as well as if a promontory were, as well as if a manor of thy friend's or thine own were..."

John Donne

Introduction

Human nature contains an element of tribalism, which we have inherited from our ancestors who lived in small generically-homogeneous tribes, competing for territory on the grasslands of Africa. But humans also have a genius for cooperation. Our enormously successful civilization has been built through the sharing of ideas and innovations. All cultural groups have contributed.

Today it is vital that the cooperative side of human nature should be supported by our educational systems, our mass media and our religious leaders.

The passions of mankind

The explosion of human knowledge

Cultural evolution depends on the non-genetic storage, transmission, diffusion and utilization of information. The development of human speech, the invention of writing, the development of paper and printing, and finally in modern times, mass media, computers and the Internet - all these have been crucial steps in society's explosive accumulation of information and knowledge. Human cultural evolution proceeds at a constantly-accelerating speed, so great in fact that it threatens to shake society to pieces.

Every species changes gradually through genetic evolution; but with humans, cultural evolution has rushed ahead with such a speed that it has completely outstripped the slow rate of genetic change. Genetically we are quite similar to our neolithic ancestors, but their world has been replaced by a world of quantum theory, relativity, supercomputers, antibiotics, genetic engineering and space telescopes - unfortunately also a world of nuclear weapons and nerve gas.

Because of the slowness of genetic evolution in comparison to the rapid and constantly-accelerating rate of cultural change, our bodies and emotions (as Malthus put it, the "passions of mankind") are not completely adapted to our new way of life. They still reflect the way of life of our hunter-gatherer ancestors.

Within rapidly-moving cultural evolution, we can observe that technical change now moves with such astonishing rapidity that neither social institutions, nor political structures, nor education, nor public opinion can keep pace. The lightning-like pace of technical progress has made many of our ideas and institutions obsolete. For example, the absolutely-sovereign nation-state and the institution of war have both become dangerous anachronisms in an era of instantaneous communication, global interdependence and all-destroying weapons.

In many respects, human cultural evolution can be regarded as an enormous success. However, at the start of the 21st century, most thoughtful observers agree that civilization is entering a period of crisis. As all curves move exponentially upward - population, production, consumption, rates of scientific discovery, and so on - one can observe signs of increasing environmental stress, while the continued existence and spread of nuclear weapons threatens civilization with destruction. Thus while the explosive growth of knowledge has brought many benefits, the problem of achieving a stable, peaceful and sustainable world remains serious, challenging and unsolved.

Tribal emotions and nationalism

In discussing conflicts, we must be very careful to distinguish between two distinct types of aggression exhibited by both humans and animals. The first is intra-group aggression, which is often seen in rank-determining struggles, for example when two wolves fight for pack leadership, or when males fight for the privilege of mating with females. Another, completely different, type of aggression is seen when a group is threatened by outsiders. Most animals, including humans, then exhibit a communal defense response - self-sacrificing and heroic combat against whatever is perceived to be an external threat. It is this second type of aggression that makes war possible.

Arthur Koestler has described inter-group aggression in an essay entitled *The Urge to Self-Destruction*,[a] where he writes: "Even a cursory glance at history should convince one that individual crimes, committed for selfish motives, play a quite insignificant role in the human tragedy compared with the numbers massacred in unselfish love of one's tribe, nation, dynasty, church or ideology... Wars are not fought for personal gain, but out of loyalty and devotion to king, country or cause..."

"We have seen on the screen the radiant love of the Führer on the faces of the Hitler Youth... They are transfixed with love, like monks in ecstasy on religious paintings. The sound of the nation's anthem, the sight of its proud flag, makes you feel part of a wonderfully loving community. The fanatic is prepared to lay down his life for the object of his worship, as the lover is prepared to die for his idol. He is, alas, also prepared to kill anybody who represents a supposed threat to the idol." The emotion described here by Koestler is the same as the communal defense mechanism ("militant enthusiasm") described below in biological terms by the Nobel Laureate ethologist Konrad Lorenz.

---

[a] In *The Place of Value in a World of Facts*, A. Tiselius and S. Nielsson editors, Wiley, New York, (1970).

In *On Aggression*, Lorenz gives the following description of the emotions of a hero preparing to risk his life for the sake of the group: "In reality, militant enthusiasm is a specialized form of communal aggression, clearly distinct from and yet functionally related to the more primitive forms of individual aggression. Every man of normally strong emotions knows, from his own experience, the subjective phenomena that go hand in hand with the response of militant enthusiasm. A shiver runs down the back and, as more exact observation shows, along the outside of both arms. One soars elated, above all the ties of everyday life, one is ready to abandon all for the call of what, in the moment of this specific emotion, seems to be a sacred duty. All obstacles in its path become unimportant; the instinctive inhibitions against hurting or killing one's fellows lose, unfortunately, much of their power. Rational considerations, criticisms, and all reasonable arguments against the behavior dictated by militant enthusiasm are silenced by an amazing reversal of all values, making them appear not only untenable, but base and dishonorable. Men may enjoy the feeling of absolute righteousness even while they commit atrocities. Conceptual thought and moral responsibility are at their lowest ebb. As the Ukrainian proverb says: 'When the banner is unfurled, all reason is in the trumpet'."

"The subjective experiences just described are correlated with the following objectively demonstrable phenomena. The tone of the striated musculature is raised, the carriage is stiffened, the arms are raised from the sides and slightly rotated inward, so that the elbows point outward. The head is proudly raised, the chin stuck out, and the facial muscles mime the 'hero face' familiar from the films. On the back and along the outer surface of the arms, the hair stands on end. This is the objectively observed aspect of the shiver!"

"Anybody who has ever seen the corresponding behavior of the male chimpanzee defending his band or family with self-sacrificing courage will doubt the purely spiritual character of human enthusiasm. The chimp, too, sticks out his chin, stiffens his body, and raises his elbows; his hair stands on end, producing a terrifying magnification of his body contours as seen from the front. The inward rotation of the arms obviously has the purpose of turning the longest-haired side outward to enhance the effect. The whole combination of body attitude and hair-raising constitutes a bluff. This is also seen when a cat humps its back, and is calculated to make the animal appear bigger and more dangerous than it really is. Our shiver, which in German poetry is called a 'Heiliger Schauer', a 'holy' shiver, turns out to be the vestige of a prehuman vegetative response for making a fur bristle which we no longer have. To the humble seeker for biological truth, there cannot be the slightest doubt that human militant enthusiasm evolved out of a communal defense response of our prehuman ancestor."

Lorenz goes on to say, "An impartial visitor from another planet, looking at man as he is today - in his hand the atom bomb, the product of his intelligence - in his heart the aggression drive, inherited from his anthropoid ancestors, which the same intelligence cannot control - such a visitor would not give mankind much chance of survival."

Members of tribe-like groups are bound together by strong bonds of altruism and loyalty. Echos of these bonds can be seen in present-day family groups, in team sports, in the fellowship of religious congregations, and in the bonds that link soldiers to their army comrades and to their nation.

Warfare involves not only a high degree of aggression, but also an extremely high degree of altruism. Soldiers kill, but they also sacrifice their own lives. Thus patriotism and duty are as essential to war as the willingness to kill.

Tribalism involves passionate attachment to one's own group, self-sacrifice for the sake of the group, willingness both to die and to kill if necessary to defend the group from its enemies, and belief that in case of a conflict, one's own group is always in the right. Unfortunately these emotions make war possible; and today a Third World War might lead to the destruction of civilization.

Population genetics

The mystery of self-sacrifice in war

At first sight, the willingness of humans to die defending their social groups seems hard to explain from the standpoint of Darwinian natural selection. After the heroic death of such a human, he or she will be unable to produce more children, or to care for those already born.Therefore one might at first suppose that natural selection would work strongly to eliminate the trait of self-sacrifice from human nature. However, the theory of population genetics and group selection can explain both the willingness of humans to sacrifice themselves for their own group, and also the terrible aggression that they sometimes exhibit towards competing groups. It can explain both intra-group altruism and inter-group aggression.

Fisher, Haldane and Hamilton

The idea of group selection in evolution was proposed in the 1930's by J.B.S. Haldane and R.A. Fischer, and more recently it has been discussed by W.D. Hamilton.

Fig. 9.1 Nikolaas Tinbergen (left) and Konrad Lorenz. They and Karl von Frisch shared the 1973 Nobel Prize in Medicine and Physiology for studies of behavior patterns in animals. (Public domain.)

If we examine altruism and aggression in humans, we notice that members of our species exhibit great altruism towards their own children. Kindness towards close relatives is also characteristic of human behavior, and the closer the biological relationship is between two humans, the greater is the altruism they tend to show towards each other. This profile of altruism is easy to explain on the basis of Darwinian natural selection since two closely related individuals share many genes and, if they cooperate, the genes will be more effectively propagated.

To explain from an evolutionary point of view the communal defense mechanism discussed by Lorenz - the willingness of humans to kill and be killed in defense of their communities - we have only to imagine that our ancestors lived in small tribes and that marriage was likely to take place within a tribe rather than across tribal boundaries. Under these circumstances, each tribe would tend to consist of genetically similar individuals. The tribe itself, rather than the individual, would be the unit on which the evolutionary forces of natural selection would act.

Fig. 9.2 Sir Ronald Aylmer Fischer (1890-1962). In his book "The Genetical Foundations of Natural Selection", published in 1930, Fischer laid the foundations of population genetics. (Public domain.)

According to the group selection model, a tribe whose members showed altruism towards each other would be more likely to survive than a tribe whose members cooperated less effectively. Since several tribes might be in competition for the same territory, successful aggression against a neighboring group could increase the chances for survival of one's own tribe. Thus, on the basis of the group selection model, one would expect humans to be kind and cooperative towards members of their own group, but at the same time to sometimes exhibit aggression towards members of other groups, especially in conflicts over territory. One would also expect intergroup conflicts to be most severe in cases where the boundaries between groups are sharpest - where marriage is forbidden across the boundaries.

Language, religion and tribal markings

In biology, a species is defined to be a group of mutually fertile organisms. Thus all humans form a single species, since mixed marriages between all known races will produce children, and subsequent generations in mixed marriages are also fertile. However, although there is never a biological barrier to marriages across ethnic and racial boundaries, there are often very severe cultural barriers.

Irenäus Eibl-Ebesfeldt, a student of Konrad Lorenz, introduced the word *pseudospeciation* to denote cases where cultural barriers between two groups of humans are so strongly marked that marriages across the boundary are difficult and infrequent. In such cases, she pointed out, the two groups function as though they were separate species, although from a biological standpoint this is nonsense. When two such groups are competing for the same land, the same water, the same resources, and the same jobs, the conflicts between them can become very bitter indeed. Each group regards the other as being "not truly human".

In his book *The Biology of War and Peace*, Eibl-Eibesfeldt discusses the "tribal markings" used by groups of humans to underline their own identity and to clearly mark the boundary between themselves and other groups. One of the illustrations in his book shows the marks left by ritual scarification on the faces of the members of certain African tribes. These scars would be hard to counterfeit, and they help to establish and strengthen tribal identity. Seeing a photograph of the marks left by ritual scarification on the faces of African tribesmen, it is impossible not to be reminded of the dueling scars that Prussian army officers once used to distinguish their caste from outsiders.

Surveying the human scene, one can find endless examples of signs that mark the bearer as a member of a particular group - signs that can be thought of as "tribal markings": tattoos; piercing; bones through the nose or ears; elongated necks or ears; filed teeth; Chinese binding of feet; circumcision, both male and female; unique hair styles; decorations of the tongue, nose, or naval; peculiarities of dress, kilts, tartans, school ties, veils, chadors, and headdresses; caste markings in India; use or non-use of perfumes; codes of honor and value systems; traditions of hospitality and manners; peculiarities of diet (certain foods forbidden, others preferred); giving traditional names to children; knowledge of dances and songs; knowledge of recipes; knowledge of common stories, literature, myths, poetry or common history; festivals, ceremonies, and rituals; burial customs, treatment of the dead and ancestor worship; methods of building and decorating homes; games and sports peculiar to a culture; relationship to animals, knowledge of horses and ability to ride; nonrational systems of belief. Even a baseball hat worn

Fig. 9.3 1908 photo of a Filipino Bontoc warrior bearing a Head hunters "Chaklag" Tattoo. Tribal markings help social groups to establish their identity and to sharply define the boundaries of the group. Within the group boundaries, humans tend to exhibit altruism, while across the boundaries, aggression is often exhibited. In modern nations, genetically dissimilar humans often use tribal markings to establish social cohesion over a larger group than would otherwise be possible. (Public domain.)

backwards or the professed ability to enjoy atonal music can mark a person as a member of a special "tribe". Undoubtedly there are many people in New York who would never think of marrying someone who could not appreciate the paintings of Jasper Johns, and many in London who would consider anyone had not read all the books of Virginia Wolfe to be entirely outside the bounds of civilization.

By far the most important mark of ethnic identity is language, and within a particular language, dialect and accent. If the only purpose of language were communication, it would be logical for the people of a small country like Denmark to stop speaking Danish and go over to a more universally-understood international language such as English. However, language has another function in addition to communication: It is also a mark of identity. It establishes the boundary of the group.

Within a particular language, dialects and accents mark the boundaries of subgroups. For example, in England, great social significance is attached to accents and diction, a tendency that George Bernard Shaw satirized in his play, *Pygmalion*, which later gained greater fame as the musical comedy, *My Fair Lady*. This being the case, we can ask why all citizens of England do not follow the example of Eliza Doolittle in Shaw's play, and improve their social positions by acquiring Oxford accents. However, to do so would be to run the risk of being laughed at by one's peers and regarded as a traitor to one's own local community and friends. School children everywhere can be very cruel to any child who does not fit into the local pattern. At Eton, an Oxford accent is compulsory; but in a Yorkshire school, a child with an Oxford accent would suffer for it.

Next after language, the most important "tribal marking" is religion. It seems probable that in the early history of our hunter-gatherer ancestors, religion evolved as a mechanism for perpetuating tribal traditions and culture. Like language, and like the innate facial expressions studied by Darwin, religion is a universal characteristic of all human societies. All known races and cultures practice some sort of religion. Thus a tendency to be religious seems to be built into human nature, or at any rate, the needs that religion satisfies seem to be a part of our inherited makeup. Otherwise, religion would not be as universal as it is.

Formation of group identity

Although humans originally lived in small, genetically homogeneous tribes, the social and political groups of the modern world are much larger, and are often multiracial and multiethnic.

There are a number of large countries that are remarkable for their diversity, for example Brazil, Argentina and the United States. Nevertheless it has been possible to establish social cohesion and group identity within each of these enormous nations. India and China too, are mosaics of diverse peoples, but nevertheless, they function as coherent societies. Thus we see that group identity is a social construction, in which artificial "tribal markings" define the boundaries of the group.

As an example of the use of tribal markings to establish social cohesion over a large group of genetically dissimilar humans, one can think of the role of baseball and football in the United States. Affection for these sports and knowledge of their intricacies is able to establish social bonds that transcend racial and religious barriers.

One gains hope for the future by observing how it has been possible to produce both internal peace and social cohesion over very large areas of the globe - areas that contain extremely diverse populations. The difference between making large, ethnically diverse countries function as coherent sociopolitical units and making the entire world function as a unit is not very great.

Since group identity is a social construction, it is not an impossible goal to think of enlarging the already-large groups of the modern world to include all of humanity.

Non-human examples of aggression and altruism

Aggression associated with mating

We must be careful not to confuse intergroup aggression with aggression associated with mating behavior. Among many species of fish, birds and animals, males fight for the privilege of mating. This type of aggression is often associated with sexual dimorphism, i.e. secondary differences in structure between males and females of the same species. For example, the large antlers of male deer are used for rank-determining fights, which confer greater reproductive success on the winner; but herds of deer do not engage in war with other herds. Thus there is a distinction between rank-determining aggression and inter-group aggression.

Chimpanzees and bonobos

The line of descent leading to humans diverged from the line leading to chimpanzees and bonobos between 5 and 6 million years ago. Chimps and bonobos look very similar, and until recent times, naturalists did not realize that they are separate species. However, modern studies have revealed the distinctness of the two species, as well as great differences in their social behavior. Chimpanzee groups are male-dominated, and far more aggressive than bonobo societies, which are female-dominated. Besides the aggression associated with mating (just discussed), chimpanzees also exhibit terrible inter-group aggression.

In his book *Before the Dawn*, Nicholas Wade describes what Jane Goodall, John Mitani, and other primatologists have discovered concerning male chimpanzees' aggression towards neighboring groups of their own species: "Chimpanzees carefully calculate the odds, and seek to minimize risk, a very necessary procedure if one fights on a regular basis. They prefer to attack an isolated individual, and then retreat into their own territory. If they encounter an opposing patrol, they will access the size of their opponents' party, and retreat if outnumbered. Researchers have confirmed this behavior by playing the call of a single male to chimp parties of various sizes. They find that the chimps will approach if they number three or more; parties of two will slink away. Three against one is the preferred odds: two to hold the victim down, and a third to batter him to death." Interestingly, the female-dominated bonobo societies do not exhibit this type of inter-group warfare, which, among chimpanzees, is conducted exclusively by the males.

The social insects

The social[b] insects, ants, bees, wasps and termites, exhibit nearly perfect altruism towards members of their own group. This extreme form of altruism towards near relations (kin altruism) is closely connected with the peculiar method of reproduction of the social insects.[c] The workers are sterile or nearly sterile, while the queen is the only reproductive female. The result of this special method of reproduction is that very nearly perfect altruism is possible within a hive or nest, since genetic changes favoring antisocial behavior would be detrimental to the hive or nest as a whole. The hive or nest can, in some sense, be regarded as a superorganism, with the individuals cooperating totally in much the same way that cells cooperate within a multicellular organism. The social insects exhibit aggression towards members of their own species from other hives or nests, and can be said to engage in wars.

The evolution of cooperation

From Thomas Huxley to Lynn Margulis and symbiosis

Charles Darwin (1809-1882) was acutely aware of close and mutually beneficial relationships between organisms. For example, in his work on the fertilization of flowers, he studied the ways in which insects and plants can become exquisitely adapted to each other's needs.

On the other hand Thomas Henry Huxley (1825-1895), although he was a strong supporter of Darwin, saw competition as the main mechanism of

---

[b]The technical term is *eusocial*.

[c]Interestingly a similar method of reproduction, associated with extreme intra-group altruism has evolved among mammals, but is represented by only two species: the naked mole rat and Damaraland mole rat.

Fig. 9.4 Thomas Henry Huxley (1825-1895), characatured in Vanity Fair. Huxley was a strong supporter of Darwin, but he placed much more emphasis on competition in evolution than Darwin did. In fact, Darwin himself was strongly aware of the great role that cooperation plays. (Public domain.)

evolution. In his essay *Struggle for Existence and its Bearing Upon Man* Huxley wrote: "From the point of view of the moralist, the animal world is about on the same level as a gladiators' show. The creatures are fairly well treated and set to fight; hereby the strongest, the swiftest, and the cunningest live to fight another day. The spectator has no need to turn his thumbs down, as no quarter is granted."

Prince Peter Kropotkin (1842-1921) argued strongly against Huxley's point of view in his book *Mutual Aid; A Factor of Evolution*. "If we ask Nature", Kropotkin wrote, "'who are the fittest: those who are continually at war with each other, or those who support one another?', we at once see that those animals that acquire habits of mutual aid are undoubtedly the fittest. They have more chances to survive, and they attain, in their respective classes, the highest development of intelligence and bodily organization."

Fig. 9.5 The biologist Lynn Margulis argued strongly that eukaryotic cells should be regarded as cooperative communities of simpler organisms that once lived independently. At first she was almost alone in this view, but today it is generally accepted. Most of the great upward steps in evolution have involved cooperation. (Public domain.)

Today, the insights of modern biology show that although competition plays an important role, most of the great upward steps in evolution have involved cooperation. The biologist Lynn Margulis (1938-) has been one of the pioneers of the modern viewpoint which recognizes symbiosis as a central mechanism in evolution.

One-celled organisms seen as examples of cooperation

The first small bacterial cells (prokaryotic cells) can be thought of as cooperative communities in which autocatalytic molecules thrived better together than they had previously done separately.

The next great upward step in evolution, the development of large and complex (eukaryotic) cells, also involved cooperation: Many of their components, for example mitochondria (small granular structures that are needed for respiration) and chloroplasts (the photosynthetic units of higher plants) are believed to have begun their existence as free-living prokaryotic cells. They now have become components of complex cells, cooperating biochemically with the other subcellular structures. Both mitochondria and chloroplasts possess their own DNA, which shows that they were once free-living bacteria-like organisms, but they have survived better in a cooperative relationship.

Cooperation between cells; multicellular organisms

Multicellular organisms evolved from cooperative communities of eukaryotic cells. Some insights into how this happened can be gained from examples which are just on the borderline between the multicellular organisms and single-celled ones. The cooperative behavior of a genus of unicellular eukaryotes called slime molds is particularly interesting because it gives us a glimpse of how multicellular organisms may have originated. The name of the slime molds is misleading, since they are not fungi, but are similar to amoebae. Under ordinary circumstances, the individual cells wander about independently searching for food, which they draw into their interiors and digest. However, when food is scarce, they send out a chemical signal of distress. (Researchers have analyzed the molecule which expresses slime mold unhappiness, and they have found it to be cyclic adenosine monophosphate.) At this signal, the cells congregate and the mass of cells begins to crawl, leaving a slimy trail. At it crawls, the community of cells gradually develops into a tall stalk, surmounted by a sphere - the "fruiting body". Inside the sphere, spores are produced by a sexual process. If a small animal, for example a mouse, passes by, the spores may adhere to its coat; and in this way they may be transported to another part of the forest where food is more plentiful. Thus slime molds represent a sort of missing link between unicellular and multicellular or organisms. Normally the cells behave as individualists, wandering about independently, but when challenged by a shortage of food, the slime mold cells join together into an entity which closely resembles a multicellular organism. The cells even seem to exhibit altruism, since those forming the stalk have little chance of survival, and yet they are willing to perform their duty, holding up the sphere at the top so that the spores will survive and carry the genes of the community into the future.

Fig. 9.6 A photo showing several types of sponges. Sponges and slime molds are on the borderline between single celled organisms and multicellular ones. The single cells of these species can live independently, but they can also function as members of a cooperating colony. (Public domain.)

Multicellular organisms often live in a symbiotic relationship with other species. For example, in both animals and humans, bacteria are essential for the digestion of food. Fungi on the roots of plants aid their absorption of water and nutrients. Communities of bacteria and other organisms living in the soil are essential for the recycling of nutrients. Insects are essential to many plants for pollination.

Fig. 9.7 A honey bee collecting pollen. The almost perfectly altruistic behavior of bees towards members of their own hive is a consequence of their special method of reproduction, which insures that all the members of the hive are more closely related to each other than they would be to a potential offspring. A hive of bees can be regarded as a superorganism, with the individuals playing roles that are analogous to the roles played by individual cells in a multicellular organism. The degree of cooperation in human society is so great that it too can to some extent be regarded as a superorganism. (Wikipedia.)

Cooperation in groups of animals and human groups

The social behavior of groups of animals, flocks of birds and communities of social insects involves cooperation as well as rudimentary forms of language. Various forms of language, including chemical signals, postures and vocal signals, are important tools for orchestrating cooperative behavior.

The highly developed language of humans made possible an entirely new form of evolution. In cultural evolution (as opposed to genetic evolution), information is passed between generations not in the form of a genetic code, but in the form of linguistic symbols. With the invention of writing, and later the invention of printing, the speed of human cultural evolution greatly increased. Cooperation is central to this new form of evolution. Cultural advances can be shared by all humans.

The evolution of human cooperation

Intertribal aggression in prehistoric humans

Fig. 9.8 Moses depicted in a painting by Rembrandt. Many of the great ethical teachers of history lived at a time when the social unit was increasing in size - when tribalism needed to be replaced by a wider ethic. (Public domain.)

In his book *War Before Civilization* (Oxford University Press, 1996), Professor Lawrence H. Keeley of the University of Illinois states that 87%

of all prehistoric tribal societies were at war at least once per year, with 65% fighting continuously with neighboring tribes. Keeley cites as an example a massacre at Crow Creek, South Dakota, where "archaeologists found the remains of more than 500 men, women and children, who had been slaughtered, scalped and mutilated a century and a half before the arrival of Columbus (ca. AD 1325)." Other examples include a 12,000 year old Nubian cemetery, where half of the bodies apparently died by violence. Also cited is a nineteenth century study of intertribal warfare among Australia's indigenous Murgin people showing that over a twenty-year period, a quarter of the men died in war. Many more examples are given by Harvard archaeologist Stephen A. LeBlanc in *Constant Battles*, (St. Martin's Press, 2003). Commenting on such studies, Nicolas Wade wrote (in *Before the Dawn*, Penguin Group, 2007), "Had the same casualty rate been suffered by the population of the twentieth century, its war deaths would have totaled two billion people." Thus, despite the terrifying effectiveness of modern weapons, the percentage of the population killed by war seems to be much smaller today than it was in prehistoric times. However, we need to abolish nuclear weapons before a catastrophic thermonuclear war changes this hopeful statistic.

Trading in primitive societies

Although primitive societies engaged in frequent wars, they also cooperated through trade. Peter Watson, an English historian of ideas, believes that long-distance trade took place as early as 150,000 before the present. There is evidence that extensive trade in obsidian and flint took place during the stone age. Evidence for wide ranging prehistoric obsidian and flint trading networks has been found in North America. Ancient burial sites in Southeast Asia show that there too, prehistoric trading took place across very large distances. Analysis of jade jewelry from the Philippines, Thailand, Malaysia and Viet Nam shows that the jade originated in Taiwan.

The invention of writing was prompted by the necessities of trade. In prehistoric Mesopotamia, clay tokens marked with simple symbols were used for accounting as early as 8,000 BC. Often these tokens were kept in clay jars, and symbols on the outside of the jars indicated the contents. About 3,500 BC, the use of such tokens and markings led to the development of pictographic writing in Mesopotamia, and this was soon followed by the cuneiform script, still using soft clay as a medium. The clay tablets were later dried and baked to ensure permanency. The invention of writing led to a great acceleration of human cultural evolution. Since ideas could now be exchanged and preserved with great ease through writing, new advances in technique could be shared by an ever larger cooperating community of humans. Our species became more and more successful as its genius for cooperation developed.

Gracilization and decreasing sexual dimorphism

Early ancestors of modern humans had a relatively heavy (robust) bone structure in relation to their height. This robust bone structure seems to have been favored by frequent combat. During their evolution, modern humans became less robust and more gracile. In other words, their skeletons became lighter in relation to their height. Simultaneously the height and weight of males became less different from the height and weight of females. These trends are generally interpreted as indicating that combat became less important as present-day humans evolved.

Ethics and growth of the social unit

Early religions tended to be centered on particular tribes, and the ethics associated with them were usually tribal in nature. However, the more cosmopolitan societies that began to form after the Neolithic agricultural revolution required a more universal code of ethics. It is interesting to notice that many of the great ethical teachers of human history, for example Moses, Socrates, Plato, Aristotle, Lao Tzu, Confucius, Buddha, and Jesus, lived at the time when the change to larger social units was taking place. Tribalism was no longer appropriate. A wider ethic was needed.

Today the size of the social unit is again being enlarged, this time enlarged to include the entire world. Narrow loyalties have become inappropriate and there is an urgent need for a new ethic - a global ethic. Loyalty to one's nation needs to be supplemented by a higher loyalty to humanity as a whole.

Interdependence in modern human society

All of the great upward steps in the evolution of life on earth have involved cooperation: Prokaryotes, the first living cells, can be thought of as cooperative communities of autocatalysts; large, complex eukaryote cells are now believed to have evolved as cooperative communities of prokaryotes; multicellular organisms are cooperative communities of eukaryotes; multicellular organisms cooperate to form societies; and different species cooperate to form ecosystems. Indeed, James Lovelock has pointed out that the earth as a whole is a complex interacting system that can be regarded as a huge organism.

The enormous success of humans as a species is due to their genius for cooperation. The success of humans is a success of cultural evolution, a new form of evolution in which information is passed between generations, not in the form of DNA sequences but in the form of speech, writing, printing and finally electronic signals. Cultural evolution is built on cooperation, and has reached great heights of success as the cooperating community has become larger and larger, ultimately including the entire world.

Without large-scale cooperation, modern science would never have evolved. It developed as a consequence of the invention of printing, which allowed painfully gained detailed knowledge to be widely shared. Science derives its great power from concentration. Attention and resources are brought to bear on a limited problem until all aspects of it are understood. It would make no sense to proceed in this way if knowledge were not permanent, and if the results of scientific research were not widely shared. But today the printed word and the electronic word spread the results of research freely to the entire world. The whole human community is the repository of shared knowledge.

The achievements of modern society are achievements of cooperation. We can fly, but no one builds an airplane alone. We can cure diseases, but only through the cooperative efforts of researchers, doctors and medicinal firms. We can photograph and understand distant galaxies, but the ability to do so is built on the efforts of many cooperating individuals.

An isolated sponge cell can survive, but an isolated human could hardly do so. Like an isolated bee, a human would quickly die without the support of the community. The comfort and well-being that we experience depends on far-away friendly hands and minds, since trade is global, and the exchange of ideas is also global.

Finally, we should be conscious of our cooperative relationships with other species. We could not live without the bacteria that help us to digest our food. We could not live without the complex communities of organisms in the soil that convert dead plant matter into fertile topsoil. We could not live without plants at the base of the food chain, but plants require pollination, and pollination frequently requires insects. An intricate cooperative network of inter-species relationships is necessary for human life, and indeed necessary for all life. Competition plays a role in evolution, but the role of cooperation is greater.

Two sides of human nature

Looking at human nature, both from the standpoint of evolution and from that of everyday experience, we see the two faces of Janus; one face shines radiantly; the other is dark and menacing. Two souls occupy the human breast, one warm and friendly, the other murderous. Humans have developed a genius for cooperation, the basis for culture and civilization; but they are also capable of genocide; they were capable of massacres during the Crusades, capable of genocidal wars against the Amerinds, capable of the Holocaust, of Hiroshima, of the killing-fields of Cambodia, of Rwanda, and of Darfur

As an example of the two sides of human nature, we can think of Scandinavia. The Vikings were once feared throughout Europe. The Book of Common Prayer in England contains the phrase "Protect us from the fury of the Northmen!". Today the same people are so peaceful and law-abiding that they can be taken as an example for how we would like a future world to look. Human nature has the possibility for both kinds of behavior depending on the circumstances. This being so, there are strong reasons to enlist the help of education and religion to make the bright side of human nature win over the dark side. Today, the mass media are an important component of education, and thus the mass media have a great responsibility for encouraging the cooperative and constructive side of human nature rather than the dark and destructive side. In the next chapter we will explore the question of how the media can better fulfill this responsibility.

Suggestions for further reading

(1) D.R. Griffin, *Animal Mind - Human Mind*, Dahlem Conferenzen 1982, Springer, Berlin, (1982).
(2) S. Savage-Rumbaugh, R. Lewin, et al., *Kanzi: The Ape at the Brink of the Human Mind*, John Wiley and Sons, New York, (1996).
(3) R. Dunbar, *Grooming, Gossip, and the Evolution of Language*, Harvard University Press, (1998).
(4) R.I.M. Dunbar, *Primate Social Systems*, Croom Helm, London, (1988).
(5) J.H. Greenberg, *Research on language universals*, Annual Review of Anthropology, **4**, 75-94 (1975).
(6) M.E. Bitterman, *The evolution of intelligence*, Scientific American, January, (1965).
(7) R. Fox, *In the beginning: Aspects of hominid behavioral evolution*, Man, **NS 2**, 415-433 (1967).
(8) M.S. Gazzaniga, *The split brain in man*, Scientific American, **217**, 24-29 (1967).
(9) D. Kimura, *The asymmetry of the human brain*, Scientific American, **228**, 70-78 (1973).

(10) R.G. Klein, *Anatomy, behavior, and modern human origins*, Journal of World Prehistory, **9 (2)**, 167-198 (1995).

(11) G. Klein, *The Human Career, Human Biological and Cultural Origins*, University of Chicago Press, (1989).

(12) N.G. Jablonski and L.C. Aiello, editors, *The Origin and Diversification of Language*, Wattis Symposium Series in Anthropology. Memoirs of the California Academy of Sciences, No. 24, The California Academy of Sciences, San Francisco, (1998).

(13) S. Pinker, *The Language Instinct: How the Mind Creates Language*, Harper-Collins Publishers, New York, (1995).

(14) S. Pinker, *Talk of genetics and visa versa*, Nature, **413**, 465-466, (2001).

(15) S. Pinker, *Words and rules in the human brain*, Nature, **387**, 547-548, (1997).

(16) J.H. Barkow, L. Cosmides and J. Tooby, editors, *The Adapted Mind: Evolutionary Psychology and the Generation of Culture*, Oxford University Press, (1995).

(17) D.R. Begun, C.V. Ward and M.D. Rose, *Function, Phylogeny and Fossils: Miocene Hominid Evolution and Adaptations*, Plenum Press, New York, (1997).

(18) R.W. Byrne and A.W. Whitten, *Machiavellian Intelligence: Social Expertise and the Evolution of Intellect in Monkeys, Apes and Humans*, Cambridge University Press, (1988),

(19) V.P. Clark, P.A. Escholz and A.F. Rosa, editors, *Language: Readings in Language and Culture*, St Martin's Press, New York, (1997).

(20) T.W. Deacon, *The Symbolic Species: The Co-evolution of Language and the Brain*, W.W. Norton and Company, New York, (1997).

(21) C. Gamble, *Timewalkers: The Prehistory of Global Colonization*, Harvard University Press, (1994).

(22) K.R. Gibson and T. Inglod, editors, *Tools, Language and Cognition in Human Evolution*, Cambridge University Press, (1993).

(23) P. Mellers, *The Emergence of Modern Humans: An Archeological Perspective*, Edinburgh University Press, (1990).

(24) P. Mellers, *The Neanderthal Legacy: An Archeological Perspective of Western Europe*, Princeton University Press, (1996).

(25) S. Mithen, *The Prehistory of the Mind*, Thames and Hudson, London, (1996).

(26) D. Haraway, *Signs of dominance: from a physiology to a cybernetics of primate biology, C.R. Carpenter, 1939-1970*, Studies in History of Biology, **6**, 129-219 (1983).

(27) D. Johanson and M. Edey, *Lucy: The Beginnings of Humankind*, Simon and Schuster, New York, (1981).

(28) B. Kurtén, *Our Earliest Ancestors*, Colombia University Press, New York, (1992).
(29) R. Lass, *Historical Linguistics and Language Change*, Cambridge University Press, (1997).
(30) R.E. Leakey and R. Lewin, *Origins Reconsidered*, Doubleday, New York, (1992).
(31) P. Lieberman, *The Biology and Evolution of Language*, Harvard University Press, (1984).
(32) C.S.L. Lai, S.E. Fisher, J.A, Hurst, F. Vargha-Khadems, and A.P. Monaco, *A forkhead-domain gene is mutated in a severe speech and language disorder*, Nature, **413**, 519-523, (2001).
(33) W. Enard, M. Przeworski, S.E. Fisher, C.S.L. Lai, V. Wiebe, T. Kitano, A.P. Monaco, and S. Pääbo, *Molecular evolution of FOXP2, a gene involved in speech and language*, Nature AOP, published online 14 August 2002.
(34) M. Gopnik and M.B. Crago, *Familial aggregation of a developmental language disorder*, Cognition, **39**, 1-50 (1991).
(35) K.E. Watkins, N.F. Dronkers, and F. Vargha-Khadem, *Behavioural analysis of an inherited speech and language disorder. Comparison with acquired aphasia*, Brain, **125**, 452-464 (2002).
(36) J.D. Wall and M. Przeworski, *When did the human population size start increasing?*, Genetics, **155**, 1865-1874 (2000).
(37) L. Aiello and C. Dean, *An Introduction to Human Evolutionary Anatomy*, Academic Press, London, (1990).
(38) F. Ikawa-Smith, ed., *Early Paleolithic in South and East Asia*, Mouton, The Hague, (1978).
(39) M. Aitken, *Science Based Dating in Archeology*, Longman, London, (1990).
(40) R.R. Baker, *Migration: Paths Through Space and Time*, Hodder and Stoughton, London, (1982).
(41) P. Bellwood, *Prehistory of the Indo-Malaysian Archipelago*, Academic Press, Sidney, (1985).
(42) P.J. Bowler, *Theories of Human Evolution: A Century of Debate, 1884-1944*, Basil Blackwell, Oxford, (1986).
(43) P.J. Bowler, *Evolution: The History of an Idea*, University of California Press, (1989).
(44) P.J. Bowler, *Fossils and Progress: Paleontology and the Idea of Progressive Evolution in the Nineteenth Century*, Science History Publications, New York, (1976).
(45) G. Isaac and M. McCown, eds., *Human Origins: Louis Leaky and the East African Evidence*, Benjamin, Menlo Park, (1976).

(46) F.J. Brown, R. Leaky, and A. Walker, *Early Homo erectus skeleton from west Lake Turkana, Kenya*, Nature, **316**, 788-92, (1985).
(47) K.W. Butzer, *Archeology as Human Ecology*, Cambridge University Press, (1982).
(48) A.T. Chamberlain and B.A. Wood, *Early hominid phylogeny*, Journal of Human Evolution, **16**, 119-33, (1987).
(49) P. Mellars and C. Stringer, eds., *The Human Revolution: Behavioural and Biological Perspectives in the Origins of Modern Humans*, Edinburgh University Press, (1989).
(50) B. Fagan, *The Great Journey: The Peopling of Ancient America*, Thames and Hudson, London, (1987).
(51) R.A. Foley, ed., *Hominid Evolution and Community Ecology*, Academic Press, New York, (1984).
(52) S.R. Binford and L.R. Binford, *Stone tools and human behavior*, Scientific American, **220**, 70-84, (1969).
(53) B.F. Skinner and N. Chomsky, *Verbal behavior*, Language, **35** 26-58 (1959).
(54) D. Bickerton, *The Roots of Language*, Karoma, Ann Arbor, Mich., (1981).
(55) E. Lenneberg in *The Structure of Language: Readings in the Philosophy of Language*, J.A. Fodor and J.A. Katz editors, Prentice-Hall, Englewood Cliffs N.J., (1964).
(56) M. Ruhelen, *The Origin of Language*, Wiley, New York, (1994).
(57) C.B. Stringer and R. McKie, *African Exodus: The Origins of Modern Humanity*, Johnathan Cape, London (1996).
(58) R. Lee and I. DeVore, editors, *Kalahari Hunter-Gatherers*, Harvard University Press, (1975).
(59) D. Schamand-Besserat, *Before Writing, Volume 1, From Counting to Cuneiform*, University of Texas Press, Austin, (1992).
(60) D. Schamandt-Besserat, *How Writing Came About*, University of Texas Press, Austin, (1992).
(61) A. Robinson, *The Story of Writing*, Thames, London, (1995).
(62) A. Robinson, *Lost Languages: The Enegma of the World's Great Undeciphered Scripts*, McGraw-Hill, (2002).
(63) D. Jackson, *The Story of Writing*, Taplinger, New York, (1981).
(64) G. Jeans, *Writing: The Story of Alphabets and Scripts*, Abrams and Thames, (1992).
(65) W.M. Senner, editor, *The Origins of Writing*, University of Nebraska Press, Lincoln and London, (1989).
(66) F. Coulmas, *The Writing Systems of the World*, Blackwell, Oxford, (1989).

(67) W.G. Bolz, *The Origin and Early Development of the Chinese Writing System*, American Oriental Society, New Haven Conn., (1994).
(68) T.F. Carter, *The Invention of Printing in China and its Spread Westward*, Ronald Press, (1925).
(69) E. Eisenstein, *The Printing Revolution in Early Modern Europe*, Cambridge University Press, (1983).
(70) M. Olmert, *The Smithsonian Book of Books*, Wing Books, New York, (1992).
(71) D.J. Futuyma, *Evolutionary Biology*, Sinauer Associates, Sunderland Mass., (1986).
(72) B. Glass, O. Temkin, and W.L. Strauss, eds., *Forerunners of Darwin: 1745-1859*, Johns Hopkins Press, Baltimore, (1959).
(73) R. Milner, *The Encyclopedia of Evolution*, an Owl Book, Henry Holt and Company, New York, (1990).
(74) T.A. Appel, *The Cuvier-Geoffroy Debate: French Biology in the Decades before Darwin*, Oxford University Press, (1987).
(75) P. Corsi, *The Age of Lamarck: Evolutionary Theories in France, 1790-1834*, University of California Press, Berkeley, (1988).
(76) M. McNeil, *Under the Banner of Science: Erasmus Darwin and his Age*, Manchester University Press, Manchester, (1987).
(77) L.G. Wilson, *Sir Charles Lyell's Scientific Journals on the Species Question*, Yale University Press, New Haven, (1970).
(78) E.O. Wilson, *Sociobiology*, Harvard University Press (1975).
(79) E.O. Wilson, *On Human Nature*, Bantham Books, New York, (1979).
(80) A.B. Adams, *Eternal Quest: The Story of the Great Naturalists*, G.P. Putnam's Sons, New York, (1969).
(81) A.S. Packard, *Lamarck Pinker, the Founder of Evolution: His Life and Work*, Longmans, Green, and Co., New York, (1901).
(82) C. Darwin, *An historical sketch of the progress of opinion on the Origin of Species, previously to the publication of this work*, Appended to third and later editions of *On the Origin of Species*, (1861).
(83) L. Eiseley, *Darwin's Century: Evolution and the Men who Discovered It*, Dobleday, New York, (1958).
(84) Francis Darwin (editor), *The Autobiography of Charles Darwin and Selected Letters*, Dover, New York (1958).
(85) Charles Darwin, *The Voyage of the Beagle*, J.M. Dent and Sons Ltd., London (1975).
(86) Charles Darwin, *The Origin of Species*, Collier MacMillan, London (1974).
(87) Charles Darwin, *The Expression of Emotions in Man and Animals*, The University of Chicago Press (1965).

(88) H.F. Osborne, *From the Greeks to Darwin: The Development of the Evolution Idea Through Twenty-Four Centuries*, Charles Scribner and Sons, New York, (1929).
(89) Sir Julian Huxley and H.B.D. Kettlewell, *Charles Darwin and his World*, Thames and Hudson, London (1965).
(90) Allan Moorehead, *Darwin and the Beagle*, Penguin Books Ltd. (1971).
(91) Ruth Moore, *Evolution*, Time-Life Books (1962).
(92) L. Barber, *The Heyday of Natural History: 1820-1870*, Doubleday and Co., Garden City, New York, (1980).
(93) A. Desmond, *Huxley*, Addison Wesley, Reading, Mass., (1994).
(94) A. Desmond and J. Moore, *Darwin*, Penguin Books, (1992).
(95) R. Owen, (P.R. Sloan editor), *The Hunterian Lectures in Comparative Anatomy, May-June, 1837*, University of Chicago Press, (1992).
(96) C. Nichols, *Darwinism and the social sciences*, Phil. Soc. Scient. **4**, 255-277 (1974).
(97) M. Ruse, *The Darwinian Revolution*, University of Chicago Press, (1979).
(98) R. Dawkins, *The Extended Phenotype*, Oxford University Press, (1982).
(99) R. Dawkins, *The Blind Watchmaker*, W.W. Norton, (1987).
(100) R. Dawkins, *River out of Eden: A Darwinian View of Life*, Harper Collins, (1995).
(101) R. Dawkins, *Climbing Mount Improbable*, W.W. Norton, (1996).
(102) R. Dawkins, *The Selfish Gene*, Oxford University Press, (1989).
(103) S.J. Gould, *Ever Since Darwin*, W.W. Norton, (1977).
(104) R.G.B. Reid, *Evolutionary Theory: The Unfinished Synthesis*, Croom Helm, (1985).
(105) M. Ho and P.T. Saunders, editors, *Beyond Neo-Darwinism: An Introduction to a New Evolutionary Paradigm*, Academic Press, London, (1984).
(106) J. Maynard Smith, *Did Darwin Get it Right? Essays on Games, Sex and Evolution*, Chapman and Hall, (1989).
(107) E. Sober, *The Nature of Selection: Evolutionary Theory in Philosophical Focus*, University of Chicago Press, (1984).
(108) B.K. Hall, *Evolutionary Developmental Biology*, Chapman and Hall, London, (1992).
(109) J. Thompson, *Interaction and Coevolution*, Wiley and Sons, (1982).
(110) R.A. Fischer, *The Genetical Theory of Natural Selection*, Clarendon, Oxford, (1930).
(111) J.B.S. Haldane, *Population genetics*, New Biology **18**, 34-51, (1955).
(112) N. Tinbergen, *The Study of Instinct*, Oxford University Press, (1951).

(113) N. Tinbergen, *The Herring Gull's World*, Collins, London, (1953).
(114) N. Tinbergen, *Social Behavior in Animals*, Methuen, London, (1953).
(115) N. Tinbergen, *Curious Naturalists*, Country Life, London, (1958).
(116) N. Tinbergen, *The Animal in its World: Explorations of an Ethologist*, Allan and Unwin, London, (1973).
(117) K. Lorenz, *On the evolution of behavior*, Scientific American, December, (1958).
(118) K. Lorenz, *Evolution and Modification of Behavior* Harvard University Press, Cambridge, MA, (1961).
(119) K. Lorenz, *Studies in Animal and Human Behavior. I and II.*, Harvard University Press, (1970) and (1971).
(120) K. Lorenz, *On Aggression*, Bantem Books, (1977).
(121) P.H. Klopfer and J.P. Hailman, *An Introduction to Animal Behavior: Ethology's First Century*, Prentice-Hall, New Jersey, (1969).
(122) J. Jaynes, *The historical origins of "Ethology" and "Comparative Psychology"*, Anim. Berhav. **17**, 601-606 (1969).
(123) W.H. Thorpe, *The Origin and Rise of Ethology: The Science of the Natural Behavior of Animals*, Heinemann, London, (1979).
(124) R.A. Hinde, *Animal Behavior: A Synthesis of Ethological and Comparative Psychology*, McGraw-Hill, New York, (1970).
(125) R.A. Hinde, *Biological Bases of Human Social Behavior*, McGraw-Hill, New York (1977).
(126) R.A. Hinde, *Individuals, Relationships and Culture: Links Between Ethology and the Social Sciences*, Cambridge University Press, (1987).
(127) R.A. Hinde, *Ethology: Its Nature and Relationship With Other Sciences*
(128) R.A. Hinde, *Non-Verbal Communication*, Cambridge University Press, (1972).
(129) R.A. Hinde, A.-N. Perret-Clermont and J. Stevenson-Hinde, editors, *Social Relationships and Cognative Development*, Clarendon, Oxford, (1985).
(130) R.A. Hinde and J. Stevenson-Hinde, editors, *Relationships Within Families: Mutual Influences*, Clarendon Press, Oxford, (1988).
(131) J.H. Crook, editor, *Social Behavior in Birds and Mammals*, Academic Press, London, (1970).
(132) P. Ekman, editor, *Darwin and Facial Expression*, Academic Press, New York, (1973).
(133) P. Ekman, W.V. Friesen and P. Ekworth, *Emotions in the Human Face*, Pergamon, New York, (1972).
(134) N. Blurton Jones, editor, *Ethological Studies of Child Behavior*, Cambridge University Press, (1975).

(135) M. von Cranach, editor, *Methods of Inference from Animals to Human Behavior*, Chicago/Mouton, Haag, (1976); Aldine, Paris, (1976).
(136) I. Eibl-Eibesfeldt, *Ethology, The Biology of Behavior*, Holt, Rinehart and Winston, New York, (1975).
(137) I. Eibl-Eibesfeldt and F.K. Salter, editors, *Indoctrinability, Ideology, and Warfare: Evolutionary Perspectives*, Berghahn Books, (1998).
(138) I. Eibl-Eibesfeldt, *Human Ethology*, Walter De Gruyter Inc., (1989).
(139) I. Eibl-Eibesfeldt, *Love and Hate*, Walter De Gruyter Inc., (1996).
(140) I. Eibl-Eibesfeldt, *The Biology of Peace and War*, Thames and Hudson, New York (1979).
(141) I. Eibl-Eibesfeldt, *Der Vorprogramiert Mensch*, Molden, Vienna, (1973).
(142) I. Eibl-Eibesfeldt, *Liebe und Hass*, Molden, Vienna, (1973).
(143) J. Bowlby, *By ethology out of psychoanalysis: An experiment in interbreeding*, Animal Behavior, **28**, 649-656 (1980).
(144) B.B. Beck, *Animal Tool Behavior*, Garland STPM Press, New York, (1980).
(145) R. Axelrod, *The Evolution of Cooperation*, Basic Books, New York, (1984).
(146) J.D. Carthy and F.L. Ebling, *The Natural History of Aggression*, Academic Press, New York, (1964)
(147) D.L. Cheney and R.M. Seyfarth, *How Monkeys See the World: Inside the Mind of Another Species*, University of Chicago Press, (1990).
(148) F. De Waal, *Chimpanzee Politics*, Cape, London, (1982).
(149) M. Edmunds, *Defense in Animals*, Longman, London, (1974).
(150) R.D. Estes, *The Behavior Guide to African Mammals*, University of California Press, Los Angeles, (1991).
(151) R.F. Ewer, *Ethology of Mammals*, Logos Press, London, (1968).
(152) E. Morgan, *The Scars of Evolution*, Oxford University Press, (1990).
(153) W.D. Hamilton, *The genetical theory of social behavior. I and II*, J. Theor. Biol. **7**, 1-52 (1964).
(154) R.W. Sussman, *The Biological Basis of Human Behavior*, Prentice Hall, Englewood Cliffs, (1997).
(155) Albert Szent-Györgyi, *The Crazy Ape*, Philosophical Library, New York (1970).
C. Zhan-Waxler, *Altruism and Aggression: Biological and Social Origins*, Cambridge University Press (1986).
(156) R. Dart, *The predatory transition from ape to man*, International Anthropological and Linguistic Review, **1**, (1953).
(157) R. Fox, *In the beginning: Aspects of hominid behavioral evolution*, Man, **NS 2**, 415-433 (1967).

(158) R.G. Klein, *Anatomy, behavior, and modern human origins*, Journal of World Prehistory, **9 (2)**, 167-198 (1995).
(159) D.R. Begun, C.V. Ward and M.D. Rose, *Function, Phylogeny and Fossils: Miocene Hominid Evolution and Adaptations*, Plenum Press, New York, (1997).
(160) P.J. Bowler, *Theories of Human Evolution: A Century of Debate, 1884-1944*, Basil Blackwell, Oxford, (1986).
(161) G.C. Conroy, *Primate Evolution*, W.W. Norton, New York, (1990).
(162) G. Klein, *The Human Career, Human Biological and Cultural Origins*, University of Chicago Press, (1989).
(163) D.P. Barash *Sociobiology and Behavior*, Elsevier, New York, (1977).
(164) N.A. Chagnon and W. Irons, eds., *Evolutionary Biology and Human Social Behavior, an Anthropological Perspective*, Duxbury Press, N. Scituate, MA, (1979).
(165) E. Danielson, *Vold, en Ond Arv?*, Gyldendal, Copenhagen, (1929).
(166) M.R. Davie, *The Evolution of War*, Yale University Press, New Haven, CT, (1929).
(167) T. Dobzhanski, *Mankind Evolving*, Yale University Press, New Haven, CT, (1962).
(168) R.L. Holloway, *Primate Aggression: Territoriality and Xenophobia*, Academic Press, New York, (1974).
(169) P. Kitcher, *Vaulting Ambition: Sociobiology and the Quest for Human Nature*, MIT Press, Cambridge, MA, (1985).
(170) S.L.W. Mellen, *The Evolution of Love*, Freeman, Oxford, (1981).
(171) A. Roe and G.G. Simpson, *Behavior and Evolution*, Yale University Press, New Haven, CT, (1958).
(172) N.J. Smelser, *The Theory of Collective Behavior*, Free Press, New York, (1963).
(173) R. Trivers, *Social Evolution*, Benjamin/Cummings, Menlo Park, CA, (1985).
(174) W. Weiser, *Konrad Lorenz und seine Kritiker*, Piper, Munich, (1976).
(175) W. Wickler, *Biologie der 10 Gebote*, Piper, Munich, (1971).
(176) J. Galtung, *A structural theory of aggression*, Journal of Peace Research, **1**, 95-119, (1964).
(177) G.E. Kang, *Exogamy and peace relations of social units: A cross-cultural test*, Ethology, **18**, 85-99, (1979).
(178) A. Montagu, *Man and Aggression*, Oxford University Press, New York, (1968).
(179) W.A. Nesbitt, *Human Nature and War*, State Education Department of New York, Albany, (1973).
(180) W. Suttles, *Subhuman and human fighting*, Anthropologica, **3**, 148-163, (1961).

(181) V. Vale and Andrea Juno, editors, *Modern Primitives: An Investigation of Contemporary Adornment and Ritual*, San Francisco Research, (1990).

(182) P.P.G. Bateson and R.A. Hinde, editors, *Growing Points in Ethology: Based on a Conference Sponsored by St. John's College and King's College, Cambridge*, Cambridge University Press, (1976).

(183) P. Bateson, editor, *The Development and Integration of Behaviour: Essays in Honour of Robert Hinde*, Cambridge University Press, (1991).

Chapter 10

Education for Peace

"We have to extend our loyalty to the whole of the human race.... A war-free world will be seen by many as Utopian. It is not Utopian. There already exist in the world large regions, for example the European Union, within which war is inconceivable. What is needed is to extend these..."

Sir Joseph Rotblat, Nobel Peace Prize Acceptance Speech, 1995

Introduction

Since modern war has become prohibitively dangerous, there is an urgent need for peace education. Why do we pay colossal sums for war, which we know is the source of so much human suffering, and which threatens to destroy human civilization? Why not instead support peace and peace education?

In this chapter, we will see that many groups and individuals are already working for this goal. With even a little more support, they would be much more effective.

The growth of global consciousness

Besides a humane, democratic and just framework of international law and governance, we urgently need a new global ethic, - an ethic where loyalty to family, community and nation will be supplemented by a strong sense of the brotherhood of all humans, regardless of race, religion or nationality. Schiller expressed this feeling in his "Ode to Joy", a part of which is the text of Beethoven's Ninth Symphony. Hearing Beethoven's music and Schiller's words, most of us experience an emotion of resonance and unity with the message: All humans are brothers and sisters - not just some - all! It is almost a national anthem of humanity. The feelings that the music and words provoke are similar to patriotism, but broader. It is this sense of a universal human family that we need to cultivate in education, in the mass media, and in religion. We already appreciate music, art and literature from the entire world, and scientific achievements are shared by all, regardless of their country of origin. We need to develop this principle of universal humanism so that it will become the cornerstone of a new ethic.

Reformed teaching of history

Educational reforms are urgently needed, particularly in the teaching of history. As it is taught today, history is a chronicle of power struggles and war, told from a biased national standpoint. Our own race or religion is superior; our own country is always heroic and in the right.

We urgently need to replace this indoctrination in chauvinism by a reformed view of history, where the slow development of human culture is described, giving adequate credit to all who have contributed. Our modern civilization is built on the achievements of many ancient cultures. China, Japan, India, Mesopotamia, Egypt, Greece, the Islamic world, Christian Europe, and the Jewish intellectual traditions all have contributed. Potatoes, corn, squash, vanilla, chocolate, chili peppers, pineapples, quinine, etc. are gifts from the American Indians. Human culture, gradually built up over thousands of years by the patient work of millions of hands and minds, should be presented as a precious heritage - far too precious to be risked in a thermonuclear war.

The teaching of history should also focus on the times and places where good government and internal peace have been achieved, and the methods by which this has been accomplished. Students should be encouraged to think about what is needed if we are to apply the same methods to the world as a whole. In particular, the histories of successful federations should be studied, for example the Hanseatic League, the Universal Postal Union, the federal governments of Australia, Brazil, Germany, Switzerland, the United States, Canada, and so on. The recent history of the European Union provides another extremely important example. Not only the successes, but also the problems of federations should be studied in the light of the principle of subsidiarity.[a] The essential features of federations should be clarified,[b] as well as the reasons why weaker forms of union have proved to be unsuccessful.

Reformed education of economists and businessmen

The education of economists and businessmen needs to face the problems of global poverty - the painful contrast between the affluence and wastefulness of the industrial North and the malnutrition, disease and illiteracy endemic in the South. Students of economics and business must look for the roots of poverty not only in population growth and war, but also in the history of colonialism and neocolonialism, and in defects in global financial institutions and trade agreements. They must be encouraged to formulate proposals for the correction of North-South economic inequality.

The economic impact of war and preparation for war should be included in the training of economists. Both direct and indirect costs should be studied. An example of an indirect cost of war is the effect of unimaginably enormous military budgets in reducing the amount of money available for solving the serious problems facing the world today.

---

[a]The principle of subsidiarity states that within a federation, decisions should be taken at the lowest level at which there are no important externalities. Thus, for example, decisions affecting air quality within Europe should be taken in Bruxelles because winds blow freely across national boundaries, but decisions affecting only the local environment should be taken locally.

[b]One of the most important of these features is that federations have the power to make and enforce laws that are binding on individuals, rather than trying to coerce their member states.

Law for a united world

Law students should be made aware of the importance of international law. They should be familiar with its history, starting with Grotius and the Law of the Sea. They should know the histories of the International Court of Justice and the Nuremberg Principles. They should study the United Nations Charter (especially the articles making war illegal) and the Universal Declaration of Human Rights, as well as the Rome Treaty and the foundation of the International Criminal Court. They should be made aware of a deficiency in the present United Nations - the lack of a legislature with the power to make laws that are binding on individuals.

Students of law should be familiar with all of the details of the World Court's historic Advisory Opinion on Nuclear Weapons, a decision that make the use or threat of use of nuclear weapons illegal. They should also study the Hague and Geneva Conventions, and the various international treaties related to nuclear, chemical and biological weapons. The relationship between the laws of the European Union and those of its member states should be given high importance. The decision by the British Parliament that the laws of the EU take precedence over British law should be a part of the curriculum.

Teaching global ethics

Professors of theology should emphasize three absolutely central components of religious ethics: the duty to love and forgive one's enemies, the prohibition against killing, and the concept of universal human brotherhood. They should make their students conscious of a responsibility to give sermons that are relevant to the major political problems of the modern world, and especially to relate the three ethical principles just mentioned to the problem of war. Students of theology should be made conscious of their responsibility to soften the boundaries between ethnic groups, to contribute to interreligious understanding, and to make marriage across racial and religious boundaries more easy and frequent.

The social responsibility of scientists

In teaching science too, reforms are needed. Graduates in science and engineering should be conscious of their responsibilities. They must resolve never to use their education in the service of war, nor for the production of weapons, nor in any way that might be harmful to society or to the environment.

Science and engineering students ought to have some knowledge of the history and social impact of science. They could be given a course on the history of scientific ideas; but in connection with modern historical developments such as the industrial revolution, the global population explosion, the development of nuclear weapons, genetic engineering, and information technology, some discussion of social impact of science could be introduced. One might hope to build up in science and engineering students an understanding of the way in which their own work is related to the general welfare of humankind, and a sense of individual social and ethical responsibility. These elements are needed in science education if rapid technological progress is to be beneficial to society rather than harmful.

The changes just mentioned in the specialized lawyers, theologians, scientists and engineers should have a counterpart in elementary education. The basic facts about peace and war should be communicated to children in simple language, and related to the everyday experiences of children. Teachers' training colleges ought to discuss with their student-teachers the methods that can be used to make peace education a part of the curriculum at various levels, and how it can be related to familiar concepts. They should also discuss the degree to which the painful realities of war can be explained to children of various ages without creating an undesirable amount of anxiety.

Peace education can be made a part of the curriculum of elementary schools through (for example) theme days or theme weeks in which the whole school participates. This method has been used successfully in many European schools. During the theme days the children have been encouraged to produce essays, poems and drawings illustrating the difference between peace and war, and between negative peace and positive peace.[c] Another activity has been to list words inspired by the concept "peace", rapidly and by free association, and to do the same for the concept "war". Drama has also been used successfully in elementary school peace education, and films have proved to be another useful teaching aid.

The problems of reducing global inequalities, of protecting human rights, and of achieving a war-free world can be introduced into grade school courses in history, geography, religion and civics. The curriculum of these courses is frequently revised, and advocates of peace education can take curriculum revisions as opportunities to introduce much-needed reforms that will make the students more international in their outlook. The argument (a true one) should be that changes in the direction of peace education will make students better prepared for a future in which peace will be a central issue and in which they will interact with people of other

---

[c] Negative peace is merely the absence of war. In positive peace, neighboring nations are actively engaged in common projects of mutual benefit, in cultural exchanges, in trade, in exchanges of students and so on.

nations to a much greater extent than was the case in previous generations. The same can be said for curriculum revisions at the university level.

Large nations compared with global government

The problem of achieving internal peace over a large geographical area is not insoluble. It has already been solved. There exist today many nations or regions within each of which there is internal peace, and some of these are so large that they are almost worlds in themselves. One thinks of China, India, Brazil, Australia, the Russian Federation, the United States, and the European Union. Many of these enormous societies contain a variety of ethnic groups, a variety of religions and a variety of languages, as well as striking contrasts between wealth and poverty. If these great land areas have been forged into peaceful and cooperative societies, cannot the same methods of government be applied globally?

But what are the methods that nations use to achieve internal peace? Firstly, every true government needs to have the power to make and enforce laws that are binding on individual citizens. Secondly the power of taxation is a necessity. These two requirements of every true government have already been mentioned; but there is a third point that still remains to be discussed:

Within their own territories, almost all nations have more military power than any of their subunits. For example, the US Army is more powerful than the State Militia of Illinois. This unbalance of power contributes to the stability of the Federal Government of the United States. When the FBI wanted to arrest Al Capone, it did not have to bomb Chicago. Agents just went into the city and arrested the gangster. Even if Capone had been enormously popular in Illinois, the government of the state would have realized in advance that it had no chance of resisting the US Federal Government, and it still would have allowed the "Feds" to make their arrest. Similar considerations hold for almost all nations within which there is internal peace. It is true that there are some nations within which subnational groups have more power than the national government, but these are frequently characterized by civil wars.

Of the large land areas within which internal peace has been achieved, the European Union differs from the others because its member states still maintain powerful armies. The EU forms a realistic model for what can be achieved globally in the near future by reforming and strengthening the United Nations. In the distant future, however, we can imagine a time when a world federal authority will have much more power than any of its member states, and when national armies will have only the size needed to maintain local order.

Today there is a pressing need to enlarge the size of the political unit from the nation-state to the entire world. The need to do so results from the terrible dangers of modern weapons and from global economic interdependence. The progress of science has created this need, but science has also given us the means to enlarge the political unit: Our almost miraculous modern communications media, if properly used, have the power to weld all of humankind into a single supportive and cooperative society.

Culture, education and human solidarity

Cultural and educational activities have a small ecological footprint, and therefore are more sustainable than pollution-producing, fossil-fuel-using jobs in industry. Furthermore, since culture and knowledge are shared among all nations, work in culture and education leads societies naturally towards internationalism and peace.

Economies based on a high level of consumption of material goods are unsustainable and will have to be abandoned by a future world that renounces the use of fossil fuels in order to avoid catastrophic climate change, a world where non-renewable resources such as metals will become increasingly rare and expensive. How then can full employment be maintained?

The creation of renewable energy infrastructure will provide work for a large number of people; but in addition, sustainable economies of the future will need to shift many workers from jobs in industry to jobs in the service sector. Within the service sector, jobs in culture and education are particularly valuable because they will help to avoid the disastrous wars that are currently producing enormous human suffering and millions of refugees, wars that threaten to escalate into an all-destroying global thermonuclear war.[d]

Human nature has two sides: It has a dark side, to which nationalism and militarism appeal; but our species also has a genius for cooperation, which we can see in the growth of culture. Our modern civilization has been built up by means of a worldwide exchange of ideas and inventions. It is built on the achievements of many ancient cultures. China, Japan, India, Mesopotamia, Egypt, Greece, the Islamic world, Christian Europe, and the Jewish intellectual traditions all have contributed. Potatoes, corn, squash, vanilla, chocolate, chilli peppers, and quinine are gifts from the American Indians.[e]

We need to reform our educational systems, particularly the teaching of history. As it is taught today, history is a chronicle of power struggles

---

[d] http://www.fredsakademiet.dk/library/need.pdf
http://eruditio.worldacademy.org/issue-5/article/urgent-need-renewable-energy
[e] http://eruditio.worldacademy.org/article/evolution-cooperation

Fig. 10.1 Malala Yousefzai, winner of the 2014 Nobel Peace Prize, says: "One child, one teacher, one book and one pen can change the world!" (Wikipedia.)

and war, told from a biased national standpoint. We are taught that our own country is always heroic and in the right. We urgently need to replace this indoctrination in chauvinism by a reformed view of history, where the slow development of human culture is described, giving credit to all who have contributed. When we teach history, it should not be about power struggles. It should be about how human culture was gradually built up over thousands of years by the patient work of millions of hands and minds. Our common global culture, the music, science, literature and art that all of us share, should be presented as a precious heritage - far too precious to be risked in a thermonuclear war.

We have to extend our loyalty to the whole of the human race, and to work for a world not only free from nuclear weapons, but free from war. A war-free world is not utopian but very practical, and not only practical but necessary. It is something that we can achieve and must achieve. Today their are large regions, such as the European Union, where war would be inconceivable. What is needed is to extend these.

Nor is a truly sustainable economic system utopian or impossible. To achieve it, we should begin by shifting jobs to the creation of renewable energy infrastructure, and to the fields of culture and education. By so

Fig. 10.2 Cultural exchanges lead to human solidarity (Public domain.)

doing we will support human solidarity and avoid the twin disasters of catastrophic war and climate change.

UNESCO and peace education

Advocates of education for peace can obtain important guidance and encouragement from UNESCO - the United Nations Educational, Scientific and Cultural Organization.[f] The Constitution of UNESCO, was written immediately after the end of the Second World War, during which education had been misused (especially in Hitler's Germany) to indoctrinate students in such a way that they became uncritical and fanatical supporters of military dictatorships. The founders of the United Nations were anxious to correct this misuse, and to make education instead one of the foundations of a peaceful world. One can see this hope in the following paragraph from UNESCO's Constitution:

"The purpose of the Organization is to contribute to peace and security by promoting collaboration among nations through education, science and culture in order to further universal respect for justice, for the rule of law

---

[f] http://www.unicef.org/education/files/PeaceEducation.pdf

and for the human rights and fundamental freedoms which are affirmed for the peoples of the world, without distinction of race, sex, language or religion, by the Charter of the United Nations."

In other words, UNESCO was given the task of promoting education for peace, and of promoting peace through international cooperation in education.

In 1946 the General Conference of UNESCO adopted a nine-point resolution concerning the improvement of textbooks in such a way as to make them support international understanding, paying particular attention to history teaching and civic education. During the next decade, UNESCO produced publications and hosted seminars to promote improvements in the teaching of history, geography and modern languages, so that these subjects could be more instrumental in developing mutual understanding between nations and between cultures. A meeting of French, German, British and American teachers was organized in 1952, with the goal of removing national prejudices from textbooks. Every two years after this date bilateral and multilateral consultations of history teachers have taken place under the auspices of UNESCO.

Here are a few voices that express the aims and ideals of UNESCO over the years:

- Ellen Wilkinson (United Kingdom) (Former UK Minister of Education, Chairwoman of the conference establishing UNESCO in 1945): *What can this organization do? Can we replace nationalist teaching by a conception of humanity that trains children to have a sense of mankind as well as of national citizenship? That means working for international understanding*
- Maria Montessori (Italy), pioneer of modern education and education for peace, Fourth Session of the General Conference of UNESCO, Florence 1950: *If one day UNESCO resolved to involve children in the reconstruction of the world and building peace, if it chose to call on them, to discuss with them, and recognize the value of all the revelations they have for us, it would find them of immense help in infusing new life into this society which must be founded on the cooperation of all.*
- Jamie Torres Bodet (Mexico), Director-General of UNESCO, 1948-1952, (The UNESCO Courier, 1951): *Knowledge and understanding of the principles of the Universal Declaration of Human Rights and their practical application must begin during childhood. Efforts to make known the rights and duties they imply will never be fully effective unless schools in all countries make teaching about the declaration a regular part of their curriculum...*

- Lionel Elvin (United Kingdom), Director of the Department of Education of UNESCO, 1950-1956 (UNESCO Courier, 1953): *If UNESCO were only an office in Paris, its task would be impossible. It is more than that: it is an association of some sixty-five countries which have pledged themselves to do all they can, not only internationally but within their own boundaries, to advance the common aim of educating for peace. The international side comes in because we shall obviously do this faster and better and with more mutual trust if we do it together.*
- Jawaharlal Nehru (India) Prime Minister, 1947-1964 (Address on a visit to UNESCO, 1962): *It is then the minds and hearts of men that have to be approached for mutual understanding, knowledge and appreciation of each other and through the proper kind of education... But we have seen that education by itself does not lead to a conversion of minds towards peaceful purposes. Something more is necessary, new standards, new values and perhaps a kind of spiritual background and a feeling of commonness of mankind.*
- James P. Grant (United States). Executive Director of UNICEF, 1980-1995, (International Conference on Education, Geneva, 1994): *Education for peace must be global, for as the communications revolution transforms the world into a single community, everyone must come to understand that they are affected by what happens elsewhere, and that their lives , too, have an impact. Solidarity is a survival strategy in the global village.*

During the time when he was Secretary-General of UNESCO, Federico Mayor Zaragoza of Spain introduced the concept of a *Culture of Peace*. He felt, as many did, that civilization was entering a period of crisis. Federico Mayor believed this crisis to be as much spiritual as it was economic and political. It was necessary, he felt, to counteract our present power-worshiping culture of violence with a Culture of Peace, a set of ethical and aesthetic values, habits and customs, attitudes towards others, forms of behavior and ways of life that express

- Respect for life and for the dignity and human rights of individuals.
- Rejection of violence.
- Recognition of equal rights for men and women.
- Upholding the principles of democracy, freedom, justice, solidarity, tolerance and the acceptance of differences.
- Understanding between nations and countries and between ethnic, religious, cultural and social groups.

Mayor and UNESCO implemented this idea by designating the year 2000 as the International Year of the Culture of Peace. In preparation for this year, a meeting of Nobel Peace Prize Laureates launched *Manifesto 2000*, a campaign in which the following pledge of the Culture of Peace was widely circulated and signed:

Recognizing my share of responsibility for the future of humanity, especially for today's children and those of future generations, I pledge - in my daily life, in my family, my work, my community, my country and my region - to:

(1) *respect the life and dignity of every person without discrimination or prejudice;*
(2) *practice active non-violence, rejecting violence in all its forms: physical, sexual, psychological, economical and social, in particular towards the most deprived and vulnerable such as children and adolescents;*
(3) *share my time and material resources in a spirit of generosity to put an end to exclusion, injustice and political and economic oppression;*
(4) *defend freedom of expression and cultural diversity, giving preference always to dialogue and listening without engaging in fanaticism, defamation and the rejection of others;*
(5) *promote consumer behavior that is responsible and development practices that respect all forms of life and preserve the balance of nature on the planet;*
(6) *contribute to the development of my community, with the full participation of women and respect for democratic principles, in order to create together new forms of solidarity.*

In addition, Federico Mayor and UNESCO initiated a Campaign for the Children of the World, and this eventually developed into the International Decade for a Culture of Peace and Non-Violence for the Children of the World (2001-2010). In support of this work, the UN General Assembly drafted a Program of Action on a Culture of Peace (53rd Session, 2000). The Program of Action obliges it signatories to "ensure that children, from an early age, benefit from education on the values, attitudes, modes of behavior and ways of life to enable them to resolve any dispute peacefully and in a spirit of respect for human dignity and of tolerance and non-discrimination", and to "encourage the revision of educational curricula, including textbooks..."

Just as this program was starting, the September 11 terrorist attacks gave an enormous present to the culture of violence and war, and almost silenced the voices speaking for a Culture of Peace. However, military solutions have never provided true security, even for the strongest countries. Expensive and technologically advanced weapons systems may enrich arms manufacturers and military lobbies, but they do not provide security - only an unbelievably expensive case of the jitters. By contrast, the Culture of Peace can give us hope for the future.

Some examples of peace education in Denmark

A book entitled "Et barn har brug for fred!" ("A Child Needs Peace!") by Nils Hartmann of the Danish UNICEF Committee provides a good example of peace education at the elementary level. Here are rough translations of a few of the paragraphs of Nils Hartmann's book:

Peace and solidarity

A more just division of the resources of the world requires that we, in our part of the world, feel more solidarity with people in the less developed countries - in other words we must feel that we have much in common with them. People who feel solidarity with each other don't fight. They are friends.

Solidarity means more than just making sacrifices for each other. If we only give others things we have too much of, something is missing. True solidarity also means that we must have respect for each other - respect for each other's culture, actions, religion and life. When we respect each other, we are also open towards each other. We need each other and learn from each other.

Peace and fundamental needs.

When people's fundamental needs are satisfied, they are able to feel secure, and the reasons for war and conflicts disappear. But it is important that every person satisfies these fundamental needs in a way that doesn't harm or exploit others.

- *If I buy a weapon in order to feel more safe, there will be others who feel threatened.*
- *If I exploit others in order to satisfy my own needs, there will be dissatisfaction and conflicts.*
- *If I use more food than I need, others will go hungry.*
- *If I dig a well and claim all the water for myself, others will go thirsty.*
- *If I buy unnecessary things, others will go without necessities*

What can we get for the money that is wasted on armaments?

In 1985 the world used about 8,000 billion (8,000,000,000,000) kroner[g] for military purposes. In other words, half a billion kroner are being wasted while this lesson is going on. Here are a few examples of things we could have bought for a fraction of that amount of money:

Health

Almost everywhere in the world there is a lack of doctors, nurses and hospitals. This is especially true in the poorest country districts and slums of developing countries. A large number of children in these countries need to be vaccinated against some of the illnesses that are already eliminated from our part of the world. Measles, whooping cough, diphtheria, polio, tuberculosis and lockjaw cost the lives of millions of children each year. Also, many children need to come to a health clinic to get medicine and vitamins. Building up even a very basic health system would do wonders. The cost of a basic health system for the whole world is estimated to be 17 billion kroner per year.

Safe drinking water

More than 2 billion people have no way of getting safe water. Impure water and lack of water lead to many diseases. Today, diarrhoea is the most common cause of death for small children in the developing countries. The United Nations has declared the period 1981-1990 to be the International Water Decade. The United Nations has calculated that by using a total of 50 billion kroner, it would be possible to give pure drinking water to all the people of the world.

Education

In developing countries, less than half of the adults have more than a year of schooling. Education is the best investment that we can make if we want to modernize a society and to create positive development. Building schools for all of the developing countries, educating teachers, and producing teaching materials would cost 55 billion kroner.

These paragraphs from Nils Hartmann's book are illustrated with photographs of children from the developing countries. The paragraphs are written in simple language, and the examples used are related to the needs of children.

---

[g] Eight Danish kroner = one US dollar.

The Danish National Group of Pugwash Conferences on Science and World Affairs

In March, 1954, the US tested a hydrogen bomb at the Bikini Atoll in the Pacific Ocean. It was 1000 times more powerful than the Hiroshima bomb. The Japanese fishing boat, Lucky Dragon, was 130 kilometers from the Bikini explosion, but radioactive fallout from the test killed one crew member and made all the others seriously ill.

Concerned about the effects of a large-scale war fought with such bombs, or even larger ones, Albert Einstein and Bertrand Russell published a manifesto containing the words: "Here then is the problem that we present to you, stark and dreadful and inescapable: Shall we put an end to the human race, or shall mankind renounce war?... There lies before us, if we choose, continual progress in happiness, knowledge and wisdom. Shall we, instead, choose death because we cannot forget our quarrels? We appeal as human beings to human beings: Remember your humanity, and forget the rest. If you can do so, the way lies open to a new Paradise; if you cannot, there lies before you the risk of universal death."

The Russell-Einstein Manifesto called for a meeting of scientists from both sides of the Cold War to try to minimize the danger of a thermonuclear conflict. The first meeting took place in 1957 at the summer home of the Canadian philanthropist Cyrus Eaton at the small village of Pugwash, Nova Scotia.

Fig. 10.3 The Russell-Einstein Manifesto: "Shall we put an end to the human race, or shall mankind renounce war?" (Pugwash Conferences.)

From this small beginning, a series of conferences developed, in which scientists, especially physicists, attempted to work for peace, and tried to address urgent problems related to science. These conferences were called Pugwash Conferences on Science and World Affairs, taking their name from the small village in Nova Scotia where the first meeting was held. From the start, the main aim of the meetings was to reduce the danger that civilization would be destroyed in a thermonuclear war.

Many countries have local Pugwash groups, and the Danish National Pugwash Group is one of these. Our activities include conferences at the Danish Parliament, aimed at influencing decision-makers, but other activities are aimed influencing public opinion. Peace education activities include the award of student peace prizes on United Nations Day.

United Nations Day Student Peace Prizes

In collaboration with the Danish Peace Academy, and with the help of the Hermod Lannung Foundation the Danish National Group of Pugwash Conferences on Science and World Affairs has offered prizes each year to students at 10 Danish gymnasiums for projects related to global problems and their solutions and to the United Nations.

These projects are essays, dramatic sketches, videos, websites, posters, etc., and they were judged on UN Day, before large audiences of students. The background for this project is as follows: In 2007, in collaboration with several other NGOs, we arranged a visit to Copenhagen by Dr. Tadatoshi Akiba, the Mayor of Hiroshima. In connection with his visit, we arranged a Peace Education Conference at the University of Copenhagen.

In connection with Dr. Akiba's visit, we also arranged a day of peace education at Copenhagen's Open Gymnasium. About 15 people from various branches of Denmark's peace movement arrived at the gymnasium at 7.00 a.m., and between 8.00 and 10.00 they talked to 15 groups of about 25-50 students about topics related to peace. At 10.30, all 500 students assembled in a large hall, where Dr. Akiba gave an address on abolition of nuclear weapons. A chorus from the gymnasium sang, and finally there was a panel discussion.

The students were extremely enthusiastic about the whole program. The success of our 2007 effort made us want to do something similar in 2008, and perhaps to broaden the scope. Therefore we wrote to the Minister of Education, and proposed that October 24, United Nations Day, should be a theme day in all Danish schools and gymnasiums, a day devoted to the discussion of global problems and their solutions. We received the very kind reply. The Minister said that he thought our idea was a good one, but that he did not have the power to dictate the curricula to schools. We needed to contact the individual schools, gymnasiums and municipalities.

In the autumn of 2008 we arranged a United Nations Day program on October 24 at Sankt Annæ Gymnasium with the cooperation of Nørre Gymnasium. We offered prizes to drama students at the two gymnasiums for the best peace-related dramatic sketch, a condition being that the sketches should be performed and judged before a large audience. Our judges were the famous actress Mia Luhne, Johan Olsen, the lead singer of a popular rock group, and the dramatist Steen Haakon Hansen. The students' sketches and the judges speeches about the meaning of peace were very strong and moving. Everyone was very enthusiastic about the day. The judges have said that they would be willing to work with us again on peace-related cultural events.

Fig. 10.4 A painting representing the work of the United Nations. It won first prize at a UN Day Student Peace Prize competition. (Danish National Pugwash Group.)

Our successes in 2007 and 2008 have made us wish to continue and possibly expand the idea of making United Nations Day a theme day in Danish schools and gymnasiums, a day for discussion of global problems and their solutions, with special emphasis on the role of the United Nations. The Hermod Lannung Foundation supported our project for extending this idea to 10 Danish gymnasiums from 2010 until 2016.

The Grundtvigian Peoples' Colleges

A unique feature of the Danish educational system is the adult education that is available at about a hundred Folkehøjskole (Peoples' Colleges). This tradition of adult education dates back to the Danish poet-bishop N.F.S. Grundtvig (1783-1872). Besides writing more than half of the hymns presently used in Danish churches, Grundtvig also introduced farmers' cooperatives into Denmark and founded a system of adult education.

At the time when Grundtvig lived, the Industrial Revolution had already transformed England into a country that exported manufactured goods but was unable to feed itself because of its large population. In this situation, Denmark began a prosperous trade, exporting high quality agricultural produce to England (for example dairy products, bacon, and so on). Grundtvig realized that it would be to the advantage of small-scale Danish farmers to process and export these products themselves, thus avoiding losing a part of their profits to large land-owners or other middlemen who might do the processing and exporting for them. He organized the small farmers into cooperatives, and in order to give the farmers enough knowledge and confidence to run the cooperatives, Grundtvig created a system of adult education: the Peoples' Colleges. The cooperatives and the adult education system contributed strongly to making Denmark a prosperous and democratic country.

Of the hundred or so Grundtvigian Peoples' Colleges exiting today, about forty offer peace education as a subject. An example of such a peace education course was the two-week summer school "Towards a Non-violent Society", held at the International College in Elsinore during the summer of 1985. Since it was supported not only by the students' fees but also by a government subsidy, the summer school was able to pay the travel and living expenses for lecturers who came from many parts of the world.

Among the stars of the summer school were former US Governor Harold Stassen, the only living person who had signed the UN Charter; the famous Cambridge University ethologist, Professor Robert Hinde; Professor Suman Khana from India, an expert on non-violence and Gandhi; Sister George, a Catholic nun from Jerusalem, who spoke 12 languages during the course of her daily work and who was an expert on the conflicts of the Middle East; and Meta Ditzel, a member of the Danish Parliament who advocated legislation to make excessively violent videos less easily available to children. Other lectures were given by representatives of Amnesty International and the Center for Rehabilitation of Torture Victims.

In discussing Danish peace education initiatives, we must not fail to mention Holger Terp's enormous and popular Danish Peace Academy website.[h] Despite serious health problems, which include almost complete loss of vision and multiple heart bypass operations, Holger Terp singlehandedly established a unique website devoted to peace education. The Danish Peace Academy website contains more than 99,000 files in Danish, English and German. The website is visited by many thousands of students from around the world.

The World Conference of Religions for Peace

Other powerful voices for peace have been raised by the World Conference of Religions for Peace, which met for the first time in October 1970 in Kyoto, Japan.[i] At this meeting, more than 1000 religious leaders gathered to discuss the grave dangers posed by modern war. Among them were representatives of the Baha'i, Mahayana and Trevada Buddhists, Protestants, Roman Catholics, Orthodox Christians, Confucians, representatives of several streams of Hinduism, a number of communities of indigenous faith, Shiite and Sunni Muslims, Jains, Reform Jews, Shintos, Sikhs, Zoroastrians, and representatives of a number of new religions.

The WCRP sponsors many projects related to conflict resolution, the world's children, development, disarmament and security, human rights, and peace education. For example, in the field of peace education, WCRP sponsors a project in Israel called "Common Values/Different Sources" which brings together Jews, Muslims and Christians to study sacred texts together in search of shared values, eventually resulting in a book for classroom use. In England and Germany, another WCRP project analyzes school textbooks' treatment of religious traditions that are foreign to the books' intended audiences.

Dr. Edy Korthals Altes, a former Ambassador of the Netherlands to Poland and Spain and an Honorary President of the World Conference of Religions for Peace, has expressed his vision of our current global situation in the following words: "We need a new concept of security. The old concept dates back to the Romans who said 'If you want peace, prepare for war.' The new concept I would propose is exactly the opposite, 'If you want peace, prepare for peace.' While this may sound simplistic, it is difficult to put into practice since the application of justice and solidarity in international political and economic relations requires sacrifices from 'those who have.' I would give three reasons why the old concept of 'security' is no longer valid: a) The extreme vulnerability of modern society; b) The tremendous

---

[h] www.fredsakademiet.dk

[i] Subsequent World Assemblies of the WCRP have been held in Louvain, Belgium, (1974); Princeton New Jersey, (1979); Nairobi, Kenya, (1984); Melbourne, Australia, (1989); Riva del Garde, Italy, (1994); and Amman, Jordan, (1999).

destructive power of modern arms and terrorism; c) The interdependence between nations. These three elements are closely interconnected. It is therefore imperative to apply justice and solidarity in our international relations. If not, disaster looms!"

Dr. Altes feels that economic reforms are needed if global peace is to be achieved. "Not only economic justice is involved", he writes, "but also political justice. A clear example of which is the current situation in the Middle East. There must also be justice in the economic world situation in which 1/5 of the world population enjoys a high standard of living while 1/5 lives in terrible poverty, millions dying every year from hunger. This 'North South gap' is increasing!"

Discussing "myths that underlie our present economic system", he points to

(1) "The notion that each person has unlimited material needs. We are told to 'consume more' which is totally contrary to any religion. What is more, it is a self-defeating program that is contrary to humanity in general. The New Testament is clear 'you shall not live on bread alone.' Our deeper needs are not for material goods but for inner growth."
(2) "Unlimited growth. The economy, my firm, my salary should all grow. In a finite planet, this is total nonsense. This maxim of growth has brought about great ecological damage."
(3) Idolatry of the Free Market. I am in favor of a free market, but one that is set in the context of social and human conditions. We need to apply means to avoid the 'law of the jungle' in the market place."

No enumeration of religious voices raised in the cause of peace would be complete without mention of the Religious Society of Friends (Quakers), all of whom refuse to give any support whatever to the institution of war. Although they are fundamentally opposed to war as being completely contrary to Christian ethics, the Quakers are active in caring for the victims of war, and in 1947 the American Friends Service Committee and the Friends Service Council were jointly awarded the Nobel Peace Prize.

The non-violence of Mahatma Gandhi, Martin Luther King and Nelson Mandela, the writings of the Dalai Lama, the messages of Pope John Paul II and other popes, the anti-war convictions of the Quakers, and the many projects of the World Conference of Religions for Peace all illustrate the potentialities of the world's religions as powerful forces for mobilizing public opinion in the cause of peace. One hopes that the voice of religion in this cause will become still more powerful in the future. Each week, all over the world, congregations assemble and are addressed by their leaders on ethical issues. But all too often there is no mention of the astonishing and shameful contradiction between the institution of war (especially the doctrine of "massive retaliation"), and the principle of universal human brotherhood, loving and forgiving one's enemies, and returning good for evil. At a moment of history when the continued survival of civilization is in doubt because of the incompatibility of war with the existence of thermonuclear weapons, our religious leaders ought to use their enormous influence to help to solve the problem of war, which is after all an ethical problem. In this way, religion can become part of the cure of a mortal social illness rather than part of the disease - part of the answer rather than of part of the problem.

Soka Gakkai

Soka Gakkai is a large Nichiren Buddhist religious group. Its 12 million members are centered primarily in Japan, but Soka Gakkai International (SGI) has groups in 192 countries. In Japanese, the words "Soka Gakkai" mean "Value-Creating Education". The organization was started by two Japanese educators, Tsunesaburo Makiguchi and Josei Toda, both of whom were imprisoned by their government during World War II because of their opposition to militarism. Makiguchi died as a result of his imprisonment, but Josei Toda went on to found a large and vigorous educational organization dedicated to culture, humanism, world peace and nuclear abolition.

The Toda Declaration and Daisaku Ikeda's Proposals

In 1957, before a cheering audience of 50,000 young Soka Gakkai members, Josei Toda declared nuclear weapons to be an absolute evil. He said that their possession is criminal under all circumstances, and he called the young people present to work untiringly to rid the world of all nuclear weapons.

Toda was the mentor of Daisaku Ikeda, the first president SGI. Every year, President Ikeda issues a Peace Proposal, calling for international understanding and dialogue, as well as nuclear abolition, and outlining practical steps by which he believes these goals may be achieved. In his 2013 Peace Proposal, Ikeda, noted that 2015 will be the 70th anniversary of the destruction of Hiroshima, and he proposed that the NPT review conference should take place in Hiroshima, rather that in New York. He proposed that this should be followed by "an expanded global summit for a nuclear-weapon-free world"

The Hiroshima Peace Committee and the last remaining hibakushas

In Japanese the survivors of injuries from the nuclear bombing of Hiroshima and Nagasaki are called "hibakushas". Over the years, the Soka Gakkai

Fig. 10.5 In 1957, before a cheering audience of 50,000 young Soka Gakkai members, Josei Toda declared nuclear weapons to be an absolute evil. He said that their possession is criminal under all circumstances, and he called on the young people present to work untiringly to rid the world of all nuclear weapons. (SGI International.)

Hiroshima Peace Committee has published many books containing their testimonies. The most recent of these books, "A Silence Broken", contains the testimonies of 14 men, now all in their late 70's or in their 80's, who are among the last few remaining hibakushas. All 14 of these men have kept silent until now because of the prejudices against hibakushas in Japan, where they and their children are thought to be unsuitable as marriage partners because of the effects of radiation. But now, for various reasons, they have chosen to break their silence. Many have chosen to speak now because of the Fukushima disaster.

The testimonies of the hibakushas give a vivid picture of the hell-like horrors of the nuclear attack on the civilian population of Hiroshima, both in the short term and in the long term. For example, Shigeru Nonoyama, who was 15 at the time of the attack, says: "People crawling out from crumbled houses started to flee. We decided to escape to a safe place on the hill. We saw people with melted ears stuck to their cheeks, chins glued to their shoulders, heads facing in awkward positions, arms stuck to bodies, five fingers joined together and grab nothing. Those were the people fleeing. Not merely a hundred or two, The whole town was in chaos."

"I saw the noodle shop's wife leg was caught under a fallen pole, and a fire was approaching. She was screaming, 'Help me!.Help me!' There were no soldiers, no firefighters. I later heard that her husband had cut off his wife's leg with a hatchet to save her."

"Each and every scene was hell itself. I couldn't tell the difference between the men and the women. Everybody had scorched hair, burned hair, and terrible burns. I thought I saw a doll floating in a fire cistern, but it was a baby. A wife trapped under her fallen house was crying, 'Dear, please help me, help me!' Her husband had no choice but to leave her in tears."

The Catholic Church

An outstanding example of religious leadership in addressing global problems was given by H.H. Pope John Paul II. In his Christmas address on 25 December, 2002, the Pope said that efforts for peace were urgently needed "in the Middle East, to extinguish the ominous smouldering of a conflict which, with the joint efforts of all, can be avoided."

Pope John Paul II was not an exception among the Roman Catholic Popes of the 20th century. All of them have spoken strongly against the institution of war. Especially notable are H.H. Pope Paul IV who made a one-day visit to the United Nations where his speech included the words "no more war, war never again", and H.H. Pope John XXIII, author of the eloquent encyclical, *Pacem in Terris*. One can think also of the Ecumenical Council Vatican II, which denounced the arms race as an "utterly treacherous trap for humanity", questioned the method of deterrence as a safe way to preserve a steady peace, and condemned war as a "crime against God and man himself".

In his Apostolic Exhortation, "Evangelii Gaudium", Pope Francis said: "In our time humanity is experiencing a turning-point in its history, as we can see from the advances being made in so many fields. We can only praise the steps being taken to improve people's welfare in areas such as health care, education and communications. At the same time we have to remember that the majority of our contemporaries are barely living from day to day, with dire consequences. A number of diseases are spreading. The hearts of many people are gripped by fear and desperation, even in the so-called rich countries. The joy of living frequently fades, lack of respect for others and violence are on the rise, and inequality is increasingly evident. It is a struggle to live and, often, to live with precious little dignity."

"This epochal change has been set in motion by the enormous qualitative, quantitative, rapid and cumulative advances occurring in the sciences and in technology, and by their instant application in different areas of nature and of life. We are in an age of knowledge and information, which has led to new and often anonymous kinds of power."

"Just as the commandment 'Thou shalt not kill' sets a clear limit in order to safeguard the value of human life, today we also have to say 'thou shalt not' to an economy of exclusion and inequality. Such an economy kills. How can it be that it is not a news item when an elderly homeless person dies of exposure, but it is news when the stock market loses two points? This is a case of exclusion. Can we continue to stand by when food is thrown away while people are starving? This is a case of inequality. Today everything comes under the laws of competition and the survival of the fittest, where the powerful feed upon the powerless. As a consequence, masses of people find themselves excluded and marginalized: without work, without possibilities, without any means of escape."

"In this context, some people continue to defend trickle-down theories which assume that economic growth, encouraged by a free market, will inevitably succeed in bringing about greater justice and inclusiveness in the world. This opinion, which has never been confirmed by the facts, expresses a crude and naive trust in the goodness of those wielding economic power and in the sacralized workings of the prevailing economic system. Meanwhile, the excluded are still waiting."

The Dalai Lama

In his excellent and highly readable book, *Ancient Wisdom, Modern World: Ethics for the New Millennium*, the Dalai Lama writes: "..At present and for the conceivable future, the UN is the only global institution capable of influencing and formulating policy on behalf of the international community. Of course, many people criticize it on the grounds that it is ineffective, and it is true that time and again we have seen its resolutions ignored, abandoned and forgotten. Nevertheless, in spite of its shortcomings, I for one continue to have the highest regard not only for the principles on which it was founded but also for the great deal that it has achieved since its inception in 1945. We need only ask ourselves whether or not it has helped to save lives by defusing potentially dangerous situations to see that it is more than the toothless bureaucracy some people say it is. We should also consider the great work of its subsidiary organizations, such as UNICEF, United Nations High Commission for Refugees, UNESCO and the World Health Organization..."

"I see the UN, developed to its full potential, as being the proper vehicle for carrying out the wishes of humanity as a whole. As yet it is not able to do this very effectively, but we are only just beginning to see the emergence of a global consciousness (which is made possible by the communications revolution). And in spite of tremendous difficulties, we have seen it in action in numerous parts of the world, even though at the moment there may be only one or two nations spearheading these initiatives. The fact that they are seeking the legitimacy conferred by a United Nations mandate suggests a felt need for justification through collective approbation. This, in turn, I believe to be indicative of a growing sense of a single, mutually dependent, human community."

Unfulfilled responsibilities of the mainstream media

Throughout history, art was commissioned by rulers to communicate, and exaggerate, their power, glory, absolute rightness etc, to the populace. The pyramids gave visual support to the power of the Pharaoh; portraits of rulers are a traditional form of propaganda supporting monarchies; and palaces were built as symbols of power. Modern powerholders are also aware of the importance of propaganda. Thus the media are a battleground where reformers struggle for attention, but are defeated with great regularity by the wealth and power of the establishment. This is a tragedy because today there is an urgent need to make public opinion aware of the serious problems facing civilization, and the steps that are needed to solve these problems. The mass media could potentially be a great force for public education, but in general their role is not only unhelpful - it is often negative. War and conflict are blatantly advertised by television and newspapers. Meanwhile the peace movement has almost no access to the mainstream media.

Today we are faced with the task of creating a new global ethic in which loyalty to family, religion and nation will be supplemented by a higher loyalty to humanity as a whole. In case of conflicts, loyalty to humanity as a whole must take precedence. In addition, our present culture of violence must be replaced by a culture of peace. To achieve these essential goals, we urgently need the cooperation of the mass media.

The predicament of humanity today has been called "a race between education and catastrophe": Human emotions have not changed much during the last 40,000 years, and human nature still contains an element of tribalism to which nationalistic politicians successfully appeal. The completely sovereign nation-state is still the basis of our global political system. The danger in this situation is due to the fact that modern science has given us incredibly destructive weapons. Because of these weapons, the tribal tendencies in human nature and the politically fragmented structure of our world have both become dangerous anachronisms.

After the tragedies of Hiroshima and Nagasaki, Albert Einstein said, "The unleashed power of the atom has changed everything except our way of thinking, and thus we drift towards unparalleled catastrophes." We have to learn to think in a new way. Will we learn this in time to prevent disaster? When we consider the almost miraculous power of our modern electronic media, we can be optimistic. Cannot our marvelous global communication network be used to change anachronistic ways of thought and anachronistic social and political institutions in time, so that the system will not self-destruct as science and technology revolutionize our world? If they were properly used, our instantaneous global communications could give us hope.

The success of our species is built on cultural evolution, the central element of which is cooperation. Thus human nature has two sides, tribal emotions are present, but they are balanced by the human genius for cooperation. The case of Scandinavia - once war-torn, now cooperative - shows that education is able to bring out either the kind and cooperative side of human nature, or the xenophobic and violent side. Which of these shall it be? It is up to our educational systems to decide, and the mass media are an extremely important part of education. Hence the great responsibility that is now in the hands of the media.

How do the media fulfill this life-or-death responsibility? Do they give us insight? No, they give us pop music. Do they give us an understanding of the sweep of evolution and history? No, they give us sport. Do they give us an understanding of need for strengthening the United Nations, and the ways that it could be strengthened? No, they give us sit-coms and soap operas. Do they give us unbiased news? No, they give us news that has been edited to conform with the interests of the military-industrial complex and other powerful lobbys. Do they present us with the need for a just system of international law that acts on individuals? On the whole, the subject is neglected. Do they tell of of the essentially genocidal nature of nuclear weapons, and the need for their complete abolition? No, they give us programs about gardening and making food.

A consumer who subscribes to the "package" of broadcasts sold by a cable company can often search through all 35 or 45 channels without finding a single program that offers insight into the various problems that are facing the world today. What the viewer finds instead is a mixture of pro-establishment propaganda and entertainment. Meanwhile the neglected global problems are becoming progressively more severe.

In general, the mass media behave as though their role is to prevent the peoples of the world from joining hands and working to change the world and to save it from thermonuclear and environmental catastrophes. The television viewer sits slumped in a chair, passive, isolated, disempowered and stupefied. The future of the world hangs in the balance, the fate of children and grandchildren hang in the balance, but the television viewer feels no impulse to work actively to change the world or to save it. The Roman emperors gave their people bread and circuses to numb them into political inactivity. The modern mass media seem to be playing a similar role.

The alternative media

Luckily, there are alternatives to the mainstream media, available primarily on the Internet, but also to a certain extent on radio and television and in films. One can think of such alternative media figures as Thom Hartmann, Leonardo DiCaprio, Amy Goodman and Oliver Stone, or Internet sites such as Common Dreams, EcoWatch, Truthout, Countercurrents, the Danish Peace Academy website and TMS Weekly Digest. Interestingly, Bob Dylan, a longtime counterculture hero, has recently been awarded the Nobel Prize in Literature.

Johan Galtung

One of the founders of Peace Studies and Conflict Resolution as academic disciplines, is Professor Johan Galtung (1930 -). He is the author of more than a thousand articles and over a hundred books in these fields. He was also the main founder of the Peace Research Institute Oslo in 1959, and he served as its first director until 1970. Prof. Galtung established the *Journal of Peace Research* in 1964. A few years later. in 1969, he was appointed to the world's first chair in peace and conflict studies at the University of Oslo. Dr. Jan Øberg, a student of Prof. Galtung, went on to found the influential Transnational Foundation for Peace and Future Research in Lund, Sweden.

Universities Offering Peace Studies Degrees

Among the American universities and colleges offering degrees in Peace Studies and Conflict Resolution,[j] one can mention the University of Notre Dame, the University of California, Berkeley, Georgetown University, Swarthmore College, Tufts University, Wellesley College. the University of North Carolina at Chapel Hill, Colgate University, Brandeis University, the University of Texas at Austin, George Washington University, DePauw University, Smith College, Syracuse University, Southern Methodist University, Saint Johns University, American University, Marquette University, College of Saint Benedict. University of San Diego, Creighton University, Willamette University, University of Denver, Duquesne University, John Caroll University, Earlham College, George Mason University, Juniata College, University of Utah and Manhattan College. A degree program in

---

[j]http://colleges.startclass.com/d/o/Peace-Studies-and-Conflict-Resolution

Peace Studies is also offered by Clark University.[k]

In Costa Rica, the University for Peace (UPEACE)[l] offers a wide variety of courses. The departments of UPEACE include Environment and Development, International Law and Human Rights, and Peace and Conflict Studies. UPEACE also offers online education.[m]

The many educational institutions founded by Soka Gakkai International offer courses in peace studies. Among these are Soka University Japan, the Toda Institute for Global Peace, and Soka University of America.

Masters courses in peace studies and conflict resolution[n] are also offered at Universitat Oberta de Catallunya, University of Malta, Durham University, Trinity College Dublin, Alice Salimon University of Applied Sciences Berlin, University of Nicosia, Australian National University, Middlebury Institute of International Studies at Monterey, Swansea University, Aarhus University, Utrecht University, University of Kent, CIFE, University of Technology Sidney, University of Bridgeport, Duquesne University, SOAS University of London, Chapman University, SIT Graduate Institute, Kings College London, Goethe University Frankfurt, Joan B. Kroc School of Peace Studies, Johns Hopkins University School of Advanced International Studies, University of Bradford Faculty of Social and International Studies, and University of East Anglia Faculty of Social Sciences.

Jakob von Uexküll and The World Future Council

Jakob von Uexküll belongs to a brilliant family. His grandfather was a famous Baltic-German physiologist who founded the discipline of Biosemiotics. Besides being a former Member of the European Parliament and a leader of the German Green Party, von Uexküll himself founded both the Right Livelihood Award (sometimes called the Alternative Nobel Prize) and also the World Future Council.[o] Here are a few excerpts from one of his speeches to the WFC:

"Today we are heading for unprecedented dangers and conflicts, up to and including the end of a habitable planet in the foreseeable future, depriving all future generations of their right to life and the lives of preceding

---

[k] https://www2.clarku.edu/departments/peacestudies/gradprograms.cfm
[l] https://www.upeace.org/academic/academic-departments/peace-and-conflict-studies/peace-education
[m] http://www.elearning.upeace.org/
[n] http://www.masterstudies.com/Masters-Degree/Political-Science/Peace-and-Conflict-Studies/
[o] http://www.rightlivelihood.org/
http://www.worldfuturecouncil.org/
http://www.worldfuturecouncil.org/gpact/

generations of meaning and purpose."

"This apocalyptic reality is the elephant in the room. Current policies threaten temperature increases triggering permafrost melting and the release of ocean methane hydrates which would make our earth unliveable, according to research presented by the British Government Met office at the Paris Climate Conference."

"The myth that climate change is conspiracy to reduce freedom is spread by a powerful and greedy elite which has largely captured governments to preserve their privileges in an increasingly unequal world."

"Long before that point, our prosperity, security, culture and identity will disintegrate. A Europe unable to cope with a few million war refugees will collapse under the weight of tens or even hundreds of millions of climate refugees."

Some of the major organizations in the peace movement

Among the many organizations working actively for peace education, one can think of the following:[p]

- The Nuclear Age Peace Foundation
- The International Peace Bureau
- International Physicians for the Prevention of Nuclear War
- Greenpeace
- Pugwash Conferences on Science and World Affairs
- Global Zero
- Abolition 2000
- Mayors for Peace
- International Campaign to Abolish Nuclear Weapons (ICAN)
- World Association of World Federalists
- Campaign for Nuclear Disarmament
- Pax Christi
- American Friends Service Committee
- The Society of Prayer for World Peace
- The Danish Peace Academy
- International Network of Engineers and Scientists for Global Responsibility (INES)
- War Resistors International
- Stockholm International Peace Research Institute (SIPRI)
- Peace Research Institute, Oslo
- Soka Gakkai International (SGI)
- Hiroshima Peace Memorial Museum

---
[p]The list is in no particular order, and is by no means complete.

- Transcend International
- Transnational Foundation for Peace and Future Research (TFF)
- Gandhi International Institute for Peace
- Bertrand Russell Peace Foundation
- Lawyers' Committee on Nuclear Policy
- Parliamentarians for Nuclear Nonproliferation and Disarmament
- Nuclear Abolition Forum
- Code Pink
- Jewish Voice for Peace
- Women's International League for Peace and Freedom
- World Beyond War
- Global Security Institute
- The Council of Canadians
- International Fellowship of Reconciliation
- Physicians for Social Responsibility
- Anglican Pacifist Fellowship
- Institute for Economics and Peace
- Veterans Against War
- The Elders
- Nobel Women's Initiative
- Peace Pledge Union
- United Nations Integrated Peacebuilding Office
- The Committee for a Sane Nuclear Policy
- Seeds of Peace
- Middle Powers Initiative

Fig. 10.6   Italian Nobel Peace Prize winner Ernesto Teodoro Moneta first adopted the motto In Varietate Concordia/In Varietate Unitas. (Public domain.)

In the world as it is, young people are indoctrinated with nationalism. History is taught in such a way that one's own nation is seen as heroic and in the right, while other nations are seen as inferior or as enemies.

In the world as it could be, young people would be taught to feel loyalty to humanity as a whole. History would be taught in such a way as to emphasize the contributions that all nations and all races have made to the common cultural heritage of humanity.

Fig. 10.7 Journalists at work in Montreal in the 1940s. Today, the mass media form an extremely important part of our total educational system. (Public domain.)

In the world as it is, modern communications media, such as television, films and newspapers, have an enormous influence on public opinion. However, this influence is only rarely used to build up international understanding and mutual respect.

In the world as it could be, mass communications media would be more fully used to bridge human differences. Emphasis would be shifted from the sensational portrayal of conflicts to programs that widen our range of sympathy and understanding.

Fig. 10.8 A painting showing the Tower of Babel. (Public domain.)

In the world as it is, international understanding is blocked by language barriers.

In the world as it could be, an international language would be selected, and every child would be taught it as a second language.

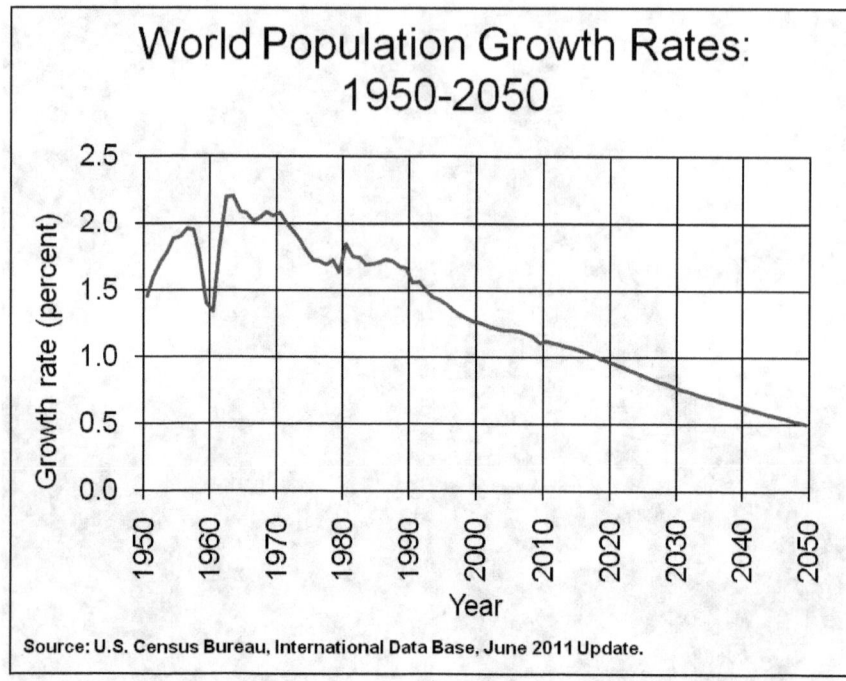

Fig. 10.9 In the sustainable society of the future, work related to culture and education will play an increasingly important role. Resource-using and pollution-producing industrial production will necessarily become less important as non-renewable resources vanish. (Public domain.)

In the world as it is, young people are often faced with the prospect of unemployment. This is true both in the developed countries, where automation and recession produce unemployment, and in the developing countries, where unemployment is produced by overpopulation and by lack of capital.

In the world as it could be, the idealism and energy of youth would be fully utilized by the world community to combat illiteracy and disease, and to develop agriculture and industry in the Third World. These projects would be financed by the UN using revenues derived from taxing international currency transactions.

Suggestions for further reading

(1) B. Broms, *United Nations*, Suomalainen Tiedeakatemia, Helsinki, (1990).
(2) S. Rosenne, *The Law and Practice at the International Court*, Dordrecht, (1985).
(3) S. Rosenne, *The World Court - What It Is and How It Works*, Dordrecht, (1995).
(4) J. D'Arcy and D. Harris, *The Procedural Aspects of International Law (Book Series), Volume 25*, Transnational Publishers, Ardsley, New York, (2001).
(5) H. Cullen, *The Collective Complaints Mechanism Under the European Social Charter*, European Law Review, Human Rights Survey, p. 18-30, (2000).
(6) S.D. Bailey, *The Procedure of the Security Council*, Oxford, (1988).
(7) R.A. Akindale, *The Organization and Promotion of World Peace: A Study of Universal-Regional Relationships*, Univ. Toronto Press, Toronto, Ont., (1976).
(8) J.S. Applegate, *The UN Peace Imperative*, Vantage Press, New York, (1992).
(9) S.E. Atkins, *Arms Control, Disarmament, International Security and Peace: An Annotated Guide to Sources, 1980-1987*, Clio Press, Santa Barbara, CA, (1988).
(10) N. Ball and T. Halevy, *Making Peace Work: The Role of the International Development Community*, Overseas Development Council, Washington DC, (1996).
(11) F. Barnaby, Ed., *The Gaia Peace Atlas: Survival into the Third Millennium*, Doubleday, New York, (1988)
(12) J.H. Barton, *The Politics of Peace: An Evaluation of Arms Control*, Stanford Univ. Press, Stanford, CA, (1981).
(13) W. Bello, *Visions of a Warless World*, Friends Committee on National Education Fund, Washington DC, (1986).
(14) A. Boserup and A. Mack, *Abolishing War: Cultures and Institutions; Dialogue with Peace Scholars Elise Boulding and Randall Forsberg*, Boston Research Center for the Twenty-first Century, Cambridge, MA, (1998).
(15) E. Boulding et al., *Bibliography on World Conflict and Peace*, Westview Press, Boulder, CO, (1979).
(16) E. Boulding et al., Eds., *Peace, Culture and Society: Transnational Research Dialogue*, Westview Press, Boulder, CO, (1991).
(17) A.T. Bryan et al., Eds., *Peace, Development and Security in the Caribean*, St. Martins Press, New York, (1988).

(18) A.L. Burns and N. Heathcote, *Peace-Keeping by UN Forces from Suez to Congo*, Praeger, New York, (1963).
(19) F. Capra and C. Spretnak, *Green Politics: The Global Promise*, E.P. Dutton, New York, (1986).
(20) N. Carstarphen, *Annotated Bibliography of Conflict Analysis and Resolution*, Inst. for Conflict Analysis and Resolution, George Mason Univ., Fairfax, VA, (1997).
(21) N. Chomsky, *Peace in the Middle East? Reflections on Justice and Nationhood*, Vintage Books, New York, (1974).
(22) G. Clark and L. Sohn, *World Peace Through World Law*, World Without War Pubs., Chicago, IL, (1984).
(23) K. Coates, *Think Globally, Act Locally: The United Nations and the Peace Movements*, Spokesman Books, Philadelphia, PA, (1988).
(24) G. De Marco and M. Bartolo, *A Second Generation United Nations: For Peace and Freedom in the 20th Century*, Colombia Univ. Press, New York, (1997).
(25) F.M. Deng and I.W. Zartman, Eds., *Conflict Resolution in Africa*, Brookings Institution, Washington, DC, (1991).
(26) W. Desan, *Let the Future Come: Perspectives for a Planetary Peace*, Georgetown Univ. Press, Washington, DC, (1987).
(27) D. Deudney, *Whole Earth Security. A Geopolitics of Peace*, Worldwatch paper 55. Worldwatch Institute, Washington, DC, (1983).
(28) A.J. Donovan, *World Peace? A Work Based on Interviews with Foreign Diplomats*, A.J. Donovan, New York, (1986).
(29) R. Duffey, *International Law of Peace*, Oceania Pubs., Dobbs Ferry, NY, (1990).
(30) L.J. Dumas, *The Socio-Economics of Conversion From War to Peace*, M.E. Sharpe, Armonk, NY, (1995).
(31) W. Durland, *The Illegality of War*, National Center on Law and Pacifism, Colorado Springs, CO, (1982).
(32) F. Esack, *Qur'an, Liberation and Pluralism: An Islamic Perspective on Interreligious Solidarity Against Oppression*, Oxford Univ. Press, London, (1997).
(33) I. Hauchler and P.M. Kennedy, Eds., *Global Trends: The World Almanac of Development and Peace*, Continuum Pubs., New York, (1995).
(34) H.B. Hollins et al., *The Conquest of War: Alternative Strategies for Global Security*, Westview Press, Boulder, CO, (1989).
(35) H.J. Morgenthau, *Peace, Security and the United Nations*, Ayer Pubs., Salem, NH, (1973).
(36) C.C. Moskos, *Peace Soldiers: The Sociology of a United Nations Military Force*, Univ. of Chicago Press, Chicago, IL, (1976).

(37) L. Pauling, *Science and World Peace*, India Council for Cultural Relations, New Delhi, India, (1967).
(38) C. Peck, *The United Nations as a Dispute Resolution System: Improving Mechanisms for the Prevention and Resolution of Conflict*, Kluwer, Law and Tax, Cambridge, MA, (1996).
(39) D. Pepper and A. Jenkins, *The Geography of Peace and War*, Basil Blackwell, New York, (1985).
(40) J. Perez de Cuellar, *Pilgrimage for Peace: A Secretary General's Memoir*, St. Martin's Press, New York, (1997).
(41) R. Pickus and R. Woito, *To End War: An Introduction to the Ideas, Books, Organizations and Work That Can Help*, World Without War Council, Berkeley, CA, (1970).
(42) S.R. Ratner, *The New UN Peacekeeping: Building Peace in Lands of Conflict after the Cold War*, St. Martins Press, New York, (1995).
(43) I.J. Rikhye and K. Skjelsbaek, Eds., *The United Nations and Peacekeeping: Results, Limitations and Prospects: The Lessons of 40 Years of Experience*, St. Martins Press, New York, (1991).
(44) J. Rotblat, Ed., *Scientists in Quest for Peace: A History of the Pugwash Conferences*, MIT Press, Cambridge, MA, (1972).
(45) J. Rotblat, Ed., *Scientists, The Arms Race, and Disarmament*, Taylor and Francis, Bristol, PA, (1982).
(46) J. Rotblat, Ed., *Striving for Peace, Security and Development in the World*, World Scientific, River Edge, NJ, (1991).
(47) J. Rotblat, Ed., *Towards a War-Free World*, World Scientific, River Edge, NJ, (1995).
(48) J. Rotblat, Ed., *Nuclear Weapons: The Road to Zero*, Westview, Boulder, CO, (1998).
(49) J. Rotblat and L. Valki, Eds., *Coexistance, Cooperation and Common Security*, St. Martins Press, New York, (1988).
(50) United Nations, *Peaceful Settlement of Disputes between States: A Select Bibliography*, United Nations, New York, (1991).
(51) United States Arms Control and Disarmament Agency, *Arms Control and Disarmament Agreements: Texts and Histories of Negotiations*, USACDA, Washington, DC, (updated annually)
(52) D. Fahrni, *An Outline History of Switzerland - From the Origins to the Present Day*, Pro Helvetia Arts Council of Switzerland, Zurich, (1994).
(53) J.M. Luck, *A History of Switzerland*, Sposs, Palo Alto, CA, (1985).

Chapter 11

The Future of International Law

With law shall our land be built up, but with lawlessness laid waste.

Njal's Saga, Iceland, c 1270

To coerce the states is one of the maddest projects that was ever devised... Can any reasonable man be well disposed towards a government which makes war and carnage the only means of supporting itself, a government that can exist only by the sword? Every such war must involve the innocent with the guilty. The single consideration should be enough to dispose every peaceable citizen against such government... What is the cure for this great evil? Nothing, but to enable the... laws to operate on individuals, in the same manner as those of states do.

Alexander Hamilton

Introduction

After the invention of agriculture, roughly 10,000 years ago, humans began to live in progressively larger groups, which were sometimes multi-ethnic. In order to make towns, cities and finally nations function without excessive injustice and violence, both ethical and legal systems were needed. Today, in an era of global economic interdependence, instantaneous worldwide communication and all-destroying thermonuclear weapons, we urgently need new global ethical principles and a just and enforcible system of international laws.

What is law?

The principles of law, ethics, politeness and kindness function in slightly different ways, but all of these behavioral rules help human societies to function in a cohesive and trouble-free way. Law is the most coarse. The mesh is made finer by ethics, while the rules of politeness and kindness fill in the remaining gaps.

Legal systems began at a time at a time when tribal life was being replaced by life in villages, towns and cities. One of the oldest legal documents that we know of is a code of laws enacted by the Babylonian king Hammurabi in about 1754 BC. It consists of 282 laws, with scaled punishments, governing household behavior, marriage, divorce, paternity, inheritance, payments for services, and so on. An ancient 2.24 meter stele inscribed with Hammurabi's Code can be seen in the Louvre. The laws are written in the Akkadian language, using cuneiform script.

Humanity's great ethical systems also began during a period when the social unit was growing very quickly. It is an interesting fact that many of history's greatest ethical teachers lived at a time when the human societies were rapidly increasing in size. One can think, for example of Moses, Confucius, Lao-Tzu, Gautama Buddha, the Greek philosophers, and Jesus. Muhammad came slightly later, but he lived and taught at a time when tribal life was being replaced by city life in the Arab world. During the period when these great teachers lived, ethical systems had become necessary to over-write raw inherited human emotional behavior patterns in such a way that increasingly large societies could function in a harmonious and cooperative way, with a minimum of conflicts.

Magna Carta, 1215

2015 marks the 800th anniversary of the Magna Carta, which is considered to be the foundation of much of our modern legal system. It was drafted

Fig. 11.1 Figures at top of stele "fingernail" above Hammurabi's code of laws. (Public domain.)

by the Archbishop of Canterbury to make peace between the unpopular Norman King John of England and a group of rebel barons. The document promised the protection of church rights, protection for the barons from illegal imprisonment, access to swift justice, and limitations feudal payments to the Crown. It was renewed by successive English sovereigns, and its protection against illegal imprisonment and provisions for swift justice were extended from the barons to ordinary citizens. It is considered to be the basis for British constitutional law, and in 1789, it influenced the drafting of the Constitution of the United States. Lord Denning described the Magna Carta as "the greatest constitutional document of all times: the foundation of the freedom of the individual against the arbitrary authority of the despot".

Fig. 11.2 King John is forced to sign the Magna Carta. (Public domain.)

The English Bill of Rights, 1689

When James II was overthrown by the Glorious Revolution the Dutch stadholder William III of Orange-Nassau and his wife, Mary II of England were invited to be joint sovereigns of England. The Bill of Rights was originally part of the invitation, informing the couple regarding the limitations that would be imposed on their powers. Later the same year, it was incorporated into English law. The Bill of Rights guaranteed the supremacy of Parliament over the monarch. It forbid cruel and unusual punishments, excessive bail and excessive fines. Freedom of speech and free elections were also guaranteed, and a standing army in peacetime was forbidden without the explicit consent of Parliament. The Bill of Rights was influenced by the writings of the Liberal philosopher, John Locke (1632-1704).

The United States Constitution and Bill of Rights, 1789

The history of the Federal Constitution of the United States is an interesting one. It was preceded by the Articles of Confederation, which were written by the Second Continental Congress between 1776 and 1777, but it soon became clear that Confederation was too weak a form of union for a collection of states.

George Mason, one of the drafters of the Federal Constitution, believed that "such a government was necessary as could directly operate on individuals, and would punish those only whose guilt required it", while another drafter, James Madison, wrote that the more he reflected on the use of force, the more he doubted "the practicality, the justice and the efficacy of it when applied to people collectively, and not individually."

Finally, Alexander Hamilton, in his Federalist Papers, discussed the Articles of Confederation with the following words: "To coerce the states is one of the maddest projects that was ever devised... Can any reasonable man be well disposed towards a government which makes war and carnage the only means of supporting itself, a government that can exist only by the sword? Every such war must involve the innocent with the guilty. The single consideration should be enough to dispose every peaceable citizen against such government... What is the cure for this great evil? Nothing, but to enable the... laws to operate on individuals, in the same manner as those of states do."

In other words, the essential difference between a confederation and a federation, both of them unions of states, is that a federation has the power to make and to enforce laws that act on individuals, rather than attempting to coerce states (in Hamilton's words, "one of the maddest projects that was ever devised.") The fact that a confederation of states was found to be far too weak a form of union is especially interesting because our present United Nations is a confederation. We are at present attempting to coerce states with sanctions that are "applied to people collectively and not individually." The International Criminal Court, which we will discuss below, is a development of enormous importance, because it acts on individuals, rather than attempting to coerce states.

There are many historical examples of successful federations; but in general, unions of states based on the principle of confederation have proved to be too weak. Probably our best hope for the future lies in gradually reforming and strengthening the United Nations, until it becomes a federation.

In the case of the Federal Constitution of the United States, there were Anti-Federalists who opposed its ratification because they feared that it would be too powerful. Therefore, on June 8, 1789, James Madison introduced in the House of Representatives a series of 39 amendments to the constitution, which would limit the government's power. Of these, only amendments 3 to12 were adopted, and these have become known collectively as the Bill of Rights.

Fig. 11.3 James Madison, wrote that the more he reflected on the use of force, the more he doubted "the practicality, the justice and the efficacy of it when applied to people collectively, and not individually." He later introduced the Constitutional amendments that became the U.S. Bill of Rights. (Public domain.)

Of the ten amendments that constitute the original Bill of Rights, we should take particular notice of the First, Fourth and Sixth, because they have been violated repeatedly and grossly by the present government of the United States.

The First Amendment requires that "Congress shall make no law respecting an establishment of religion, or prohibiting the free exercise thereof; or abridging the freedom of speech, or of the press; or the right of the people peaceably to assemble, and to petition the Government for a redress of grievances." The right to freedom of speech and freedom of the press has been violated by the punishment of whistleblowers. The right to assemble peaceably has also been violated repeatedly and brutally by the present government's militarized police.

The Fourth Amendment states that "The right of the people to be secure in their persons, houses, papers, and effects, against unreasonable searches and seizures, shall not be violated, and no Warrants shall issue, but upon probable cause, supported by Oath or affirmation, and particularly describing the place to be searched, and the persons or things to be seized." It is hardly necessary to elaborate on the U.S. Government's massive violations of the Fourth Amendment. Edward Snowden's testimony has revealed a huge secret industry carrying out illegal and unwarranted searches and seizures of private data, not only in the United States, but also throughout the world. This data can be used to gain power over citizens and leaders through blackmail. True democracy and dissent are thereby eliminated.

The Sixth Amendment requires that "In all criminal prosecutions, the accused shall enjoy the right to a speedy and public trial, by an impartial jury of the State and district wherein the crime shall have been committed, which district shall have been previously ascertained by law, and to be informed of the nature and cause of the accusation; to be confronted with the witnesses against him; to have compulsory process for obtaining witnesses in his favor, and to have the Assistance of Counsel for his defense." This constitutional amendment has also been grossly violated.

In the context of federal unions of states, the Tenth Amendment is also interesting. This amendment states that "The powers not delegated to the United States by the Constitution, nor prohibited by it to the States, are reserved to the States respectively, or to the people." We mentioned above that historically, federations have been very successful. However, if we take the European Union as an example, it has had some problems connected with the principle of subsidiarity, according to which as few powers as possible should be decided centrally, and as many issues as possible should be decided locally. The European Union was originally designed as a free trade area, and because of its history commercial considerations have trumped environmental ones. The principle of subsidiarity has not been followed, and enlightened environmental laws of member states have been declared to be illegal by the EU because they conflicted with free trade. These are difficulties from which we can learn as we contemplate the conversion of the United Nations into a federation.

The United States Bill of Rights was influenced by John Locke and by the French philosophers of the Enlightenment. The French Declaration of the Rights of Man (August, 1789) was almost simultaneous with the U.S. Bill of Rights.

We can also see the influence of Enlightenment philosophy in the wording of the U.S. Declaration of independence (1776): "We hold these truths to be self-evident, that all men are created equal, that they are endowed by their Creator with certain unalienable Rights, that among these are Life, Liberty and the pursuit of Happiness.–That to secure these rights, Governments are instituted among Men, deriving their just powers from the consent of the governed..." Another criticism that can be leveled against the present government of the United States is that its actions seem to have nothing whatever to do with the consent of the governed, not to mention the violations of the rights to life, liberty and the pursuit of happiness implicit in extrajudicial killings.

Kellogg-Briand Pact, 1928

World War I was a catastrophe that still casts a dark shadow over the future of humanity. It produced enormous suffering, brutalization of values, irreparable cultural loss, and a total of more than 37 million casualties, military and civilian. Far from being the "war to end war", the conflict prepared the way for World War II, during which nuclear weapons were developed; and these now threaten the existence the of human species and much of the biosphere.

After the horrors of World War I, the League of Nations was set up in the hope of ending the institution of war forever. However, many powerful nations refused to join the League, and it withered. Another attempt to outlaw war was made in 1928. in the form of a pact named after its authors, U.S. Secretary of State, Frank B. Kellogg and French Foreign Minister Astrid Briand. The Kellogg-Briand Pact is formally called the General Treaty for the Renunciation of War as an Instrument of National Policy. It was ultimately ratified by 62 Nations, including the United States (by a Senate vote of 85 to 1). Although frequently violated, the Pact remains in force today, establishing a norm which legally outlaws war.

United Nations Charter, 1945

The Second World War was even more disastrous than the First. Estimates of the total number of people who died as a result of the war range between 50 million and 80 million. With the unspeakable suffering caused by the war fresh in their minds, representatives of the victorious allied countries assembled in San Francisco to draft the charter of a global organization which they hoped would end the institution of war once and for all.

The Preamble to the United Nations Charter starts with the words: "We, the peoples of the United Nations, determined to save succeeding generations from the scourge of war, which twice in our lifetime has brought untold sorrow to mankind; and to unite our strength to maintain international peace and security; and to ensure, by the acceptance of principles and the institution of methods, that armed force shall not be used, save in the common interest; and to employ international machinery for the promotion of the economic and social advancement of all peoples, have resolved to combine our efforts to accomplish these aims."

Article 2 of the UN Charter requires that "All members shall refrain in their international relations from the threat or use of force against the territorial integrity or political independence of any state." This requirement is somewhat qualified by Article 51, which says that "Nothing in the present Charter shall impair the inherent right of individual or collective self-defense if an armed attack occurs against a Member of the United Nations, until the Security Council has taken measures necessary to maintain international peace and security." Thus, in general, war is illegal under the UN Charter. Self-defense against an armed attack is permitted, but only for a limited time, until the Security Council has had time to act. The United Nations Charter does not permit the threat or use of force in preemptive wars, or to produce regime changes, or for so-called "democratization", or for the domination of regions that are rich in oil.[a]

Clearly, the United Nations Charter aims at abolishing the institution of war once and for all; but the present Charter has proved to be much too weak to accomplish this purpose, since it is a confederation of the member states rather than a federation. This does not mean that that our present United Nations is a failure. Far from it! The UN has achieved almost universal membership, which the League of Nations failed to do. The Preamble to the Charter speaks of " the promotion of the economic and social advancement of all peoples", and UN agencies, such as the World Health Organization, the Food and Agricultural Organization and UNESCO, have worked very effectively to improve the lives of people throughout the world. Furthermore, the UN has served as a meeting place for diplomats from all countries, and many potentially serious conflicts have been resolved by informal conversations behind the scenes at the UN. Finally, although often unenforceable, resolutions of the UN General Assembly and declarations by the Secretary General have great normative value.

---

[a] http://www.un.org/en/documents/charter/preamble.shtml

Fig. 11.4 Clearly, the United Nations Charter aims at abolishing the institution of war once and for all. (Public domain.)

When we think of strengthening and reforming the UN, then besides giving it the power to make and enforce laws that are binding on individuals, we should also consider giving it an independent and reliable source of income. As it is, rich and powerful nations seek to control the UN by means of its purse strings: They give financial support only to those actions that are in their own interests.

A promising solution to this problem is the so-called "Tobin tax", named after the Nobel-laureate economist James Tobin of Yale University. Tobin proposed that international currency exchanges should be taxed at a rate between 0.1 and 0.25 percent. He believed that even this extremely low rate of taxation would have the beneficial effect of damping speculative transactions, thus stabilizing the rates of exchange between currencies. When asked what should be done with the proceeds of the tax, Tobin said, almost as an afterthought, "Let the United Nations have it."

The volume of money involved in international currency transactions is so enormous that even the tiny tax proposed by Tobin would provide the United Nations with between 100 billion and 300 billion dollars annually. By strengthening the activities of various UN agencies, the additional income would add to the prestige of the United Nations and thus make the organization more effective when it is called upon to resolve international political conflicts. The budgets of UN agencies, such as the World Health Organization, the Food and Agricultural Organization, UNESCO and the UN Development Programme, should not just be doubled but should be multiplied by a factor of at least twenty.

With increased budgets the UN agencies could sponsor research and other actions aimed at solving the world's most pressing problems: AIDS, drug-resistant infections diseases, tropical diseases, food insufficiencies, pollution, climate change, alternative energy strategies, population stabilization, peace education, as well as combating poverty, malnutrition, illiteracy, lack of safe water and so on. Scientists would would be less tempted to find jobs with arms-related industries if offered the chance to work on idealistic projects. The United Nations could be given its own television channel, with unbiased news programs, cultural programs, and "State of the World" addresses by the UN Secretary General.

In addition, the voting system of the United Nations General Assembly needs to be reformed, and the veto power in the Security Council needs to be abolished.

International Court of Justice, 1946

The International Court of Justice (ICJ) is the judicial arm of the United Nations. It was established by the UN Charter in 1945, and it began to function in 1946. The ICJ is housed in the Peace Palace in the Hague, a beautiful building constructed with funds donated by Andrew Carnegie. Since 1946, the ICJ has dealt with only 161 cases. The reason for this low number is that only disputes between nations are judged, and both the countries involved in a dispute have to agree to abide by the Court's jurisdiction before the case can be accepted.

Besides acting as an arbitrator in disputes between nations, the ICJ also gives advisory opinions to the United Nations and its agencies. An extremely important judgment of this kind was given in 1996: In response to questions put to it by WHO and the UN General Assembly, the Court ruled that "the threat and use of nuclear weapons would generally be contrary to the rules of international law applicable in armed conflict, and particularly the principles and rules of humanitarian law." The only possible exception to this general rule might be "an extreme circumstance of self-defense, in which the very survival of a state would be at stake". But the Court refused to say that even in this extreme circumstance the threat or use of nuclear weapons would be legal. It left the exceptional case undecided. In addition, the World Court added unanimously that "there exists an obligation to pursue in good faith and bring to a conclusion negotiations leading to nuclear disarmament in all its aspects under strict international control."

This landmark decision has been criticized by the nuclear weapon states as being decided "by a narrow margin", but the structuring of the vote made the margin seem more narrow than it actually was. Seven judges voted against Paragraph 2E of the decision (the paragraph which states that the threat or use of nuclear weapons would be generally illegal, but which mentions as a possible exception the case where a nation might be defending itself from an attack that threatened its very existence.) Seven judges voted for the paragraph, with the President of the Court, Muhammad Bedjaoui of Algeria casting the deciding vote. Thus the Court adopted it, seemingly by a narrow margin. But three of the judges who voted against 2E did so because they believed that no possible exception should be mentioned! Thus, if the vote had been slightly differently structured, the result would have be ten to four.

Of the remaining four judges who cast dissenting votes, three represented nuclear weapons states, while the fourth thought that the Court ought not to have accepted the questions from WHO and the UN. However Judge Schwebel from the United States, who voted against Paragraph 2E, nevertheless added, in a separate opinion, "It cannot be accepted that the use of nuclear weapons on a scale which would, or could, result in the deaths of many millions in indiscriminate inferno and by far-reaching fallout, have pernicious effects in space and time, and render uninhabitable much of the earth, could be lawful."

Judge Higgins from the UK, the first woman judge in the history of the Court, had problems with the word "generally" in Paragraph 2E and therefore voted against it, but she thought that a more profound analysis might have led the Court to conclude in favor of illegality in all circumstances.

Judge Fleischhauer of Germany said, in his separate opinion, "The nuclear weapon is, in many ways, the negation of the humanitarian considerations underlying the law applicable in armed conflict and the principle of neutrality. The nuclear weapon cannot distinguish between civilian and military targets. It causes immeasurable suffering. The radiation released by it is unable to respect the territorial integrity of neutral States."

President Bedjaoui, summarizing the majority opinion, called nuclear weapons "the ultimate evil", and said "By its nature, the nuclear weapon, this blind weapon, destabilizes humanitarian law, the law of discrimination in the use of weapons... The ultimate aim of every action in the field of nuclear arms will always be nuclear disarmament, an aim which is no longer Utopian and which all have a duty to pursue more actively than ever."

Nuremberg Principles, 1947

In 1946, the United Nations General Assembly unanimously affirmed "the principles of international law recognized by the Charter of the Nuremberg Tribunal and the judgment of the Tribunal". The General Assembly also established an International Law Commission to formalize the Nuremberg Principles. The result was a list that included Principles VI, which is particularly important in the context of the illegality of NATO:

Principle VI: The crimes hereinafter set out are punishable as crimes under international law:

(a) Crimes against peace: (I) Planning, preparation, initiation or waging of a war of aggression or a war in violation of international treaties, agreements or assurances; (II) Participation in a common plan or conspiracy for accomplishment of any of the acts mentioned under (I).

Robert H. Jackson, who was the chief United States prosecutor at the Nuremberg trials, said that "To initiate a war of aggression is therefore not only an international crime; it is the supreme international crime, differing from other war crimes in that it contains within itself the accumulated evil of the whole." Furthermore, the Nuremberg principles state that "The fact that a person acted pursuant to order of his Government or of a superior does not relieve him from responsibility under international law, provided a moral choice was in fact possible to him." The training of soldiers is designed to make the trainees into automatons, who have surrendered all powers of moral judgment to their superiors. The Nuremberg Principles put the the burden of moral responsibility squarely back where it ought to be: on the shoulders of the individual.

The Universal Declaration of Human Rights, 1948

On December 10, 1948, the General Assembly of the United Nations adopted a Universal Declaration of Human Rights. 48 nations voted for adoption, while 8 nations abstained from voting. Not a single state voted

Fig. 11.5 In 1946, the United Nations General Assembly unanimously affirmed "the principles of international law recognized by the Charter of the Nuremberg Tribunal and the judgment of the Tribunal". The General Assembly also established an International Law Commission to formalize the Nuremberg Principles. The photo shows defendants, including Hermann Göring, at the Nuremberg trials. (Public domain.)

against the Declaration. In addition, the General Assembly decided to continue work on the problem of implementing the Declaration. The Preamble to the document stated that it was intended "as a common standard of achievement for all peoples and nations, to the end that every individual and every organ of society, keeping this Declaration constantly in mind, shall strive by teaching and education to promote respect for these rights and freedoms."

Articles 1 and 2 of the Declaration state that "all human beings are born free and equal in dignity and in rights", and that everyone is entitled to the rights and freedoms mentioned in the Declaration without distinctions of any kind. Neither race color, sex, language, religion, political or other opinion, national or social origin, property or social origin must make a difference. The Declaration states that everyone has a right to life, liberty and security of person and property. Slavery and the slave trade are prohibited, as well as torture and cruel, inhuman or degrading punishments. All people must be equal before the law, and no person must be subject to arbitrary arrest, detention or exile. In criminal proceedings an accused person must be presumed innocent until proven guilty by an impartial public hearing where all necessary provisions have been made for the defense of the accused.

No one shall be subjected to interference with his privacy, family, home or correspondence. Attacks on an individual's honor are also forbidden. Everyone has the right of freedom of movement and residence within the borders of a state, the right to leave any country, including his own, as well as the right to return to his own country. Every person has the right to a nationality and cannot be arbitrarily deprived of his or her nationality.

All people of full age have a right to marry and to establish a family. Men and women have equal rights within a marriage and at its dissolution, if this takes place. Marriage must require the full consent of both parties.

The Declaration also guarantees freedom of religion, of conscience, and of opinion and expression, as well as freedom of peaceful assembly and association. Everyone is entitled to participate in his or her own government, either directly or through democratically chosen representatives. Governments must be based on the will of the people, expressed in periodic and genuine elections with universal and equal suffrage. Voting must be secret.

Everyone has the right to the economic, social and cultural conditions needed for dignity and free development of personality. The right to work is affirmed. The job shall be of a person's own choosing, with favorable conditions of work, and remuneration consistent with human dignity, supplemented if necessary with social support. All workers have the right to form and to join trade unions.

Article 25 of the Declaration states that everyone has the right to an adequate standard of living, including food, clothing, housing and medical care, together with social services. All people have the right to security in the event of unemployment, sickness, disability, widowhood or old age. Expectant mothers are promised special care and assistance, and children, whether born in or out of wedlock, shall enjoy the same social protection. Everyone has the right to education, which shall be free in the elementary stages. Higher education shall be accessible to all on the basis of merit. Education must be directed towards the full development of the human personality and to strengthening respect for human rights and fundamental freedoms. Education must promote understanding, tolerance, and friendship among all nations, racial and religious groups, and it must further the activities of the United Nations for the maintenance of peace.

A supplementary document, the Convention on the Rights of the Child, was adopted by the United Nations General Assembly on the 12th of December, 1989. Furthermore, in July 2010, the General Assembly passed a resolution affirming that everyone has the right to clean drinking water and proper sanitation.

Many provisions of the Universal Declaration of Human Rights, for example Article 25, might be accused of being wishful thinking. In fact, Jean Kirkpatrick, former US Ambassador to the UN, cynically called the Declaration "a letter to Santa Claus". Nevertheless, like the Millennium Development Goals, the Universal Declaration of Human Rights has great value in defining the norms towards which the world ought to be striving.

It is easy to find many examples of gross violations of basic human rights that have taken place in recent years. Apart from human rights violations connected with interventions of powerful industrial states in the internal affairs of third world countries, there are many cases where governmental forces in the less developed countries have violated the human rights of their own citizens. Often minority groups have been killed or driven off their land by those who coveted the land, as was the case in Guatemala in 1979, when 1.5 million poor Indian farmers were forced to abandon their villages and farms and to flee to the mountains of Mexico in order to escape murderous attacks by government soldiers. The blockade of Gaza and extrajudicial killing by governments must also be regarded as blatant human rights violations, and there are many recent examples of genocide.

Wars in general, and in particular, the use of nuclear weapons, must be regarded as gross violations of human rights. The most basic human right is the right to life; but this is right routinely violated in wars. Most of the victims of recent wars have been civilians, very often children and women. The use of nuclear weapons must be regarded as a form of genocide, since they kill people indiscriminately, babies, children, young adults in their prime, and old people, without any regard for guilt or innocence.

Geneva Conventions, 1949

According to Wikipedia, "The Geneva Conventions comprise four treaties, and three additional protocols, that establish the standards if international law for the humanitarian treatment of war. The singular term, Geneva Convention, usually denotes the agreements of 1949, negotiated in the aftermath of the Second World War (1939-1945), which updated the terms of the first three treaties (1864, 1906, 1929) and added a fourth. The Geneva Conventions extensively defined the basic rights of wartime prisoners (civilians and military personnel); established protection for the wounded; and established protections for civilians in and around a war-zone. The treaties if 1949 were ratified, in whole or with reservations, by 196 countries."

In a way, one might say that the Geneva Conventions are an admission of defeat by the international community. We tried to abolish war entirely through the UN Charter, but failed because the Charter was too weak.

Under the Fourth Geneva Convention, collective punishment is war crime. Article 33 states that "No protected person may be punished for an offense that he or she did not personally commit." Articles 47-78 also impose substantial obligations on occupying powers, with numerous provisions for the general welfare of the inhabitants of an occupied territory. Thus Israel violated the Geneva Conventions by its collective punishment of the civilian population of Gaza in retaliation for largely ineffective Hamas rocket attacks. The larger issue, however, is the urgent need for lifting of Israel's brutal blockade of Gaza, which has created what Noam Chomsky calls the "the world's largest open-air prison". This blockade violates the Geneva conventions because Israel, as an occupying power, has the duty of providing for the welfare of the people of Gaza.

Nuclear Non-Proliferation Treaty, 1968

In the 1960's, negotiations were started between countries that possessed nuclear weapons, and others that did not possess them, to establish a treaty that would prevent the spread of these highly dangerous weapons, but which would at the same time encourage cooperation in the peaceful uses of nuclear energy. The resulting treaty has the formal title Treaty on the Non-Proliferation of Nuclear Weapons (abbreviated as the NPT). The treaty also aimed at achieving general and complete disarmament. It was opened for signature in 1968, and it entered into force on the 11th of May, 1970.

190 parties have joined the NPT, and more countries have ratified it than any other arms limitation agreement, an indication of the Treaty's great importance. Four countries outside the NPT have nuclear weapons: India, Pakistan, North Korea and Israel. North Korea had originally joined the NPT, but it withdrew in 2003. The NPT has three main parts or "pillars", 1) non-proliferation, 2) disarmament, and 3) the right to peaceful use of nuclear technology. The central bargain of the Treaty is that "the NPT non-nuclear weapon states agree never to acquire nuclear weapons and the NPT nuclear weapon states agree to share the benefits of peaceful use of nuclear technology and to pursue nuclear disarmament aimed at the ultimate elimination of their nuclear arsenals".

Articles I and II of the NPT forbid states that have nuclear weapons to help other nations to acquire them. These Articles were violated, for example, by France, which helped Israel to acquire nuclear weapons, and by China, which helped Pakistan to do the same. They are also violated by the "nuclear sharing" agreements, through which US tactical nuclear weapons will be transferred to several countries in Europe in a crisis situation. It is sometimes argued that in the event of a crisis, the NPT would no longer be valid, but there is nothing in the NPT itself that indicates that it would not hold in all situations.

The most blatantly violated provision of the NPT is Article VI. It requires the member states to pursue "negotiations in good faith on effective measures relating to cessation of the nuclear arms race at an early date and to nuclear disarmament", and negotiations towards a "Treaty on general and complete disarmament". In other words, the states that possess nuclear weapons agreed to get rid of them. However, during the 47 years that have passed since the NPT went into force, the nuclear weapon states have shown absolutely no sign of complying with Article VI. There is a danger that the NPT will break down entirely because of the majority of countries in the world are so dissatisfied with this long-continued non-compliance. Looking at the NPT with the benefit of hindsight, we can see the third "pillar", the "right to peaceful use of nuclear technology" as a fatal flaw of the treaty. In practice, it has meant encouragement of nuclear power generation, with all the many dangers that go with it.

The enrichment of uranium is linked to reactor use. Many reactors of modern design make use of low enriched uranium as a fuel. Nations operating such a reactor may claim that they need a program for uranium enrichment in order to produce fuel rods. However, by operating their ultracentrifuge a little longer, they can easily produce highly enriched (weapons-usable) uranium.

The difficulty of distinguishing between a civilian nuclear power generation program and a military nuclear program is illustrated by the case of Iran. In discussing Iran, it should be mentioned that Iran is fully in compliance with the NPT. It is very strange to see states that are long-time blatant violators of the NPT threaten Iran because of a nuclear program that fully complies with the Treaty. I believe that civilian nuclear power generation is always a mistake because of the many dangers that it entails, and because of the problem of disposing of nuclear waste. However, a military attack on Iran would be both criminal and insane. Why criminal? Because such an attack would violate the UN Charter and the Nuremberg Principles. Why insane? Because it would initiate a conflict that might escalate uncontrollably into World War III.

Biological Weapons Convention, 1972

During World War II, British and American scientists investigated the possibility of using smallpox as a biological weapon. However, it was never used, and in 1969 President Nixon officially ended the American biological weapons program, bowing to the pressure of outraged public opinion. In 1972, the United States, the United Kingdom and the Soviet Union signed a Convention on the Prohibition of the Development, Production and Stockpiling of Bacteriological (Biological) and Toxin Weapons and on their Destruction. Usually this treaty is known as the Biological Weapons Convention (BWC), and it has now been signed by virtually all of the countries of the world.

However, consider the case of smallpox: A World Health Organization team led by D.A. Henderson devised a strategy in which cases of smallpox were isolated and all their contacts vaccinated, so that the disease had no way of reaching new victims. Descriptions of the disease were circulated, and rewards offered for reporting cases. The strategy proved to be successful, and finally, in 1977, the last natural case of smallpox was isolated in Somalia. After a two-year waiting period, during which no new cases were reported, WHO announced in 1979 that smallpox, one of the most frightful diseases of humankind, had been totally eliminated from the world. This was the first instance of the complete eradication of a disease, and it was a demonstration of what could be achieved by the enlightened use of science combined with international cooperation. The eradication of smallpox was a milestone in human history.

It seems that our species is not really completely wise and rational; we do not really deserve to be called "Homo sapiens". Stone-age emotions and stone-age politics are alas still with us. Samples of smallpox virus were taken to "carefully controlled" laboratories in the United States and the Soviet Union. Why? Probably because these two Cold War opponents did not trust each other, although both had signed the Biological Weapons Convention. Each feared that the other side might intend to use smallpox as a biological weapon. There were also rumors that unofficial samples of the virus had been saved by a number of other countries, including North Korea, Iraq, China, Cuba, India, Iran, Israel, Pakistan and Yugoslavia.

Chemical Weapons Convention, 1997

On the 3rd of September, 1992, the Conference on Disarmament in Geneva adopted a Convention on the Prohibition of Development, Production, Stockpiling, and Use of Chemical Weapons and on their Destruction. This agreement, which is usually called the Chemical Weapons Convention (CWC), attempted to remedy some of the shortcomings of the Geneva Protocol of 1925. The CWC went into force in 1997, after Hungary deposited the 65th instrument of ratification.

The provisions of Article I of the CWC are as follows: 1. Each State Party to this convention undertakes never under any circumstances: (a) To develop, produce, otherwise acquire, stockpile or retain chemical weapons, or transfer, directly or indirectly, chemical weapons to anyone; (b) To use chemical weapons; (c) To engage in any military preparation to use chemical weapons; (d) To assist, encourage or induce, in any way, anyone to engage in any activity prohibited to a State Party in accordance with the provisions of this Convention. 2. Each State Party undertakes to destroy chemical weapons it owns or possesses, or that are located any place under its jurisdiction or control, in accordance with the provisions of this Convention. 3. Each State Party undertakes to destroy all chemical weapons it abandoned on the territory of another State Party, in accordance with the provisions of this Convention. 4. Each State Party undertakes to destroy any chemical weapons production facilities it owns or possesses, or that are located in any place under its jurisdiction or control, in accordance with the provisions of this Convention. 5. Each State Party undertakes not to use riot control agents as a method of warfare.

The CWC also makes provision for verification by teams of inspectors, and by 2004, 1,600 such inspections had been carried out in 59 countries. It also established an Organization for the Prevention of Chemical Warfare. All of the declared chemical weapons production facilities have now been inactivated, and all declared chemical weapons have been inventoried. However of the world's declared stockpile of chemical warfare agents (70,000 metric tons), only 12 percent have been destroyed. One hopes that in the future the CWC will be ratified by all the nations of the world and that the destruction of stockpiled chemical warfare agents will become complete.

Mine Ban Treaty, 1999

In 1991, six NGOs organized the International Campaign to Ban Landmines, and in 1996, the Canadian government launched the Ottawa process to ban landmines by hosting a meeting among like-minded anti-landmine states. A year later, in 1997, the Mine Ban Treaty was adopted and opened for signatures. In the same year, Jody Williams and the International Campaign to ban Landmines were jointly awarded the Nobel Peace Prize. After the 40th ratification of the Mine Ban Treaty in 1998, the treaty became binding international law on the 1st of March, 1999. The Ottawa Treaty functions imperfectly because of the opposition of several militarily powerful nations, but nevertheless it establishes a valuable norm, and it represents an important forward step in the development of international law.

International Criminal Court, 2002

In 1998, in Rome, representatives of 120 countries signed a statute establishing an International Criminal Court (ICC), with jurisdiction over the crime of genocide, crimes against humanity, war crimes and the crime of aggression.

Four years were to pass before the necessary ratifications were gathered, but by Thursday, April 11, 2002, 66 nations had ratified the Rome agreement, 6 more than the 60 needed to make the court permanent. It would be impossible to overstate the importance of the ICC. At last, international law acting on individuals has become a reality! The only effective and just way that international laws can act is to make individuals responsible and punishable, since (in the words of Alexander Hamilton) "To coerce states is one of the maddest projects that was ever devised."

At present, the ICC functions very imperfectly because of the bitter opposition of several powerful countries, notable the United States. U.S. President George W. Bush signed into law the American Servicemembers Protection Act of 2002, which is intended to intimidate countries that ratify the treaty for the ICC. The new law authorizes the use of military force to liberate any American or citizen of a U.S.-allied country being held by the court, which is located in The Hague. This provision, dubbed the "Hague invasion clause," has caused a strong reaction from U.S. allies around the world, particularly in the Netherlands.[b]

Despite the fact that the ICC now functions so imperfectly, it is a great step forward in the development of international law. It is there and functioning. We have the opportunity to make it progressively more impartial and to expand its powers.

Arms Trade Treaty, 2013

On April 2, 2013, a historic victory was won at the United Nations, and the world achieved its first treaty limiting international trade in arms. Work towards the Arms Trade Treaty (ATT) began in the Conference on Disarmament in Geneva, which requires a consensus for the adoption of any measure. Over the years, the consensus requirement has meant that no real progress in arms control measures has been made in Geneva, since a consensus among 193 nations is impossible to achieve.

To get around the blockade, British U.N. Ambassador Mark Lyall Grant sent the draft treaty to Secretary-General Ban Ki-moon and asked him on behalf of Mexico, Australia and a number of others to put the ATT to a swift vote in the General Assembly, and on Tuesday, April 3, 2013, it

---

[b] http://www.hrw.org/news/2002/08/03/us-hague-invasion-act-becomes-law

was adopted by a massive majority. Among the people who have worked hardest for the ATT is Anna Macdonald, Head of Arms Control at Oxfam. The reason why Oxfam works so hard on this issue is that trade in small arms is a major cause of poverty and famine in the developing countries. On April 9, Anna Macdonald wrote: "Thanks to the democratic process, international law will for the first time regulate the 70 billion dollar global arms trade. Had the process been launched in the consensus-bound Conference on Disarmament in Geneva, currently in its 12th year of meeting without even being able to agree on an agenda, chances are it would never have left the starting blocks..."

The passage of the Arms Trade Treaty by a majority vote in the UN General Assembly opens new possibilities for progress on other seemingly-intractable issues. In particular, it gives hope that a Nuclear Weapons Convention might be adopted by a direct vote on the floor of the General Assembly. The adoption of the NWC, even if achieved against the bitter opposition of the nuclear weapon states, would make it clear that the world's peoples consider the threat of an all-destroying nuclear war to be completely unacceptable.

We can pass a Nuclear Weapons Convention in the UN General Assembly

A convention banning nuclear weapons could be adopted by a majority vote on the floor of the UN General Assembly, following the precedent set by the Arms Trade Treaty. Indeed, this is the path forward advocated by the International Campaign to Abolish Nuclear Weapons (ICAN). In the case of a Nuclear Weapons Convention, world public opinion would have especially great force. It is generally agreed that a full-scale nuclear war would have disastrous effects, not only on belligerent nations but also on neutral countries. Mr. Javier Perez de Cuellar, former Secretary-General of the United Nations, emphasized this point in one of his speeches:

"I feel", he said, "That the question may justifiably be put to the leading nuclear powers: by what right do they decide the fate of humanity? From Scandinavia to Latin America, from Europe and Africa to the Far East, the destiny of every man and woman is affected by their actions. No one can expect to escape from the catastrophic consequences of a nuclear war on the fragile structure of this planet..."

"Like supreme arbiters, with our disputes of the moment, we threaten to cut off the future and to extinguish the lives of innocent millions yet unborn. There can be no greater arrogance. At the same time, the lives of all those who lived before us may be rendered meaningless; for we have the power to dissolve in a conflict of hours or minutes the entire work of civilization, with the brilliant cultural heritage of humankind."

Racism, Colonialism and Exceptionalism

A just system of laws must apply equally and without exception to everyone. If a person, or, in the case of international law, a nation, claims to be outside the law, or above the law, then there is something fundamentally wrong. For example, when U.S. President Obama said in a 2013 speech, "What makes America different, what makes us exceptional, is that we are dedicated to act", then thoughtful people could immediately see that something was terribly wrong with the system. If we look closely, we find that there is a link between racism, colonialism and exceptionalism. The racist and colonialist concept of "the white man's burden" is linked to the Neo-Conservative self-image of benevolent (and violent) interference in the internal affairs of other countries.[c]

The Oslo Principles on Climate Change Obligation, 2015

The future of human civilization and the biosphere is not only threatened by thermonuclear war: It is also threatened by catastrophic climate change. If prompt action is not taken to curb the use of fossil fuels: if the presently known reserves of fossil fuels are not left in the ground, then there is a great danger that we will pass a tipping point beyond which human efforts to stop a catastrophic increase in global temperatures will be useless because feedback loops will have taken over. There is a danger of a human-initiated 6th geological extinction event, comparable with the Permian-Triassic event, during which 96 percent of marine species and 70 percent of terrestrial vertebrates became extinct.

Recently there have been a number of initiatives which aim at making the human obligation to avert threatened environmental mega-catastrophes a part of international law. One of these initiatives can be seen in the proposal of the Oslo Principles on Climate Change Obligations; another is the Universal Declaration of the Rights of Mother Earth; and a third can be found in the concept of Biocultural Rights. These are extremely important and hopeful initiatives, and they point to towards the future development of international law for which we must strive.[d]

---

[c] http://www.countercurrents.org/avery101013.htm
https://www.youtube.com/watch?v=efI6T8lovqY
https://www.youtube.com/watch?v=IdBDRbjx9jo

[d] https://www.transcend.org/tms/2015/04/oslo-principles-on-global-climate-change-obligations/
https://www.transcend.org/tms/2015/04/climate-change-at-last-a-breakthrough-to-our-catastrophic-political-impasse/
http://www.commondreams.org/news/2015/04/14/lawsuit-out-love-unprecedented-legal-action-accuses-dutch-government-failing-climate
http://www.elgaronline.com/view/journals/jhre/6-1/jhre.2015.01.01.xml

Hope for the future, and responsibility for the future

Can we abolish the institution of war? Can we hope and work for a time when the terrible suffering inflicted by wars will exist only as a dark memory fading into the past? I believe that this is really possible. The problem of achieving internal peace over a large geographical area is not insoluble. It has already been solved. There exist today many nations or regions within each of which there is internal peace, and some of these are so large that they are almost worlds in themselves. One thinks of China, India, Brazil, the Russian Federation, the United States, and the European Union. Many of these enormous societies contain a variety of ethnic groups, a variety of religions and a variety of languages, as well as striking contrasts between wealth and poverty. If these great land areas have been forged into peaceful and cooperative societies, cannot the same methods of government be applied globally?

Today, there is a pressing need to enlarge the size of the political unit from the nation-state to the entire world. The need to do so results from the terrible dangers of modern weapons and from global economic interdependence. The progress of science has created this need, but science has also given us the means to enlarge the political unit: Our almost miraculous modern communications media, if properly used, have the power to weld all of humankind into a single supportive and cooperative society.

We live at a critical time for human civilization, a time of crisis. Each of us must accept his or her individual responsibility for solving the problems that are facing the world today. We cannot leave this to the politicians. That is what we have been doing until now, and the results have been disastrous. Nor can we trust the mass media to give us adequate public discussion of the challenges that we are facing. We have a responsibility towards future generations to take matters into our own hands, to join hands and make our own alternative media, to work actively and fearlessly for better government and for a better society.

We, the people of the world, not only have the facts on our side; we also have numbers on our side. The vast majority of the world's peoples long for peace. The vast majority long for abolition of nuclear weapons, and for a world of kindness and cooperation, a world of respect for the environment. No one can make these changes alone, but together we can do it.

Together, we have the power to choose a future where international anarchy, chronic war and institutionalized injustice will be replaced by democratic and humane global governance, a future where the madness and immorality of war will be replaced by the rule of law.

---

http://therightsofnature.org/universal-declaration/

We need a sense of the unity of all mankind to save the future, a new global ethic for a united world. We need politeness and kindness to save the future, politeness and kindness not only within nations but also between nations. To save the future, we need a just and democratic system of international law; for with law shall our land be built up, but with lawlessness laid waste.

Fig. 11.6 Although the Universal Declaration of Human Rights was cynically described as "a letter to Santa Claus" by Jean Kirkpatrik, the Declaration has great normative value. It shows the goals towards which we must aim. (Public domain.)

In the world as it is, gross violations of human rights are common. These include genocide, torture, summary execution, and imprisonment without trial.

In the world as it could be, the International Human Rights Commission would have far greater power to protect individuals against violations of human rights.

Fig. 11.7 We must listen to the uniquely life-oriented voices of women. (Public domain.)

In the world as it is, women form more than half of the population, but they are not proportionately represented in positions of political and economic power or in the arts and sciences. In many societies, women are confined to the traditional roles of childbearing and housekeeping.

In the world as it could be, women in all cultures would take their place beside men in positions of importance in government and industry, and in the arts and sciences. The reduced emphasis on childbearing would help to slow the population explosion.

Fig. 11.8 Both biological diversity and cultural diversity are of immense value. (Public domain.)

In the world as it is, there are no enforcible laws to prevent threatened species from being hunted to extinction. Many indigenous human cultures are also threatened.

In the world as it could be, an enforcible system of international laws would protect threatened species. Indigenous human cultures would also be protected.

Suggestions for further reading

(1) B. Broms, *United Nations*, Suomalainen Tiedeakatemia, Helsinki, (1990).
(2) S. Rosenne, *The Law and Practice at the International Court*, Dordrecht, (1985).
(3) S. Rosenne, *The World Court - What It Is and How It Works*, Dordrecht, (1995).
(4) J. D'Arcy and D. Harris, *The Procedural Aspects of International Law (Book Series), Volume 25*, Transnational Publishers, Ardsley, New York, (2001).
(5) H. Cullen, *The Collective Complaints Mechanism Under the European Social Charter*, European Law Review, Human Rights Survey, p. 18-30, (2000).
(6) S.D. Bailey, *The Procedure of the Security Council*, Oxford, (1988).
(7) R.A. Akindale, *The Organization and Promotion of World Peace: A Study of Universal-Regional Relationships*, Univ. Toronto Press, Toronto, Ont., (1976).
(8) J.S. Applegate, *The UN Peace Imperative*, Vantage Press, New York, (1992).
(9) S.E. Atkins, *Arms Control, Disarmament, International Security and Peace: An Annotated Guide to Sources, 1980-1987*, Clio Press, Santa Barbara, CA, (1988).
(10) N. Ball and T. Halevy, *Making Peace Work: The Role of the International Development Community*, Overseas Development Council, Washington DC, (1996).
(11) F. Barnaby, Ed., *The Gaia Peace Atlas: Survival into the Third Millennium*, Doubleday, New York, (1988)
(12) J.H. Barton, *The Politics of Peace: An Evaluation of Arms Control*, Stanford Univ. Press, Stanford, CA, (1981).
(13) W. Bello, *Visions of a Warless World*, Friends Committee on National Education Fund, Washington DC, (1986).
(14) A. Boserup and A. Mack, *Abolishing War: Cultures and Institutions; Dialogue with Peace Scholars Elise Boulding and Randall Forsberg*, Boston Research Center for the Twenty-first Century, Cambridge, MA, (1998).
(15) E. Boulding et al., *Bibliography on World Conflict and Peace*, Westview Press, Boulder, CO, (1979).
(16) E. Boulding et al., Eds., *Peace, Culture and Society: Transnational Research Dialogue*, Westview Press, Boulder, CO, (1991).
(17) A.T. Bryan et al., Eds., *Peace, Development and Security in the Caribean*, St. Martins Press, New York, (1988).

(18) A.L. Burns and N. Heathcote, *Peace-Keeping by UN Forces from Suez to Congo*, Praeger, New York, (1963).
(19) F. Capra and C. Spretnak, *Green Politics: The Global Promise*, E.P. Dutton, New York, (1986).
(20) N. Carstarphen, *Annotated Bibliography of Conflict Analysis and Resolution*, Inst. for Conflict Analysis and Resolution, George Mason Univ., Fairfax, VA, (1997).
(21) N. Chomsky, *Peace in the Middle East? Reflections on Justice and Nationhood*, Vintage Books, New York, (1974).
(22) G. Clark and L. Sohn, *World Peace Through World Law*, World Without War Pubs., Chicago, IL, (1984).
(23) K. Coates, *Think Globally, Act Locally: The United Nations and the Peace Movements*, Spokesman Books, Philadelphia, PA, (1988).
(24) G. De Marco and M. Bartolo, *A Second Generation United Nations: For Peace and Freedom in the 20th Century*, Colombia Univ. Press, New York, (1997).
(25) F.M. Deng and I.W. Zartman, Eds., *Conflict Resolution in Africa*, Brookings Institution, Washington, DC, (1991).
(26) W. Desan, *Let the Future Come: Perspectives for a Planetary Peace*, Georgetown Univ. Press, Washington, DC, (1987).
(27) D. Deudney, *Whole Earth Security. A Geopolitics of Peace*, Worldwatch paper 55. Worldwatch Institute, Washington, DC, (1983).
(28) A.J. Donovan, *World Peace? A Work Based on Interviews with Foreign Diplomats*, A.J. Donovan, New York, (1986).
(29) R. Duffey, *International Law of Peace*, Oceania Pubs., Dobbs Ferry, NY, (1990).
(30) L.J. Dumas, *The Socio-Economics of Conversion From War to Peace*, M.E. Sharpe, Armonk, NY, (1995).
(31) W. Durland, *The Illegality of War*, National Center on Law and Pacifism, Colorado Springs, CO, (1982).
(32) F. Esack, *Qur'an, Liberation and Pluralism: An Islamic Perspective on Interreligious Solidarity Against Oppression*, Oxford Univ. Press, London, (1997).
(33) I. Hauchler and P.M. Kennedy, Eds., *Global Trends: The World Almanac of Development and Peace*, Continuum Pubs., New York, (1995).
(34) H.B. Hollins et al., *The Conquest of War: Alternative Strategies for Global Security*, Westview Press, Boulder, CO, (1989).
(35) H.J. Morgenthau, *Peace, Security and the United Nations*, Ayer Pubs., Salem, NH, (1973).
(36) C.C. Moskos, *Peace Soldiers: The Sociology of a United Nations Military Force*, Univ. of Chicago Press, Chicago, IL, (1976).

(37) L. Pauling, *Science and World Peace*, India Council for Cultural Relations, New Delhi, India, (1967).
(38) C. Peck, *The United Nations as a Dispute Resolution System: Improving Mechanisms for the Prevention and Resolution of Conflict*, Kluwer, Law and Tax, Cambridge, MA, (1996).
(39) D. Pepper and A. Jenkins, *The Geography of Peace and War*, Basil Blackwell, New York, (1985).
(40) J. Perez de Cuellar, *Pilgrimage for Peace: A Secretary General's Memoir*, St. Martin's Press, New York, (1997).
(41) R. Pickus and R. Woito, *To End War: An Introduction to the Ideas, Books, Organizations and Work That Can Help*, World Without War Council, Berkeley, CA, (1970).
(42) S.R. Ratner, *The New UN Peacekeeping: Building Peace in Lands of Conflict after the Cold War*, St. Martins Press, New York, (1995).
(43) I.J. Rikhye and K. Skjelsbaek, Eds., *The United Nations and Peacekeeping: Results, Limitations and Prospects: The Lessons of 40 Years of Experience*, St. Martins Press, New York, (1991).
(44) J. Rotblat, Ed., *Scientists in Quest for Peace: A History of the Pugwash Conferences*, MIT Press, Cambridge, MA, (1972).
(45) J. Rotblat, Ed., *Scientists, The Arms Race, and Disarmament*, Taylor and Francis, Bristol, PA, (1982).
(46) J. Rotblat, Ed., *Striving for Peace, Security and Development in the World*, World Scientific, River Edge, NJ, (1991).
(47) J. Rotblat, Ed., *Towards a War-Free World*, World Scientific, River Edge, NJ, (1995).
(48) J. Rotblat, Ed., *Nuclear Weapons: The Road to Zero*, Westview, Boulder, CO, (1998).
(49) J. Rotblat and L. Valki, Eds., *Coexistance, Cooperation and Common Security*, St. Martins Press, New York, (1988).
(50) United Nations, *Peaceful Settlement of Disputes between States: A Select Bibliography*, United Nations, New York, (1991).
(51) United States Arms Control and Disarmament Agency, *Arms Control and Disarmament Agreements: Texts and Histories of Negotiations*, USACDA, Washington, DC, (updated annually)
(52) D. Fahrni, *An Outline History of Switzerland - From the Origins to the Present Day*, Pro Helvetia Arts Council of Switzerland, Zurich, (1994).
(53) J.M. Luck, *A History of Switzerland*, Sposs, Palo Alto, CA, (1985).

Chapter 12

The Choice is Ours to Make

"Man lives in a new cosmic world for which he was not made. His survival depends on how well and how fast he can adapt himself to it, rebuilding all his ideas, all his social and political institutions. ...Modern science has abolished time and distance as factors separating nations. On our shrunken globe today, there is room for one group only - the family of man."

Albert Szent-Györgyi

"There lies before us, if we choose, continual progress in happiness, knowledge, and wisdom. Shall we, instead, choose death, because we cannot forget our quarrels? We appeal as human beings to human beings: Remember your humanity, and forget the rest. "

Bertrand Russell and Albert Einstein

Introduction

Because of the serious and linked challenges that we face today, fatalism, despair and inaction are not options. The peoples of the world must join hands to save the future. We can think of Helen Keller's words: "Alone we can do so little; together we can do so much!", or Shelly's: "You are many, they are few!"

The future is ours to shape. The fate of all future generations, is in our hands. The choice is ours to make.

The fragility of our complex civilization

The rapid growth of knowledge

Cultural evolution depends on the non-genetic storage, transmission, diffusion and utilization of information. The development of human speech, the invention of writing, the development of paper and printing, and finally, in modern times, mass media, computers and the Internet: all these have been crucial steps in society's explosive accumulation of information and knowledge. Human cultural evolution proceeds at a constantly-accelerating speed, so great in fact that it threatens to shake society to pieces.

In many respects, our cultural evolution can be regarded as an enormous success. However, at the start of the 21st century, most thoughtful observers agree that civilization is entering a period of crisis. As all curves move exponentially upward, population, production, consumption, rates of scientific discovery, and so on, one can observe signs of increasing environmental stress, while the continued existence and spread of nuclear weapons threaten civilization with destruction. Thus, while the explosive growth of knowledge has brought many benefits, the problem of achieving a stable, peaceful and sustainable world remains serious, challenging and unsolved.

Our modern civilization has been built up by means of a worldwide exchange of ideas and inventions. It is built on the achievements of many ancient cultures. China, Japan, India, Mesopotamia, Egypt, Greece, the Islamic world, Christian Europe, and the Jewish intellectual traditions, all have contributed. Potatoes, corn, squash, vanilla, chocolate, chili peppers, and quinine are gifts from the American Indians.

The sharing of scientific and technological knowledge is essential to modern civilization. The great power of science is derived from an enormous concentration of attention and resources on the understanding of a tiny fragment of nature. It would make no sense to proceed in this way if knowledge were not permanent, and if it were not shared by the entire world.

Science is not competitive. It is cooperative. It is a great monument built by many thousands of hands, each adding a stone to the cairn. This is true not only of scientific knowledge but also of every aspect of our culture, history, art and literature, as well as the skills that produce everyday objects upon which our lives depend. Civilization is cooperative. It is not competitive.

Our cultural heritage is not only immensely valuable; it is also so great that no individual comprehends all of it. We are all specialists, who understand only a tiny fragment of the enormous edifice. No scientist understands all of science. Perhaps Leonardo da Vinci could come close in his day, but today it is impossible. Nor do the vast majority people who use cell phones, personal computers and television sets every day understand in detail how they work. Our health is preserved by medicines, which are made by processes that most of us do not understand, and we travel to work in automobiles and buses that we would be completely unable to construct.

The fragility of modern society

As our civilization has become more and more complex, it has become increasingly vulnerable to disasters. We see this whenever there are power cuts or transportation failures due to severe storms. If electricity should fail for a very long period of time, our complex society would cease to function. The population of the world is now so large that it is completely dependent on the high efficiency of modern agriculture. We are also very dependent on the stability of our economic system.

The fragility of modern society is particularly worrying, because, with a little thought, we can predict several future threats which will stress our civilization very severely. We will need much wisdom and solidarity to get safely through the difficulties that now loom ahead of us.

We can already see the the problem of famine in vulnerable parts of the world. Climate change will make this problem more severe by bringing aridity to parts of the world that are now large producers of grain, for example the Middle West of the United States. Climate change has caused the melting of glaciers in the Himalayas and the Andes. When these glaciers are completely melted, China, India and several countries in South America will be deprived of their summer water supply. Water for irrigation will also become increasingly problematic because of falling water tables. Rising sea levels will drown many rice-growing areas in South-East Asia. Finally, modern agriculture is very dependent on fossil fuels for the production of fertilizer and for driving farm machinery. In the future, high-yield agriculture will be dealt a severe blow by the rising price of fossil fuels.

Economic collapse is another threat that we will have to face in the future. Our present fractional reserve banking system is dependent on economic growth. But perpetual growth of industry on a finite planet is a logical impossibility. Thus we are faced with a period of stress, where reform of our growth-based economic system and great changes of lifestyle will both become necessary.

How will we get through the difficult period ahead? I believe that solutions to the difficult problems of the future are possible, but only if we face the problems honestly and make the adjustments which they demand. Above all, we must maintain our human solidarity.

Who is my neighbor?

Are we losing the human solidarity that will be needed if our global society is to solve the pressing problems that are facing us today? Among the symptoms of loss of solidarity is the drift towards violence, racism and war that can be seen in some countries.[a] To worried observers it seems reminiscent of Germany and Italy in the 1930s. Another warning symptom is the inhospitable reception that refugees have received in Europe and elsewhere.[b]

Tribalism

Human emotional nature evolved over the long prehistory of our species, when our remote ancestors lived small tribes, competing for territory on the grasslands of Africa. Since marriage within a tribe was much more frequent than marriage outside it, each tribe was genetically homogeneous, and the tribe itself, rather than the individual, was the unit upon which the forces of natural selection acted. Those tribes that exhibited internal solidarity, combined with aggression towards competing tribes, survived best. Over a long period of time, tribalism became a hard-wired part of human nature.

---

[a] http://www.commondreams.org/news/2016/03/01/after-latest-display-bigotry-trump-again-faces-charges-racism   http://www.truth-out.org/opinion/item/35053-an-open-letter-to-evangelical-trump-voters
http://www.theguardian.com/commentisfree/2016/feb/29/donald-trump-us-election-2016-demagogue
http://harpers.org/archive/1964/11/the-paranoid-style-in-american-politics/2/
http://www.countercurrents.org/boyle020316.htm
http://www.commondreams.org/views/2016/03/01/dont-cry-me-america
https://www.washingtonpost.com/2016-election-results/super-tuesday/
http://www.commondreams.org/views/2016/03/01/yemen-humanitarian-pause-urgently-needed

[b] http://www.commondreams.org/news/2016/03/01/new-low-europe-police-bulldoze-camps-tear-gas-asylum-seekers-and-shutter-borders

We can see tribalism today in the emotions involved in football matches, in nationalism, and in war.

The birth of ethics

When humans began to live in larger and more cosmopolitan groups, it was necessary to overwrite some elements of raw human emotional nature. Tribalism became especially inappropriate, unless the scope of the perceived tribe could be extended to include everyone in the enlarged societies. Thus ethical principles were born. It is not just a coincidence that the greatest ethical teachers of history lived at a time when the size of cooperating human societies was being enlarged.

All of the major religions of humanity contain some form of the Golden Rule. Christianity offers an especially clear statement of this central ethical principle: According to the Gospel of Luke, after being told that he must love his neighbor as much as he loves himself, a man asks Jesus, "Who is my neighbor?". Jesus then replies with the Parable of the Good Samaritan, in which we are told that our neighbor need not be a member of our own tribe, but can live far away and can belong to a completely different nation or ethnic group. Nevertheless, that person is still our neighbor, and deserves our love and care.

The central ethical principle which is stated so clearly in the Parable of the Good Samaritan is exactly what we need today to avoid disaster. We must enlarge our loyalties to include the whole of humanity. We must develop a global ethic of comprehensive human solidarity, or else perish from a combination of advanced technology combined with primitive tribalism. Space-age science is exceedingly dangerous when it is combined with stone-age politics.[c]

The need for global solidarity comes from the instantaneous worldwide communication and economic interdependence that has resulted from advanced science and technology. Advanced technology, our almost miraculous ability to communicate through the Internet, Skype and smartphones, could weld the world into a single peaceful and cooperative unit. But we must learn to use global communication as a tool for developing worldwide human solidarity.

Each week, all over the world, congregations assemble and are addressed by their religious leaders on ethical issues. But all too often there is no mention of the astonishing and shameful contradiction between the institution of war (especially the doctrine of "massive retaliation"), and the principle of universal human brotherhood, loving and forgiving one's enemies, and returning good for evil.

---

[c] http://www.learndev.org/dl/SpaceAgeScienceStoneAgePolitics-Avery.pdf

Fig. 12.1 Who is my neighbor? The Parable of the Good Samaritan tells us that our neighbor may belong to an entirely different ethnic group. Nevertheless, he or she deserves our love and care. (Wikipedia.)

At a moment of history, when the continued survival of civilization is in doubt because of the incompatibility of war with the existence of thermonuclear weapons, our religious leaders ought to use their enormous influence to help to solve the problem of war, which is after all an ethical problem.

This is how Bertrand Russell expressed the need for human solidarity: "All who are not lunatics agree about certain things. That it is better to be alive than dead, better to be adequately fed than starved, better to be free than a slave. Many people desire those things only for themselves and their friends; they are quite content that their enemies should suffer. These people can be refuted by science: Humankind has become so much one family that we cannot insure our own prosperity except by insuring that of everyone else. If you wish to be happy yourself, you must resign yourself to seeing others also happy."

New ethics to match new technology

Modern science has, for the first time in history, offered humankind the possibility of a life of comfort, free from hunger and cold, and free from the constant threat of death through infectious disease. At the same time, science has given humans the power to obliterate their civilization with nuclear weapons, or to make the earth uninhabitable through overpopulation and pollution. The question of which of these paths we choose is literally a matter of life or death for ourselves and our children.

Will we use the discoveries of modern science constructively, and thus choose the path leading towards life? Or will we use science to produce more and more lethal weapons, which sooner or later, through a technical or human failure, may result in a catastrophic nuclear war? Will we thoughtlessly destroy our beautiful planet through unlimited growth of population and industry? The choice among these alternatives is ours to make. We live at a critical moment of history - a moment of crisis for civilization.

No one living today asked to be born at such a moment, but by an accident of birth, history has given us an enormous responsibility, and two daunting tasks: If civilization is to survive, we must not only stabilize the global population but also, even more importantly, we must eliminate the institution of war. We face these difficult tasks with an inherited emotional nature that has not changed much during the last 40,000 years. Furthermore, we face the challenges of the 21st century with an international political system based on the anachronistic concept of the absolutely sovereign nation-state. However, the human brain has shown itself to be capable of solving even the most profound and complex problems. The mind that has seen into the heart of the atom must not fail when confronted with paradoxes of the human heart.

The problem of building a stable, just, and war-free world is difficult, but it is not impossible. The large regions of our present-day world within which war has been eliminated can serve as models. There are a number of large countries with heterogeneous populations within which it has been possible to achieve internal peace and social cohesion, and if this is possible within such extremely large regions, it must also be possible globally.

We must replace the old world of international anarchy, chronic war and institutionalized injustice, by a new world of law. The United Nations Charter, the Universal Declaration of Human Rights and the International Criminal Court are steps in the right direction, but these institutions need to be greatly strengthened and reformed.

We also need a new global ethic, where loyalty to one's family and nation will be supplemented by a higher loyalty to humanity as a whole.

Creating the future

The chorus of a popular song repeats a message of comforting (but irresponsible) fatalism: "Que Sera Sera. Whatever will be, will be. The future's not ours to see. Que Sera Sera. What will be will be." But can we allow ourselves the luxury of fatalism, especially today, when our future is darkened by the twin threats of catastrophic climate change and thermonuclear war?

Must we not accept our responsibility for both the near future and the distant future. We must do all that is within our power to make our world one in which our children and their descendants can survive? We must save the environment. We must save plants and animals from extinction.

What has happened to the global environment is a human creation. Its very name, the anthhropocene, indicates that we made it. What will happen in the future will also be our creation, the sum of the choices that we make.

War is a human creation. Just as we abolished slavery, we can also abolish the institution of war. It is our responsibility to do so.

The tribal tendencies of human nature are not inevitable. Racism is not inevitable. Nationalist Chauvinism is not inevitable. The dark side of human nature can be overwritten by education and ethics. It is our responsibility to create a global ethical system that matches our advanced technology. We must create an ethic of universal human solidarity.

Global anarchy is not inevitable. We can extend the methods used to avoid war within nations to the entire world. We can reform the United Nations and create a global federation capable of effectively achieving the goals that we desire.

Our economic system is a human creation. The laws of the market are not really laws: They are choices. If we choose we could maximize human happiness, rather than maximizing production and profits.

The population explosion is not inevitable. It is a result of human choices. The threat of an extremely severe worldwide famine resulting from climate change, exploding populations and end of the fossil fuel era is not inevitable. If such a famine comes, it will be the result of human choices.

The decay of democracy is not inevitable. Oligarchy is not inevitable. These evils are the result of neglect and political irresponsibility. As citizens, we must have the courage to restore democracy in countries where it has disappeared, and to create it in countries where it never existed.

We live in a special time, a time of crisis. Here are the responsibilities that history has given to our generation:

- We need system change, not climate change!
- We need a new economic system, a new society, a new social contract, a new way of life.

- We must achieve a steady-state economic system. Limitless growth on a finite planet is a logical absurdity.
- We must restore democracy in countries where it has decayed, and create it in countries where it never existed.
- We must decrease economic inequality.
- We must break the power of corporate greed.
- We must leave fossil fuels in the ground.
- We must stabilize and ultimately reduce global population to a level that can be supported by sustainable agriculture.
- We must abolish the institution of war before modern weapons destroy us.
- And finally, we must develop a mature ethical system to match our new technology.

No one is exempt from these responsibilities. No one can achieve these goals alone; but together we can create the future that we choose.

The Nobel laureate biochemist Albert Szent-Györgyi once wrote:

"The story of man consists of two parts, divided by the appearance of modern science.... In the first period, man lived in the world in which his species was born and to which his senses were adapted. In the second, man stepped into a new, cosmic world to which he was a complete stranger.... The forces at man's disposal were no longer terrestrial forces, of human dimension, but were cosmic forces, the forces which shaped the universe. The few hundred Fahrenheit degrees of our flimsy terrestrial fires were exchanged for the ten million degrees of the atomic reactions which heat the sun."

"This is but a beginning, with endless possibilities in both directions - a building of a human life of undreamt of wealth and dignity, or a sudden end in utmost misery. Man lives in a new cosmic world for which he was not made. His survival depends on how well and how fast he can adapt himself to it, rebuilding all his ideas, all his social and political institutions."

"...Modern science has abolished time and distance as factors separating nations. On our shrunken globe today, there is room for one group only - the family of man."

Fig. 12.2 Cultural activities strengthen internationalism, use very few resources, and produce almost no waste. (Wikipedia.)

In the world as it is, power and material goods are valued more highly than they deserve to be. "Civilized" life often degenerates into a struggle of all against all for power and possessions. However, the industrial complex on which the production of goods depends cannot be made to run faster and faster, because we will soon encounter shortages of energy and raw materials.

In the world as it could be, nonmaterial human qualities, such as kindness, politeness, and knowledge, and musical, artistic or literary ability would be valued more highly, and people would derive a larger part of their pleasure from conversation, and from the appreciation of unspoiled nature.

Fig. 12.3 In his poem "London", William Blake used the words "mind-forged manacles". We must rid ourselves of the mind-forged manacles that make us slaves to the institution of war. (Listverse.)

In the world as it is, the institution of slavery existed for so many millennia that it seemed to be a permanent part of human society. Slavery has now been abolished in almost every part of the world. However war, an even greater evil than slavery, still exists as an established human institution.

In the world as it could be, we would take courage from the abolition of slavery, and we would turn with energy and resolution to the great task of abolishing war.

Fig. 12.4 As Helen Keller said, "Alone we can do so little; together we can do so much!" (Wikipedia.)

In the world as it is, people feel anxious about the future, but unable to influence it. They feel that as individuals they have no influence on the large-scale course of events.

In the world as it could be, ordinary citizens would realize that collectively they can shape the future. They would join hands and work together for a better world. They would give as much of themselves to peace as peace is worth.

Fig. 12.5 George Bernard Shaw once said, "Most people look at the world as it is and ask 'Why?'. We should look at the world as it could be and ask, 'Why not?'" (Public domain.)

Fig. 12.6 Gandhi said "There is enough for every man's need, but not for every man's greed." (Public domain.)

Fig. 12.7 Dr. Martin Luther King, Jr., speaking in Washington D.C.. In his book "Strength yo Love", Dr. King wrote: "Wisdom born of experience should tell us that war is obsolete... I am convinced that the Church cannot be silent while mankind faces the threat of nuclear annihilation. If the Church is true to her mission, she must call for an end to the nuclear arms race." (Public domain.)

Fig. 12.8 Sir Joseph Rotblat, leader of Pugwash Conferences on Science and World Affairs, meeting Daisaku Ikeda, President of the 12-million strong Buddhist organization Soka Gakkai International, which aims at value-promoting education, international dialogue and peace. The 96-year-old Rotblat encouraged President Ikeda to continue their mutual struggle for the complete abolition of nuclear weapons. (SGI.)

Fig. 12.9 Nelson Mandela said: "It always seems impossible until it's done!" (Wikipedia.)

Suggestions for further reading

(1) Herman Daly, *Steady-State Economics: Second Edition with New Essays*, Island Press, (1991).
(2) Herman Daly, *Economics in a Full World*, Scientific American, Vol. 293, Issue 3, September, (2005).
(3) Herman Daly and John Cobb, *For the Common Good*, Beacon Press, Boston, (1989).
(4) E.O. Wilson, *The Diversity of Life*, Allen Lane, The Penguin Press, (1992).
(5) Lester R. Brown et. al.,*Saving the Planet. How to Shape an Environmentally Sustainable Global Economy*, W.W. Norton, New York, (1991).
(6) Muhammad Yunus, *Banker to the Poor; Microcredit and the Battle Against World Poverty*, (2003).
(7) UN Global Compact, http://www.unglobalcompact.org (2007).
(8) UN Millennium Development Goals http://www.un.org/millenniumgoals/ (2007).
(9) Amartya Sen, *Poverty and Famine; An Essay on Entitlement and Deprivation*, Oxford Univeersity Press, (1981).
(10) Amartya Sen, *Development as Freedon*, Oxford University Press, (1999).
(11) Amartya Sen, *Inequality Reexamined*, Harvard University Press, (1992).
(12) Paul F. Knitter and Chandra Muzaffar, editors, *Subverting Greed; Religious Perspectives on the Global Economy*, Orbis Books, Maryknoll, New York, (2002).
(13) Edy Korthals Altes, *The Contribution of Religions to a Just and Sustainable Economic Development*, in F. David Peat, editor, *The Pari Dialogues, Volume 1*, Pari Publishing, (2007).
(14) Hendrik Opdebeeck, *Globalization Between Market and Democracy*, in F. David Peat, editor, *The Pari Dialogues, Volume 1*, Pari Publishing, (2007).
(15) Paul Hawken *The Ecology of Commerce; A Declaration of Sustainability*, Collins Business, (2005).
(16) Luther Standing Bear, *Land of the Spotted Eagle*, Houghton Mifflin, (1933).
(17) T. Gyatso, HH the Dalai Lama, *Ancient Wisdom, Modern World: Ethics for the New Millennium*, Abacus, London, (1999).
(18) T. Gyatso, HH the Dalai Lama, *How to Expand Love: Widening the Circle of Loving Relationships*, Atria Books, (2005).
(19) J. Rotblat and D. Ikeda, *A Quest for Global Peace*, I.B. Tauris, London, (2007).

(20) M. Gorbachev and D. Ikeda, *Moral Lessons of the Twentieth Century*, I.B. Tauris, London, (2005).
(21) D. Krieger and D. Ikeda, *Choose Hope*, Middleway Press, Santa Monica CA 90401, (2002).
(22) P.F. Knitter and C. Muzaffar, eds., *Subverting Greed: Religious Perspectives on the Global Economy*, Orbis Books, Maryknoll, New York, (2002).
(23) S. du Boulay, *Tutu: Voice of the Voiceless*, Eerdmans, (1988).
(24) Earth Charter Initiative *The Earth Charter*, www.earthcharter.org
(25) P.B. Corcoran, ed., *The Earth Charter in Action*, KIT Publishers, Amsterdam, (2005).
(26) R. Costannza, ed., *Ecological Economics: The Science and Management of Sustainability*, Colombia University Press, New York, (1991).
(27) A. Peccei, *The Human Quality*, Pergamon Press, Oxford, (1977).
(28) A. Peccei, *One Hundred Pages for the Future*, Pergamon Press, New York, (1977).
(29) E. Pestel, *Beyond the Limits to Growth*, Universe Books, New York, (1989).
(30) Pope Francis I, Laudato si', https://laudatosi.com/watch
(31) John S. Avery, *The Need for a New Economic System*, Irene Publishing, Sparsnäs Sweden, (2016).
(32) John S. Avery, *Collected Essays*, Volumes 1-3, Irene Publishing, Sparsnäs Sweden, (2016).
(33) John S. Avery, *Space-Age Science and Stone-Age Politics*, Irene Publishing, Sparsnäs Sweden, (2016).
(34) John S. Avery, *Science and Society*, World Scientific, (2016).

Index

ABM Treaty, 179, 182
Abolition 2000, 321
abolition of nuclear weapons, 169
abolition of war, 47
abortion, 31
absolute limits, 47, 48
accents, 270
accident waiting to happen, 186
accidental nuclear war, 185
acts pursuant to orders, 229
Adam Smith, 29
administration of property, 23
advertising, 63
aesthetic aspects, 99
Africa, 94, 106, 120, 122, 127, 129, 148, 152, 214, 217
Agent Orange, 211
aggression, 230, 264–266
aggression and mating, 271
aggression, intergroup, 271
agricultural development, 157
agricultural land lost, 211
agricultural methods, 248
agricultural monocultures, 59
agricultural output, 48
agricultural societies, 21
agricultural yields, 124
agriculture, 16, 124, 187, 326
AIDS, 159, 219, 220
air travel, 149
Alaska, 106
Alberta, Canada, 91

Albury, 12
Aleutian Islands, 106
algae, 102
Altes, Dr. Edy Korthals, 312
altruism, 15, 265–267, 272, 276
aluminum, production of, 105
aluminum-covered plastics, 94
Amazon forest dieback, 54
American Friends Service Committee, 312
American Indians, 19
Amerinds, 148
amoebae, 275
anachronistic institutions, 317
anachronistic political system, 369
anaerobic digestion, 103
anarchy, 251
ancestor worship, 269
Anglican Church, 12
Anglican Pacifist Fellowship, 322
animal feed, 125
Annan, Kofi, 188, 219
anode, 108
Antarctic icecap, 48
anti-Catholic laws, 27
anti-corruption, 219
antibiotics, 262
anxiety about the future, 374
apprentices, 4
aquifers, 122, 126, 209
arable land, 120, 217, 220
Arctic icecap, 48

area of cropland, 122
area under food production, 125
Argentina, 119, 270
arid grasslands, 122
aridity, 51, 126, 209
armaments, 210
armaments, cost of, 306
armed attack, 228
arms industries, 221
arms industry, 230
arms manufacturers, 232
Article VI, 180, 182, 185
artificial needs, 63
Arts and Crafts movement, 34
asteroids, 185
Athabasca oil sands, 91
atmosphere, 82
atomic bomb, 264
atrocities, 230, 263
Attlee, Clement, 34
Australia, 119, 295
Austria, 104
autocatalysts, 280
autocatylitic molecules, 275
automated agriculture, 12, 42

Börjesson, Pål, 101
bacterial cells, 275
Baha'i, 311
Bahr, Egon, 190
balance of nature, 304
Bangladesh, 52, 90, 96, 217
Bank of the Villages, 217
bargaining powers, 148
Baring, Alexander, 29
Baruch Plan, 171
battleground series, 317
Bedjaoui, Muhammad, 184
bees, 276
Beloc, Hilaire, 143
belt of tar, 92
benevolence, 23
Benn, Tony, 34
Berlin Wall, 179
Besant, Annie, 33, 34
Bhutto, Zulfiquar Ali, 181

bilateral tax agreements, 148
binary plants, 107
biodiesel, 103
biodiversity, 48, 216
bioethanol, 103
biogas, 103
biological diversity, 59
biological diversity, 59, 81
biological weapons, 296
biology, 47, 220
Biology of War and Peace, 268
biomass, 45, 49, 89, 94, 95, 101, 102, 110
biotas, 59
birds killed by rotors, 99
birth abnormalities, 211
birth control, 22, 31, 33, 43, 46, 135, 208
birth control programs, 135
birth defects, 213
birth rate, 12
birth rates, 43
black rat, 31
Blair, Bruce G., 186
Blake, William, 143
blight, 27
blocks, 252
Bodet, Jamie Torres, 302
Boer War, 145
bombings, 214
bonobos, 271
Borlaug, Norman, 120
Boterro, 10
boundaries of groups, 270
bounties of nature, 23
Bradlaugh, Charles, 33
Brandenburg Gate, 179
Brazil, 104, 120, 270, 295
bread and circuses, 318
bribery, 219
Britain, 166
British Labour Party, 34
brown rat, 31
Brown, Des, 189
Brown, Lester R., 64, 127, 215
Brundtland Report, 125

Brundtland, Gro Harlem, 190
bubonic plague, 31
burial customs, 269
burning of rainforests, 48
business administration, 220

Cairo population conference, 136
Calogero, Francesco, 188
Cambridge University, 12
Campaign for Nuclear Disarmament, 321
Camus, Albert, 170
Canada, 90, 91, 93, 104, 107, 119, 295
cancer and Agent Orange, 211
cancer and uranium dust, 211
cancer from fallout, 211
cannibalism, 19
canola oil, 101, 102
Canton, bombing of, 166
Canute, 24
capital, 47, 217, 326
capital, growth of, 29
Capone, Al, 298
Captain Cook, 19
carbon dioxide, 48, 50, 81
carbon emissions, 51, 95
carbon fibers, 112
Carlyle, Thomas, 31, 34
Carnegie, Andrew, 217
carrying capacity, 23, 45, 47, 48, 63, 119, 217, 220
Carter, Jimmy, 190
caste markings, 269
catastrophic damage, 45
catastrophic famine, 46
catastrophic mistake, 186
catastrophic nuclear war, 211
cathode, 108
celibacy, 18
census, 25
Central Asia, 19
Central Atlantic region, 106
ceremonies, 269
Cerrado, 120
Chad, 148
chadors, 269

chain of causes, 13
Chamberlain, Neville, 166
charge acceptors, 94
charge donors, 94
chauvinism, 295
checks to population growth, 18, 21, 43
chemical bonds, 95
chemical energy, 48
chemical signals, 277
chemical weapons, 296
Cheney Report, 90
Chicago University, 170
child labor, 4, 20, 31, 32, 160, 208, 219
child labor laws, 34
childbearing, 357
childhood deaths in Africa, 152
children, 157, 158
children killed by wars, 183
chimpanzees, 271
China, 53, 104, 111, 122, 127, 136, 151, 270
China, coal consumption, 93
China, oil consumption, 91
Chinese nuclear weapons, 177
chloroplasts, 275
cholera, 159
Christian Bomb, 181
Christian ethics, 312
cinema, 324
cinturon de la brea, 92
circumcision, 269
civics, 298
civil society, 189
civil wars, 148, 298
civilian victims of war, 183
civilians as targets, 189
civilization, 281
classical economics, 217
climate change, 48, 51, 52, 54, 59, 81, 88, 95, 123, 126, 209, 211, 220
climatic trends, 55
closed world, 217
closed world economics, 217
cloud cover, 95, 101, 102

Club of Rome, 62
cluster bombs, 211
coal, 89
coal and liquid fuels, 88
coal consumption, 88, 93
coal reserves, 49, 93
coal-burning plants, 98
Code Pink, 322
coherent sociopolitical units, 271
Cold War, 148, 179, 185
Coleridge, William, 25
collective action, 374
collective bargaining, 219
collective paranoia, 178
colonialism, 145, 295
Common Dreams, 319
common values, 311
Commoner, Barry, 66
communal aggression, 263
communal defense response, 263
communications media, 299
communications revolution, 303
competition, 7, 268
competition for territory, 21
competitive market, 7
complete nuclear abolition, 191
complex cells, 275
complicity, 230
composite materials, 112
Comprehensive Test Ban Treaty, 182
compressed hydrogen gas, 108
compulsory labor, 219
computer software, 61
computers, 262
concentrating photovoltaics, 93, 94
conception, 33
concubinage, 22
Condorcet, Marquis de, 15, 22, 42
conflicts, 305, 324
conscience, 25, 254
conscription, 254
conservation, 64
conspicuous consumption, 221
construction and maintenance, 124
construction energy, 98
consumer behavior, 304

consumption, 263
consumption of goods, 66
consumption of meat, 209
contaminated water, 42
convection currents, 106
conventional armaments, 180
conventional oil, 91, 93
conventional petroleum, 49
conversion, 221
cooking, 103, 124
cooperation, 213, 266, 318
cooperation between cells, 275
cooperation, evolution of, 280
cooperative behavior, 277
cooperative communities, 275
Cooperative Movement, 217
COP15, 54, 208
Copenhagen Accord, 56
copper reserves, 111
Corn Laws, 29
corn silk, 126
Corporate Social Responsibility, 219
corpses, 167
corruption of morals, 26
cosmopolitan societies, 280
Council of Canadians, 322
countercurrents, 319
courage, 264
creativity, 61
crime, 14, 23, 42, 161
crimes against humanity, 228–230
crimes against peace, 228–230
crippled for life, 214
crisis, 263, 369
crisis predicted, 45
crofters, 2, 10
crop failures, 51, 126
crop wastes, 89, 101
cropland, 45, 129
cropland per capita, 130, 136
cropland per person, 46
cropland, area of, 122
cropland, limitations on, 121
CRS, 219
cruelty by children, 270
CTBT, 182, 191

Cuba, 180
Cuban Missile Crisis, 176
cubic relationship, 97
cultural barriers to marriage, 268
cultural diversity, 304
cultural evolution, 213, 262, 277, 281
cultural heritage of humanity, 323
cultural history, 295
culturally-driven growth, 44
culture, 281, 302, 305
culture of peace, 301, 303, 304, 317
culture of violence, 148, 303, 305, 317
currents of molten material, 106
curriculum revisions, 298
cyclic adenosine monophosphate (AMP), 276

Dalai Lama, 313
dances and songs, 269
dangerous climate change, 51, 56
Danish economy, 96
Danish Peace Academy, 310, 319, 321
Darrieus wind turbine, 98
Darwin, Charles, 15, 148, 270, 272
death penalty for unionism, 34
death rate, 12
decreasing sexual dimorphism, 280
deep wells, 107
defamation, 304
defoliants, 211, 214
deforestation, 59, 129, 130
degradation of topsoil, 130
demand, 123
demand for men, 8
democratic governments, 42
democratic process endangered, 230
Democratic Republic of Congo, 148
demographic studies, 22
demographic trap, 135, 208
demography, 20
demoralizing effects, 6
Denmark, 96
dense population, 42
depleted uranium shells, 211
depletion of minerals in soil, 122
deportation, 230

depression, 65
depression of 1929, 64
desert areas, 94, 102, 108
desertification, 122
destruction of cities, 230
destruction of forests, 129
deterrence, flaws in concept, 183
devastation, 230
developing countries, 148, 151, 208
development, 134, 214
devil's dynamo, 232
devotion, 263, 265
Dhanapala, Jayantha, 190
dialects, 270
dialogue, 304
diarrhoea, 158, 306
DiCaprio, Leonardo, 319
Dickens, Charles, 26
diction, 270
diet, 269
digestion of food, 276
dignity, respect for, 304
diminishing returns, 29
dirt huts, 12
disappointment, 10
disarmament, 184, 311
discrimination, 304
disease, 15, 18, 20, 21, 42, 45, 46, 148, 295, 326, 369
disease and malnutrition, 45
disease-resistant strains, 120
Dismal Science, 31
Dissenting Academy, 12
distribution problems, 123
diverse populations, 271
diversity, ethnic and racial, 270
division of labor, 7
DNA sequences, 281
doctors, lack of, 306
domestic industry, 7
domestication of animals, 21
Donne, John, 261
draft animals, 45
drainage, 6
drama, 297
Dresden firebombing, 164

drought, 122
drug addiction, 161
drug-resistant tuberculosis, 210
dry steam, 107
dry-season water supply, 53
dual use power plants, 95
dung, 89, 101
Dylan, Bob, 319
dysentery, 133

earth's crust, 106
earth's rotation, 106
earthquake activity, 106
East German refugees, 179
East India Company, 28
Eastern Asia, 121
ecological catastrophe, 46
ecological considerations, 35
ecological constraints, 47
ecological damage, 210
ecological degradation, 220
ecological principles, 220
ecology, 10, 214, 217, 220
economic growth, 29, 48, 60
economic Inequality, 215
economic inequality, 15
economic interdependence, 299
economic justice, 312
economic laws, 3, 10
economic oppression, 304
economic reforms, 312
economics, 2, 217
economy of exclusion, 155, 316
ecosystems, 280
EcoWatch, 319
ecstasy, 263
education, 13, 66, 160, 213, 282, 306, 318
education for peace, 294, 301, 311
education for women, 136, 137
education of economists, 219, 220
educational curricula, 304
educational systems, 192
Ehrlich, Paul R., 66
Eibl-Eibesfeldt, Irenäus, 268
Einstein, Albert, 163, 169, 317

Eisenhower's farewell address, 230
Eisenhower, Dwight David, 227
ElBaradei, Mohamed, 188
elderly homeless persons, 155, 316
electrical power, 94, 95
electrical power costs, 98
electrical power generation, 95
electrical power grids, 99
electricity generation, 94, 96, 103, 107, 109, 212
electrochemical reactions, 108
electrode material, 108
electrolysis of water, 94, 108
electronic media, 317
electronic signals, 281
elementary peace education, 297
eliminating war, 214
elimination of poverty, 42
Elizabeth I, Queen, 24
Elvin, Lionel, 303
emotions, 369
empty-world economics, 47, 63
Enclosure Acts, 2
end of fossil fuel era, 49, 56
endemic conflict, 148, 214
endemic disease, 149
energy, 372
energy conservation, 49
energy consumption per capita, 90
energy crisis, 105
energy efficiency, 64, 96
energy inputs of agriculture, 124
energy payback ratio, 98
energy problems, 220
energy scarcity, 209
energy shortages, 220
energy storage, 104, 108
energy use, 89
energy-intensive agriculture, 119, 125
enforcement of international laws, 254
engineering students, 297
engineers, responsibility of, 296
Eniwetok Atoll, 173
Enlightenment, 10
enrichment, 188
enthalpy, 106

environment, 13, 90, 219
environmental catastrophe, 35, 318
environmental changes, 59
environmental cost, 216
environmental degradation, 46–48, 210
environmental holocaust, 211
environmental impact, 61
environmental Kuznets curve, 66
environmental stress, 263
envy, 13
equal rights for women, 303
equal status for women, 42
equilibrium economics, 64, 217, 220
equilibrium with the environment, 63
equity, 212
Eritiria, 148
erosion, 129
erosion of ethical principles, 166
escalatory cycles of violence, 183
Esquisse, 15
Essay on Government, 12
Essay on Population, 10
Essay on Population, 2nd Ed., 22, 25, 43
ethical considerations, 35
ethical principles, 183, 220, 296
ethical responsibility, 297
ethics, 2, 10, 13
Ethiopia, 148, 151
ethnic conflicts, 42
ethnic groups, 296, 298
ethnicity, 268
eukarkyotes, 280
eukaryotic cells, 275
Europe, 96, 145
European Parliament, 191
European Union, 215, 295, 296, 298
Evabgelii Gaudium, 154, 315
evil, 23
evolution, 15, 275, 318
evolution of cooperation, 272
excess population, 31
exchange of currencies, 253
execution, 356
exploitation, 305

exponential growth, 17, 60, 63
exponential increase, 43
export of armaments, 250
extermination camps, 228–230
external circuit, 108, 109
externalities, 215
extinction, 48
extinction of species, 59, 220
extinctions, 51
extortion, 219
extraction costs, 49, 93
extreme weather conditions, 51

Fabian Society, 34
Fabians, 34, 145
factories, 8
fair trial, 229
fallout, 184, 211
family planning, 46, 135, 221
family size, 18
famine, 16, 18, 21, 42, 45, 46, 80, 119, 127, 136, 187, 209, 211, 312
famine relief, 157
fanaticism, 263, 304
FAO, 121
farm buildings, 124
farm wastes, 102, 103
fast breeder reactors, 212
fast neutrons, 212
fatal accident, 186
favelas, 133
FBI, 298
federations, 295
feedbacks and climate change, 54
feedstocks for fertilizer, 124
feedstocks for pesticides, 124
Fellowship of Reconciliation, 322
Fermi, Enrico, 172
fertile land, 29
fertility of mixed marriages, 268
fertilization of flowers, 272
fertilizers, 90, 124
fiberglass, 112
field machinery, 124
Fielden, John, 4
filed teeth, 269

films, 324
filth, 6
Final Document, 182
financial institutions, 295
finite food supply, 60
Finland, 102
firebombing of Dresden, 183
firebombing of Hamburg, 166
Fischer, R.A., 265
Fischer-Tropsch catalysts, 88
fish ladders, 104
Fissile Materials Cutoff Treaty, 182
fission, 169
fission bomb, 172
flags, 263
flocks of birds, 277
flogging, 4
flood control, 104
Florida, Richard, 61, 213
focal axis, 94, 95
food and biofuels, 103
food calorie outputs, 124
food calories per capita, 125
food per capita, 130
food production, 42, 44, 120
food shortages, 24
food supplies, 80
food supply, 16, 19, 43, 47
food, imported, 31
food-exporting countries, 119
foreign industry, 7
forest loss, 129
forests, 47
former Soviet Republic, 120
fossil fuel energy inputs, 124
fossil fuel era, 88
fossil fuels, 46–48, 50, 54, 65, 98, 110, 124
fossil fuels and synthesis, 88
Franklin, Benjamin, 10
fraud, 13
free association, 297
free market, 2, 7
free market mechanisms, 35
freedom of expression, 304
French nuclear weapons, 177

French Revolution, 10
Friends Service Committee, 321
fruiting body, 276
Fruits of Philosophy, 33
fuel cells, 94, 103, 108, 109
fuelwood, 45, 89
full-world economics, 47, 63
fungi on roots, 276
fusion energy, 89
future, 374
future generations, 217, 220, 304

Galtung, Johan, 319
Gandhi International Institute for Peace, 322
Gandhi, Mahatma, 169, 222, 313
Gaskell, Dr. Peter, 4
Gates, Bill, 217
Gavin, General, 174
general good, 13
genes, 266
genetic code, 277
genetic diversity, 59
genetic engineering, 262, 297
genetic evolution, 262, 277
genetically homogeneous tribes, 270
Geneva Conventions, 296
genocide, 148, 183, 230, 281, 356
Genscher, Hans-Dietrich, 190
geography, 298, 302
geometrical growth, 17
geothermal energy, 89, 106
geothermal power, 106
geothermal power plants., 99
Germanic tribes, 20
Germany, 166, 295
Giampietro, Mario, 124
glacial epochs, 59
glacial periods, 127
glaciers, 209
glaciers, melting of, 51, 53
gladiators, 272
Glasnost, 179
global citizenship, 213
Global Compact, 219
global energy resources, 89

global energy use, 89
global environment, 63, 90
global ethic, 213, 280, 294, 317
global food bank, 209
global governance, 214
global inequalities, 298
global inequality, 214
Global Security Institute, 322
global warming, 48, 51, 54, 95, 123
Global Zero, 321
Global Zero Campaign, 190
globalization, 220
Godwin's Utopia, 23
Godwin, William, 11–13, 15–17, 21–23, 26, 31, 42
Goering, Herman, 227
Goethe's Faust, 261
Goodall, Jane, 271
Goodman, Amy, 319
goods, 61
goods per capita, 66
Gorbachev, Mikhail, 179, 189, 190
government intervention, 49
government officials, 229
governmental commitment, 110
governmental interference, 7
governmental intervention, 2
governmental regulation, 30, 35
governmental responsibility, 64
governmental support, 110
gracilization, 280
grain production, 130
grain, price of, 29
Grameen Bank, 217
Grant, James P., 303
graphite electrodes, 108
grasslands, 122
Greece, 106
Green Revolution, 120, 124
green taxes, 215
greenhouse effect, 50, 54
greenhouse gases, 50, 54, 103
greenhouses, 95
Greenland icecap melting, 51
Greenpeace, 321
Grey, Colin S., 186

grids, 99
gross national product, 60
Grotius, 296
groundwater, 126
group identity, 270, 271
group selection, 265
Groupe Dannone, 217
groups of animals, 277
growth, 8, 60, 90, 217, 372
growth of population, 25, 42
growth of wind power, 96
growth-oriented economics, 64
Guernica, 163
guilt, 170, 183
Gulf War, 214
Gulf War oil spills, 211

Hague Conventions, 296
Hahn, Otto, 169
Haileybury, 28
hair standing on end, 264
hair-trigger alert, 185
Haldane, J.B.S., 265
half-reactions, 108
half-starved child workers, 10
Hamburg firebombing, 166
Hamilton, W.D., 265
hanging, 13
Hanseatic League, 295
happiness, 10
hardwood trees killed, 211
Hartmann, Nils, 305, 306
Hartmann, Thom, 319
harvest failures, 24
harvesting, 101
Hawken, Paul, 215
Heads of State, 229
health care, 208
heat engines, 109
heat flow, 106
heat pumps, 100
heat waves, 51
heaters, 103
heating of houses, 107
heavy armaments, 254
heavy oil, 91

Heiliger Schauer, 264
heliostatic reflectors, 95
hepatitis, 133
herbicides, 211
hereditary transmission of power, 15
hero face, 264
heroic behavior, 263
HEU, 212
hibakushas, 314
high birth rates, 42
high educational level, 42
high enthalpy resources, 106
high population density, 21
high unemployment levels, 42
high-speed ultracentrafuges, 181
high-yield grain varieties, 124
high-yield strains, 120
higher loyalty, 192, 280
higher plants, 275
higher status for women, 137
highland clearances, 2
highly enriched uranium, 187, 188, 212
highway development, 130
hillsides, 130
Himalayas, 53
Hindu Kush, 53
Hinduism, 311
Hiroshima, 167, 168, 170, 171, 173, 182, 263, 317
Hiroshima Peace Committee, 313
Hiroshima Peace Museum, 190, 321
Hiroshima-Nagasaki Protocol, 191
history, 269, 298, 302, 318, 323
history of science, 297
history teaching, 295
history, teaching, 294
Hitler Youth, 263
Hitler, Adolf, 166, 301
HIV/AIDS, 149, 151, 210, 219
Hobson, John Atkinson, 145
Holdren, John P., 66
Hong Kong, 135
honge oil, 102
Hoodbhoy, Pervez, 181
Hoover Institution, 189

hospitality, 269
hospitals, lack of, 306
hot dry rock method, 107
hot water flotation, 91
houseman, 21
housing, 208
Hubbert peak(s), 46, 49, 62, 90, 91, 111
Hubbert peak for oil, 111
Hubbert, M. King, 49, 90
human dignity, 303
human economy, 48
human failings, 185
human groups, 277
human institutions, 23
human misery, 120
human nature, 270, 281
human rights, 208, 219, 296, 298, 302, 303, 311, 356
human suffering, 45
humanitarian law, 184, 185
Hume, David, 11
hunger, 13, 25
hunter-gatherer societies, 19, 21, 44
Hurd, Lord Douglas, 189
Huxley, Thomas, 272
hydroelectric power, 89, 104
hydroelectric storage, 99
hydroflurocarbons, 50
hydrogen, 94, 108, 109
hydrogen from algae, 103
hydrogen fuel cells, 109
hydrogen technologies, 108
hydrogen technology, 99, 103
hydropower, 49, 89, 95, 104, 110
hyperbolic trajectory, 44, 45

I=PAT, 66
ICAN, 321
Iceland, 104, 106
Idolatry of the Free Market, 312
ihgnorance, 13–15
Ikeda, Daisaku, 192, 313
ill-treatment of prisoners, 230
illegality of nuclear weapons, 184, 189
illegality of war, 228, 230

illiteracy, 295, 326
immigration, 80
imperialism, 155
Imperialism: A Study, 145
imported oil, 101
imports of grain, 30
imprisonment, 230
imprisonment without trial, 356
improvement of society, 16
inadequate housing, 6
inadequate sanitation, 42
income taxes, 215
increased consumption, 88
India, 52, 111, 120, 151, 180, 189, 270
India's nuclear weapons, 180
India, coal consumption, 93
India, oil consumption, 91
Indian State Railway, 102
indigenous cultures, 358
indigenous faiths, 311
indigenous peoples, 148, 211, 213
indirect costs of war, 295
indiscriminate mass slaughter, 183
individual citizens, 298
individual responsibility, 229
Indonesia, 106, 135
Industrial Revolution, 2, 28, 44, 145, 297
industrial sector, 61
industrial workers, 10, 61
industrialized countries, 46, 90
industrialized nations, 42
industry, 16
inequality, 154, 295, 315
inequality between men and women, 15
INES, 321
infant mortality, 16, 26
infanticide, 26, 31
infectious disease, 42
information, 262, 277
information-related work, 61
infrared radiation, 50
infrastructure, 133, 208, 214, 221
inhumane acts, 230
initial investment, 94

injustice, 10
inland rainfall, 129
inorganic fertilizer, 124
input/output ratio, 124
insects, 276
inspection teams, 191
Institute for Economics and Peace, 322
institution of war, 189, 232
intellectual improvement, 13
inter-group aggression, 265
interconnected problems, 208
interdependence, 280, 312
intergroup aggression, 271
intermittency, 99, 104, 108
internal law, 229
internal peace, 214, 298
international business transactions, 253
international control, 180, 211
international cooperation, 302
International Court of Justice, 296
International Criminal Court, 230, 249, 296
international language, 325
international law, 214, 229, 249, 296, 318
International Law Commission, 229
international laws, 82
International Military Tribunal, 229
International Peace Bureau, 321
international understanding, 302, 324
International Water Decade, 306
Internet, 262
interreligious understanding, 296
intertribal aggression, 267
intertribal wars, 21
intra-group altruism, 265
invasion of Poland, 166
invention of printing, 277
invention of writing, 277, 279
investment in research, 110
invisible hand, 7
IPCC, 51, 123
IPPNW, 321
Iran, 181, 189

Iraq, 211, 214
Irish Potato Famine, 27, 59, 124
Iron Law of Wages, 29–31
irreversability, 182
irreversible climatic shifts, 55
irrigation, 104, 124, 130
Islamic Bomb, 181
Islamic fundamentalists, 181
Israel, 180
Italy, 106
Itapú Dam, 104
Ito, Hisato, 167

Jackson, Tim, 223
Jainists, 311
Japan, 106, 135, 145, 166, 228
Jatropha, 102
Jesus College, 12
Jewish Bomb, 181
Jewish Voice for Peace, 322
job security, 66
John Atkins Hobson, 155

Kamchatka Peninsula, 106
Kayfetz, J.L., 208
Keeley, Lawrence H., 278
Kennedy, John F., 176
Keynes, John Maynard, 34, 64
Khan, Abdul Qadeer, 181
killing of hostages, 230
kindness, 266
kinetic energy, 97
King, Rev. Martin Luther, 313
Kissinger, Henry, 189
Klare, Michael T., 210
knowledge, 13
Knowlton, Dr. Charles, 33
Koestler, Arthur, 263
Kristensen, Thorkil, 48, 62
Kropotkin, Prince Peter, 274
Kubrick, Stanley, 172
Kuril Island chain, 106
Kuznets curve, 66

La Grande complex, 104
labor, 47

labor standards, 219
land mines, 211, 214
landfills, 103
language, 277
language and ethnic identity, 270
language barriers, 325
Lapham, Robert J., 135
large nations, 214, 270, 298
Laski, Harold, 34
last frontier, 120
late marriage, 20, 22, 27
Laterite, 130
laterization of soil, 130
Latin America, 121
latitude, 101
Law of the Sea, 296
laws acting on individuals, 214, 249
laws binding on individuals, 298
laws of nature, 23
Lawyers' Committee on Nuclear
    Policy, 322
LeBlanc, Stephen A., 278
leeching by rain, 130
legal responsibility, 230
legislature, 249
LeMay, General Curtis E., 186
Lerma Rojo, 120
less developed countries, 305
LEU, 212
Libya, 127
lifestyle changes, 90
lifestyles, 96
limitations on cropland, 121
limits to growth, 62, 221, 372
linear progression, 17
linguistic groups, 15
linguistic symbols, 277
liquid fuels, 49
liquid fuels from coal, 88, 93
literature, 269
lithium-6 isotope, 180
livestock feed, 124
living standards, 66
lobbys, 318
local communities, 270
London Blitz, 164

London School of Economics, 34
long human childhood, 15
Lorenz, Konrad, 264, 265, 268
Los Alamos Laboratory, 170
Lovelock, James, 280
Lovens, Amory, 215
low death rate(s), 21, 42
low enriched uranium, 212
low enthalpy resources, 106
low status for women, 42
low wages, 32
lowest social class, 16
loyalty, 263, 265, 323
loyalty to humanity, 317
lubrication, 90
lust, 25
luxuries, 12
luxury goods, 221
Lybia, 181

MacDonald, Ramsay, 34
machinery, 8
magnesium, 105
magnetic bottles, 89
Mahayana Buddhists, 311
Mahler, Halfdan, 133
malaria, 149, 152, 159, 210, 220
malice, 13
malnutrition, 20, 157, 210, 295
Malthus revisited, 42
Malthus, Daniel, 11, 15, 16
Malthus, Thomas Robert, 10, 11, 16, 21–28, 31, 42, 43, 46, 60
Malthusian forces, 42, 45
Manchester Guardian, 145
manifesto 2000, 304
mantle of the earth, 106
manufactured goods, 30, 145
manufacturers, 8
marginal land, 29, 46, 121, 130
Margulis, Lynn, 272, 275
marine air, 97
market economies, 179
market failures, 215
market forces, 66
market forces in ecology, 215

market mechanisms, 64
markets, 145
marriage, 33, 267
Marshall Islands, 213
Marshall, Alfred, 27
Martineau, Harriet, 26
mass media, 192, 262, 282, 294, 317, 324
massive nuclear retaliation, 183
massive retaliation, 313
material goods, 372
material possessions, 12
Maudlin, W. Parker, 135
maximum efficiency, 109
maximum natural fertility, 43
Mayor, Federico, 303
Mayors for Peace, 321
McNamara, Robert, 176
meat consumption, 125
medicine, 59
Medvedev, President, 191
melting of glaciers, 51, 53
melting of polar ice, 51
metal ore reserves, 111
metals, 111
methane, 54, 103
methane, anthropogenic, 50
microcredit, 217
Microsoft, 217
Middle East, 94, 127, 312
Middle Powers Initiative, 322
migration to cities, 133
militant enthusiasm, 263
military budgets, 220, 295
military ecological damage, 210
military establishments, 248
military lobbies, 305
military spending, 306
military strength, 145
military training, 230
military-industrial complex, 230, 318
milk and potato diet, 27
Mill, James, 31
Mill, John Stuart, 31, 118
minerals, 47
mines, 145

minimum wage law, 31, 34
miscalculation, 185
Miscanthus, 102
misery, 16, 21, 23, 32, 33, 43, 208, 371
misplaced power, 230
missile defense system, 188
missiles, 179
mistaken for a missile strike, 185
Mitani, John, 271
mitochondria, 275
modern agriculture, 44
modern arms, 312
modern languages, 302
modern weapons, 42, 299
modernism, 189
molten lava of volcanoes, 108
molten salt, 95
monocultures, 59
monopolies, 7
monsoon, 52
Montessori, Maria, 302
moral choice, 229
moral improvement, 12
moral responsibility, 230
moral restraint, 21, 22
morality, 15
morals, 25
Morris, William, 34, 222
mortality, 18
Moscow Treaty, 182
mountain passes, 97
multicellular organisms, 272, 275, 280
multiethnic groups, 270
multinationals, 220
multiracial groups, 270
murder, 230
Muslim era, 181
mutual aid, 274

Nagasaki, 167, 173, 182, 263, 317
Nanjing, bombing of, 166
nation-state(s), 262, 299, 317
national armies, 254, 298
National Energy Policy, 90
National Health Service, 34
national pride, 189

nationalism, 181, 294, 317, 323
nationalist teaching, 302
NATO, 182
natural capital, 48
Natural Capitalism, 215
natural gas, 49, 88, 89, 111, 124
natural laws, 13
natural resources, 47
natural selection, 265, 266
Near East, 121
necessity, 25
negative peace, 297
Nehru, Jawaharlal, 34, 303
neoclassical economics, 27
neocolonialism, 145, 295
neolithic agricultural revolution, 280
net primary product, 48
neutral countries, 183
neutron bombs, 180
New Agenda Resolution, 185
New Statesman, 34
New Zealand, 106
newspapers, 317, 324
Newton's solar system, 13
NGOs, 219
Nigeria, 151
nitrous oxide, 50
Nkrumah, Kwami, 145
Nobel Peace Prize, 217
Nobel Prize in Literature, 319
Nobel Women's Initiative, 322
nomadic societies, 19
non-discrimination, 304
non-nuclear-weapon states, 180
Non-Proliferation Treaty, 180
non-renewable resources, 49, 63
non-violence, 304, 313
nonrenewable resources, 217, 220
normative value, 219
North America, 96, 106
North Korea, 180, 189
North-South contrasts, 215
Northern Africa, 121
Norway, 20, 21, 43, 104
NPT, 180
NPT Article VI, 182

NPT Review Conference(s), 182
nuclear abolition, 189
Nuclear Abolition Forum, 322
Nuclear Age Peace Foundation, 321
nuclear arms race, 169, 170
nuclear black market, 187
nuclear darkness, 187, 211
nuclear deterrence, flaws, 187
nuclear disarmament, 180
nuclear energy, 110
nuclear fusion, 89
nuclear industry, support for, 110
nuclear insanity, 167, 178
nuclear nationalism, 181
Nuclear Non-Proliferation Treaty, 180
nuclear opacity, 180
nuclear power generation, 88, 211
nuclear proliferation, 200, 211, 212
nuclear sharing, 183
nuclear terrorism, 181, 187, 188
nuclear tests, 211, 213
nuclear war, 176, 211
nuclear weapon states, 180
nuclear weapons, 200, 262, 296, 297, 369
Nuclear Weapons Convention, 191
nuclear-weapon-free world, 189
Nunn, Sam, 189
Nuremberg Principles, 228–230, 296
nurses, lack of, 306
nutrient-poor soils, 130

Oakwood Chapel, 12, 16
Obama, Barack, 183, 190, 191
obscenity, 33
ocean currents, 51
ocean level rises, 51
oceans, 82
offshore winds, 97
Ogallala aquifer, 122, 127
oil, 49, 89
oil content, 102
oil industry, 110
oil, US domestic production, 90
oilsands, 91
Oklahoma, 122

Oliver Twist, 26
one-celled organisms, 275
onshore winds, 97
open world, 217
open world economics, 217
opium, 161
oppression, 13
optimism, 10
optimum global population, 46, 111, 118
ores, 111
organic wastes, 103
organized crime, 200
Orthodox Christians, 311
Osada, Arata, 167
Our Common Future, 125
outlawing war, 228
output per hectare, 46
overcrowding, 42
overgrazing, 47, 122
overpopulation, 326, 369
overshoot and crash, 45
Owen, Lord David, 189, 190
Owen, Robert, 217
oxygen, 81, 94, 108, 109
ozone, 50
ozone layer, 187, 211

Pacific Ocean, 106
packaging and retailing, 124
paint, 90
Pakistan, 52, 120, 122, 180, 189
Pakistan's nuclear weapons, 181
palm oil, 103
pandemics, 210
Pankhurst, Emaline, 34
paper, 262
paper industry, 102
Paraguay, 104, 135
Paris Climate Conference, 320
parish assistance, 25, 26
Parr, Samuel, 18
participation of women, 304
passions of mankind, 23, 25, 262
pastoral societies, 21, 44
pasturage, 45, 121, 129

paupers, 26
Pax Christi, 321
peace, 374
peace and basic needs, 305
peace and solidarity, 305
peace education, 294, 305, 311
peace movement, 317
Peace Pledge Union, 322
Peace Research Institute, Oslo, 321
peak demand, 104
peak solar power, 94
Pecci, Aurelio, 62
penal system, 13
per capita energy use, 89, 96
per capita food calories, 125
Perestroika, 179
perfectibility, 15, 22
perfluorocarbons, 50
periodical misery, 22
permafrost, melting, 54
permission to marry, 20
Perry, William, 189
Persson, Göran, 101
pesticides, 124
petroleum, 124
petroleum price, 119
petroleum reserves, 94
petroleum, conventional, 49
petroleum-derived fibers, 45
petroleum-driven tractors, 45
phagocytosis, 276
pharmaceutical companies, 149, 219
pharmaceuticals, 90
philanthropy(ies), 13, 217
Philippine Islands, 106
photosynthesis, 48, 101, 275
photovoltaic efficiency, 94
photovoltaic panels, 93
photovoltaic production costs, 95
photovoltaics, 49, 96, 110
photovoltaics, cost of, 94
photovoltaics, global market, 95
Physicians for Social Responsibility, 322
Picasso, Pablo, 163
Pigou, Arthur Cecil, 215

Pigovian taxes, 215
Pimental, David, 124, 130
Pitt, William, 8, 25
Place, Francis, 31, 32
plague, bubonic, 31
Plan B, 215
planning wars of aggression, 230
plant disease(s), 59, 124
plant energy, 125
plant genetics, 120
plantations, 145
plastics, 90
platinum electrodes, 108
plutonium, 182, 187, 188
PNND, 322
poetry, 269
poison gas, 214
polar ice, melting, 51
Political Economy, 28
political inactivity, 318
political independence, 228
Political Justice, 12, 13, 17
political justice, 312
political oppression, 304
political structures, 262
politically unstable countries, 200
politics, 13
pollination, 276
polluted air, 42
pollution, 49, 82, 369
Pongamia pinnata, 102
Poor Laws, 24, 25
Pope Francis I, 154, 315
Pope John Paul II, 313, 315
Pope John XXIII, 315
population, 10–12, 16, 60, 66, 80, 89, 220
population crash, 119
population density, 94, 102, 111, 118
population explosion, 160, 297, 357
population genetics, 265, 266
population growth, 10, 16, 17, 22, 23, 30, 31, 46, 122, 208, 209, 220, 295
population growth and poverty, 133
population oscillations, 48
population pressure, 19, 21, 27, 47

population pressure and poverty, 28
population pressure and war, 28
population stabilization, 43, 64, 134
population/cropland ratio, 45
populations of animals, 47
positive checks, 18, 43, 46
positive peace, 297
post-fossil-fuel era, 125
postures, 277
postwar nuclear arms race, 169
potato, 31
potato blight, 59
poverty, 10, 14, 16, 18, 20, 25, 28, 31, 33, 42, 46, 63, 149, 208, 210, 214, 295, 312
poverty-related deaths, 220
power, 357
power plants, 214
power reactors, 211
power unbalance, 298
power-worshiping culture, 303
powers of government, 66
Prayer for World Peace, 321
preemptive strike, 178
prehistoric wars, 278
prejudice(s), 22, 304
preventable disease, 210
preventive checks, 18, 20, 43, 46
price of grain, 24, 29
price of petroleum, 119
priest, eunuch and tyrant, 26
Priestly, Joseph, 12
primary energy, 96
primary fuels, 95
Principle of Population, 18
Principle of population, 22
printing, 262, 277, 281
private judgement, 23
processing, 101
production, 263
productivity, 7
profane purposes, 8
profits, 4, 29, 60, 220
progress, 15, 16
progress of science, 299
prokaryotes, 280

prokaryotic cells, 275
proliferation, 211
proliferation risks, 212
promiscuity, 22
promoting peace, 302
propaganda, 317
propeller-like design, 98
property, 13, 23
prosperity, 7
prosperity without growth, 223
prostitution, 31
protein-rich residues, 102
Protestants, 311
prudence, 24, 26
Prussian army officers, 268
pseudospeciation, 268
Pu-239, 171, 180, 212
public health, 135
public opinion, 191, 324
public transport, 221
public transportation, 65
Pugwash Conferences, 169, 321
Putin, Vladimir, 190

Qaddafi, Muammar, 127
Quakers, 312, 313
Queen Noor of Jordan, 190

radar, 179, 185
radiation sickness, 169
radical change, 25
radioactive fallout, 211
radioactive nuclei, 106
rainfall, 52, 94, 102, 123, 126
rainforests, 48, 50, 216
rank-determining aggression, 271
rank-determining fights, 263
rape, 230
rapeseed, 103
rapeseed oil, 95, 101, 102
rats, 167
raw materials, 217, 372
reactor provided by France, 180
Reagan, Ronald, 179
real needs, 63
reason, 23, 25

recession, 221, 326
reciprocal links, 208, 214
recycling, 111, 112
recycling of nutrients, 276
recycling resources, 64
rededication of land, 111
reflectors, 95
reforestation, 64
Reform Act of 1832, 34
Reform Jews, 311
reformed economics, 217
reformers, 35
Reign of Terror, 17
reinvestment, 8, 145
rejection of violence, 303
religion, 192, 282, 298, 305, 317
religion and culture, 270
religion and ethnicity, 270
religious bigotry, 15
religious ethics, 296
relocation of people, 104
renewable energy, 64, 88, 89, 94, 100, 110, 221, 248
renewable energy resources, 54
renewable energy sources, 46, 49
renewable energy systems, 108
renewable energy technologies, 219
renewable natural gas, 103
rent, theory of, 28
replacement fertility, 43
replies to Malthus, 25
Reply to Parr, 18
reprocessing, 188
reprocessing of fuel rods, 211
reserve indices of metals, 111
reserves of coal, 93
reserves of metals, 111
reserves of uranium, 212
reservoirs, 104
resource curse, 147
resource wars, 209
resources, 80, 220
respect for life, 303, 304
respiration, 275
responsibility, 229, 230
revenge and counter-revenge, 183

reverse transition, 45
Reykjavik Revisited, 189
Ricardo's Iron Law of Wages, 33
Ricardo, David, 28, 31
Rifkind, Sir Malcolm, 189, 190
Rift Valley, 106
right to sustenance, 24
righteousness, 263
rights of individuals, 303
Ring of Fire, 106
rising energy prices, 95
ritual scarification, 268
rituals, 269
river blindness, 149, 159
Robertson, Lord George, 189
Robespierre, 10
Robinson, Mary, 190
Robock, A., 187
Rockefeller Foundation, 120
Roman Catholics, 311
Roman Empire, 20
Rome Treaty, 296
Rookery, 11
Roosevelt, Franklin D., 64, 65, 166
Rotblat, Sir Joseph, 169, 293
Rousseau, Henry, 11
Rudd, Kevin, 190
rule of law, 302
Ruskin, John, 34, 222
Russell Peace Foundation, 322
Russell, Bertrand, 34
Russell, Lord Bertrand, 169
Russia, 106, 151, 187
Rwanda, 230

sacred duty, 263
safe drinking water, 306
Sahel, 122
Saigon, 211
Saint Paul's Cathedral, 164
salination, 122, 130
Salix viminalis, 101
SALT Treaties, 182
Samsø, 100
sanctity of the family, 15
Sane Nuclear Policy, 322

sanitation, 158, 208
satellites, 179
Saudi Arabia, 127
Scandinavia, 20, 42, 282
Scandinavians, 20
schistosomiasis, 149, 159
Schmidt, Helmut, 190
schools, 208
schools, cost of, 306
Schultz, George, 189
science, 10, 281, 302
science education, 297
scientific discovery, 263
scientific progress, 12, 42
scientific revolution, 44
scientists, responsibility of, 296
sea level rise, 48, 51
Seager, Joni, 210
Second Essay, 18
Second World War, 10, 228, 229
secret oaths, 34
secure jobs, 66
security, 311
Security Council, 148, 228, 230
seed, 125
Seeds of Peace, 322
self-defense, 228
self-destruction, 263
self-fulfillment, 12
self-help groups, 217
self-interest, 2, 10
self-love, 23
self-sacrifice in war, 265
self-sacrificing courage, 264
selfish motives, 263, 265
selfishness, 10, 13, 23
semiconducting materials, 94
semipermeable membrane, 109
sensibility, 15
September 11 attacks, 305
sequestered carbon, 129
sequestering carbon, 216
service sector, 61
servility, 13
several hundred million deaths, 175, 183

sewage, 6
sexual dimorphism, 271
shadows on the pavement, 167
shared knowledge, 281
sharing time and resources, 304
Shaw, George Bernard, 34
Shaw, Pamela, 64
Shelley, Percy Bysshe, 26
Shiite Muslims, 311
Shintos, 311
shiver, 263
short-rotation forests, 101
Shrinivasa, Udishi, 102
Sikhs, 311
silicon, 94
Singapore, 135
SIPRI, 321
slave-like working conditions, 5
slavery, 15, 373
slaves, 32
sleeping sickness, 159
slime molds, 275
slums, 133
small arms, 148, 214
smaller families, 49, 217
Smith, Adam, 7, 8, 28, 30, 34, 60, 61, 63
social businesses, 217
social classes, 24
social cohesion, 214, 271
social conscience, 30
social construction, 271
social cost, 216
social disruption, 56
social epidemiology, 155
social impact of science, 297
social inequality, 124
social insects, 272, 277
social institutions, 262
social legislation, 33
social responsibility, 217, 219
social security, 208
soil conservation, 64, 209, 221
soil erosion, 122, 130
Soka Gakkai, 192
Soka Gakkai International, 313, 321

Soka University of America, 320
Soka University Japan, 320
solar constant, 101
solar cooking, 93
solar design in architecture, 93
solar energy, 48, 89, 93, 101, 108
solar parabolic troughs, 95
solar thermal power, 49, 93, 96, 110
solar water heating, 93
soldiers, training of, 230
solidarity, 217, 303–305
Somalia, 148
Sonora 64, 120
South Africa, 49
South African nuclear weapons, 180
South America, 106
Southern Africa, 121
Southern Asia, 121
Southey, Robert, 25
sovereignty, 251
Soviet bomb, 171
Soviet thermonuclear bomb, 173
Soviet Union, 170, 179
soybean oil, 103
spawning grounds, 104
species, 268
speech, 281
sponge method, 32
spores, 276
Støre, Jonas Gahr, 189
stabilization of population, 47
stable family structure, 15
stable population, 42
Star Wars, 179
starvation, 3, 5, 26, 27, 30, 45, 148, 157
State Militia, 298
steady-state economics, 61
steady-state economy, 65
steam engine, 2
Stern Report, 52, 54, 123, 126, 129
Stern, Sir Nicholas, 52
Stockholm, 61
Stone, Oliver, 319
Strangelove, Dr., 172
Strategic Defense Initiative, 179

strong world government, 214
struggle for existence, 272
subcellular structures, 275
submarginal land, 121
subnational organizations, 181
subprime mortgage crisis, 221
subsidies, 95
subsistence, 16, 22
suffering, 15
sugar beets, 101
suicide, 4, 169, 228
suicide, collective, 170
sulfur hexafluoride, 50
sunlight, 93, 94, 101
Sunni Muslims, 311
superbombs, 172
superheavy oil, 92
superorganisms, 272
superstition, 22, 31
support from governments, 110
Surrey, 12
survival, 16, 267
survival of the fittest, 155, 316
sustainability, 10, 43, 48, 61, 63, 64, 213, 215, 217, 220, 221
Sustainable Development Commission, 223
sustainable goals, 49
sustainable limit, 45
Swaminathan, M.S., 120
swamps, 103
Sweden, 94, 101, 102
Switzerland, 295
symbiosis, 272
Symington Committee, 174
syngas, 88
synthetic fertilizers, 45
Systems of Equality, 22–24
Szent-Györgyi, Albert, 363, 371

tactical nuclear weapons, 182, 183
tar, belt of, 92
targeting civilians, 166
tarsands, 91
Tartar tribes, 19
tattoos, 269

tax agreements, bilateral, 148
tax changes, 110
tax structure, 49
tax-gatherer, 26
taxation, 49, 65, 253
taxation, power of, 298
teacher's training colleges, 297
teaching materials, 306
teaching of economics, 217
team-spirit, 263
technical defects, 185
technology, 10, 46, 155, 315
technology, transfer of, 133
tectonic plates, 106
television, 232, 317, 324
Teller, Edward, 172
temperature and agriculture, 126
temperature equilibrium, 51
temperature increase, 51
tenant farmers, 2
terawatt, definition, 49, 89
Terp, Holger, 310
territorial integrity, 228
terror bombing, 166
terrorism, 166, 312
terrorism, nuclear, 187
terrorist organizations, 181
terrorists, 200
textbooks, 302, 304
textile mills, 3
TFF, 319, 322
The Elders, 322
The Geysers, 107
theme days, 297
theology studies, 296
theoretical maximum efficiency, 109
thermal conductivity, 106
thermal expansion of ocean, 51
thermal expansion of oceans, 48
thermonuclear bombs, 180
thermonuclear catastrophe, 318
thermonuclear device, 173
thermonuclear reactions, 89, 172
thermonuclear weapons, 263, 313
third world debt, 214
Third World War, 179

third-world cities, 42
Thirteen Practical Steps, 182
thorium, 88
thou shalt not kill, 155, 316
threatened species, 358
Three Gorges Dam, 104
tidal power, 89
Tierra del Fuego, 106
time factor, 111
Tobin tax, 253
Toda Institute of Peace, 320
Toda, Josei, 313
Tokyo, firebombing, 183
tolerance, 303, 304
Toon, O., 187
topsoil, 129
topsoil, loss of, 122
torture, 4, 10, 230, 356
total output of a society, 155
total reaction, 108
trachoma, 159
trade in arms, 250
trade unions, 34, 217
trading, 279
traditional rights and duties, 2
training of soldiers, 230
Transcend International, 321
transition towns, 222
transportation, 101, 124, 208
trees, destruction of, 122
trench warfare, 263
Trevada Buddhists, 311
trial, fair, 229
tribal markings, 268, 270
tribal religions, 280
tribalism, 263, 265, 267, 280, 317
trickle-down theories, 155, 316
tropical cyclones, 51
tropical diseases, 149, 219, 248
tropical forests, 216
tropical rainforests, 50, 59, 81, 121
Truman, Harry, 172
Truthout, 319
tuberculosis, 151, 159, 220
Turco, R., 187
Turkey, 106, 122

Turner, Ted, 217
Tutu, Archbishop Desmond, 190
TW, definition, 89
twentieth century, 10
twenty children, 8
typhoid fever, 133, 159
typhus, 159

U-238, 212
Uexküll, Jakob von, 320
ultracentrafuges, 212
UN General Assembly, 184, 185, 229, 249, 252, 304
UN Global Compact, 219
UN Peacebuilding Office, 322
UN Security Council, 148, 230
unauthorized act, 186
unconventional oil, 93
underground tests, 180
understanding between cultures, 303
understanding between religions, 303
undiscovered resources, 90
unemployment, 63, 64, 134, 221, 326
unequal distribution of incomes, 145
UNESCO, 248, 301–304
UNESCO Courier, 303
UNICEF, 303, 305
unions, 31
United Kingdoms bomb, 171
United Nations, 318
United Nations Charter, 228, 296, 302
United Nations finances, 253
United States, 43, 90, 91, 107, 119, 122, 145, 166, 170, 187, 270, 295
universal brotherhood, 280
universal code of ethics, 280
universal education, 24
universal human brotherhood, 296
universality of religion, 270
University for Peace, 320
University of Copenhagen, 54
university peace education, 320
unlimited economic growth, 2
unlimited growth, 312
unlimited material needs, 312
unsanitary housing, 20

uranium, 88
uranium enrichment, 181
uranium mining, 213
uranium reserves, 212
urban growth, 130
urbanization, 122, 133
US Federal Government, 298
US food system, 124
US grain belt, 126
US imports of oil, 90
US nuclear weapons in Europe, 182
use of force, 228
USSR, 122
utilitarians, 31
utility, 25
utopian societies, 22

vaccination, 149
vaccinations for children, 306
value systems, 269
value-creating education, 313
vanishing resources, 88
veils, 269
Venezuela, 92, 93
vertical shaft design, 98
Veterans Against War, 322
vice, 13–16, 21, 23, 43
vicious circle, 208
Vietnam War, 211, 214
Vikings, 282
violence, 304
violence, rejection of, 303
Vitousek *et al.*, 48
vocal signals, 277
volcanic activity, 106
volcanic regions, 106
Von Weizäcker, Richard, 190
voting system, UN, 252
vulnerability of modern society, 312

Wade, Nicholas, 271
Wali, Mohan, 126
Wallace, Robert, 10
war, 15, 18–21, 28, 42, 45, 46, 148, 210, 295, 313, 373
war as a business, 232

war casualties, 214
war crimes, 228–230
War Resistors International, 321
war-based economy, 230
war-free world, 298, 369
Ware, Alyn, 213
wars of aggression, 230
wars, prehistoric, 278
Warsaw, bombing of, 166
wastefulness, 295
water, 108
water availability, 123
water closets, 6
water conservation, 221
water erosion, 130
water purification plants, 214
water resources, 126
water supplies, 122, 158
water vapor, 50
water, safe, 149
wave power, 49, 110
WCED, 125
Wealth and Welfare, 215
Wealth of Nations, 7, 28
weapons and threats, 305
weapons industry, 232
weapons of mass destruction, 200
weapons-usable materials, 212
weapons-usable Pu-239, 211
webb, Sydney and Beatrice, 34
Wells, H.G., 34
wheat farms, 122
wheat varieties, 120
whistle blowers, 191
Wilkinson, Ellen, 302
Wilkinson, Richard, 155
will, 25
Wilson, E.O., 59
Wilson, Harold, 34
wind electrical power costs, 98
wind energy, 89, 96, 108
wind erosion, 122
wind parks, 96

wind power, 49, 95, 99, 110
wind turbines, 96
wind velocity, 97
windmill parks, 100
women, 357
women of childbearing age, 151
Women's International League, 322
women, education for, 136
women, higher status for, 136
women, participation of, 304
wood, 95, 101
Woolf, Leonard, 34
workhouses, 4, 26
working conditions, 3
working week, 221
World Beyond War, 322
world citizenship, 294
world community, 249
World Conference of Religions for
    Peace, 311, 313
world federal authority, 298
World Federalists, 321
World Future Council, 320
world government, 298
world public opinion, 170
World Trade Center, 2001, 188
World War I, 215
World War II, 215
Worldwatch Institute, 64, 127
worship of growth, 217
writing, 277, 281
writing, invention of, 279

xenophobia, 318

yields per hectare, 125
young population, 136
Yugoslavia, 230
Yunus, Muhammad, 190, 217

Zaragoza, Federico Mayor, 303
Zimbabwe, 148
Zoroastrians, 311

www.ingramcontent.com/pod-product-compliance
Lightning Source LLC
Chambersburg PA
CBHW070307230426
43664CB00015B/2653